CREATIVE ECONOMIES IN POST-INDUSTRIAL CITIES

Re-materialising Cultural Geography

Dr Mark Boyle, Department of Geography, University of Strathclyde, UK,
Professor Donald Mitchell, Maxwell School, Syracuse University, USA and
Dr David Pinder, Queen Mary University of London, UK

Nearly 25 years have elapsed since Peter Jackson's seminal call to integrate cultural geography back into the heart of social geography. During this time, a wealth of research has been published which has improved our understanding of how culture both plays a part in, and in turn, is shaped by social relations based on class, gender, race, ethnicity, nationality, disability, age, sexuality and so on. In spite of the achievements of this mountain of scholarship, the task of grounding culture in its proper social contexts remains in its infancy. This series therefore seeks to promote the continued significance of exploring the dialectical relations which exist between culture, social relations and space and place. Its overall aim is to make a contribution to the consolidation, development and promotion of the ongoing project of re-materialising cultural geography.

Also in the series

From the Ground Up
Community Gardens in New York City and the
Politics of Spatial Transformation
Efrat Eizenberg
ISBN 978 1 4094 2909 8

Sanctuaries of the City
Lessons from Tokyo
Anni Greve
ISBN 978 0 7546 7764 2

Cities and Fascination
Beyond the Surplus of Meaning
Edited by Heiko Schmid, Wolf-Dietrich Sahr and John Urry
ISBN 978 1 4094 1853 5

Swinging City
A Cultural Geography of London 1950–1974
Simon Rycroft
ISBN 978 0 7546 4830 7

Remembering, Forgetting and City Builders
Edited by Tovi Fenster and Haim Yacobi
ISBN 978 1 4094 0667 9

Creative Economies in Post-Industrial Cities

Manufacturing a (Different) Scene

MYRNA MARGULIES BREITBART
Hampshire College, USA

LONDON AND NEW YORK

First published 2013 by Ashgate Publishing

Published 2016 by Routledge
2 Park Square, Milton Park, Abingdon, Oxfordshire OX14 4RN
711 Third Avenue, New York, NY 10017, USA

First issued in paperback 2016

Routledge is an imprint of the Taylor & Francis Group, an informa business

British Library Cataloguing in Publication Data
Creative economies in post-industrial cities :
manufacturing a (different) scene. -- (Re-materialising cultural geography)
1. Urban renewal. 2. Urban renewal--Case studies. 3. Deindustrialization--Social aspects. 4. Cultural policy--Social aspects. 5. Cultural policy--Economic aspects. 6. Community development, Urban.
I. Series II. Breitbart, Myrna Margulies.
307.3'416-dc23

The Library of Congress has cataloged the printed edition as follows:
Breitbart, Myrna Margulies.
Creative economies in post-industrial cities : manufacturing a (different) scene / by Myrna Margulies Breitbart.
p. cm. -- (Re-materialising cultural geography)
Includes bibliographical references and index.
ISBN 978-1-4094-1084-3 (hardback)
1. Urban renewal. 2. Urban renewal--Case studies. 3. Deindustri-alization.
4. Community development, Urban. 5. Cultural policy--Economic aspects. 6. Cultural policy--Social aspects. I. Title.
HT170.B743 2013
307.3'416--dc23

2012041819

ISBN 13: 978-1-138-27708-3 (pbk)
ISBN 13: 978-1-4094-1084-3 (hbk)

For Billy

Contents

PART II: MOVING BEYOND NEOLIBERAL METHODOLOGIES IN THE STUDY AND PRACTICE OF CREATIVE ECONOMY PLANNING

List of Figures

List of Contributors

Alison Bain is an Associate Professor of Geography at York University in Toronto who studies contemporary Canadian urban and suburban culture. Her work examines the relationships between artists, cities, and suburbs with attention to questions of identity formation and urban change. Her current research program focuses on cultural production and creative practice on the margins of Canada's largest metropolitan areas.

Myrna Margulies Breitbart, is a Professor of Geography and Urban Studies in the School of Critical Social Inquiry, Hampshire College, Amherst, Massachusetts. She is also Academic Director of a curricular program that supports community engagement and collaborative learning outside the classroom. Her interdisciplinary interests focus on urban social and economic inequality, community development and the spatial dimensions of social change. She has a strong commitment to, and experience with, participatory action research, especially involving young people in urban exploration and local planning, and has written widely on these subjects. Her current research is on the new creative economy and its role in urban regeneration in smaller post-industrial cities.

Liza Fior was born in London where she continues to practice as founding partner of muf architecture/art. The work of the practice negotiates between the built and social fabric, and between public and private in projects that have been mainly focused in East London but not exclusively so. Projects range from urban design schemes to small-scale temporary interventions, landscapes and buildings – a continual dialogue between detail and strategy. Liza is co-author of, *This is What We Do: A muf Manual* (London: 2001). Unsolicited research continues to be entwined into every project. Liza has taught at the Architectural Association, The Royal College of Art and most recently at Yale.

Susan Fitzpatrick is a Geographer currently based at Strathclyde University in Glasgow where she teaches Economic Geography. Her research focuses of the ethics of representation and power within urban regeneration schemes in post-industrial economies. Susan's research on creativity and its role in urban gentrification has covered the Liverpool City of Culture in 2008; a two year public art program in the English town of West Bromwich called Black Country Creative Advantage; and most recently the Glasgow International Art Biennial. She is currently authoring a paper with Iain Mackenzie on how to define the "event" in "event-led regeneration".

Gabriel N. Gee is Assistant Professor of Art History at Franklin College, Switzerland. His research interests include British painting in the 20th century, forms and discourses in the visual arts in Northern Ireland in the late 20th century, and the interaction between aesthetics and industry in the 19th and 20th centuries. He is a co-founder of the research group on Textures and Experiences of Trans-Industriality (TETI group)

Sophie Handler is an urban researcher/practitioner and initiator of Ageing Facilities an alternative urban research initiative that actively explores new ways of "making space" for older age. Funded by the RIBA and the AHRC, this practice-led research develops live projects and focused research that facilitate other ways of imagining and using the urban environment in older age. Having completed a practice-led Ph.D. on alternative duties of elderly care (at the Bartlett, UCL) Sophie has gone on to develop research practice in the charitable and public sectors: working on the evaluation of a specialist care unit for dementia patients with the Royal Holloway University, the South London and Maudsley NHS Trust and the arts charity Age Exchange, and developing research around "age-friendly cities" with Keele University and Manchester City Council. Her work on "Resistant Sitting" is currently on display at the Canadian Centre for Architecture as part of its Imperfect Health Exhibition.

Dylann McLean is a Ph.D. candidate in Geography at York University in Toronto who studies cultural producers and their potential for transforming space. Her work examines the geographical and historical specificity of portfolio career profiles in the cultural and artistic sectors. Her current research focuses on the transformation of physical and conceptual space through transgressive artistic intervention.

Arturo E. Osorio, Rutgers University Business School, New Jersey. Arturo Osorio is an Assistant Professor of Management and Global Business. He is currently a fellow at the Center for Urban Entrepreneurship and Economic Development (CUEED). Before coming to Rutgers he spent five years researching and documenting the socioeconomic renaissance of a former mill town in Western Massachusetts. This work, focused on members of the creative class (in particular artists and artisans), explored the influence and reach of individuals' lifestyle choices in the community's everyday life. His research interests include urban entrepreneurship, the creative class, grassroots movements, and socioeconomic development of communities.

Micah Salkind is a Providence, Rhode Island-based cultural historian interested in queer histories of technology and the social impacts of Afro-diasporic cultural production on local, national and transnational economies. His scholarly research on house music and the spatial diffusion of media artifacts such as 12" records and cassette tapes complements his writerly activism on behalf of local artists and art institutions. Salkind is a Ph.D. candidate in American Studies at Brown University.

Acknowledgments

My interest in industrial cities was sparked long ago in an economic geography seminar at Clark University. Back then, our discussions of "creativity" focused on the technological and social innovations that fueled the early industrial revolution and dramatically transformed people's lives. There was no talk of cultural quarters, live/work space, innovation districts or heritage tourism. There was an appreciation for a "creative class", but this referred to skilled workers who spent long grueling hours as cotton spinners or weavers, not young software engineers and filmmakers. As students, we learned of the dominant role that manufacturing played in the economic and social life of the cities and regions we studied. Once our "Geography of Manufacturing" texts were supplemented with the works of Engels and Dickens, we also understood the contradictions embedded in places that were simultaneously famous as centers of immense mechanical ingenuity yet infamous as the sites of horrific living and working conditions for workers whose labor fueled the prosperity of a small few. Much of the motivation for this book comes from a continuing fascination with these former industrial cities, and from a desire to explore how they are working to invent a new and sustainable economy in the face of the on-going problems wrought by deindustrialization and global economic shifts.

While cliché, it is clear that this book would not have been possible without the contributions of many. First and foremost I am grateful to the authors who contributed case studies and reworked their chapters several times over. Each author was a delight to work with, and their examples, taken as a whole, provide a rich, if complicated, picture of how smaller post-industrial cities and neighborhoods are reinterpreting creative economy strategies to fit their unique circumstances. I am especially grateful to Valerie Rose and Caroline Spender at Ashgate Publishing who guided me through the publishing process.

Many educational institutions and foundations enabled my research. I am grateful to my home base, Hampshire College, for several vital summer faculty development grants and sabbaticals to support this project. The Massachusetts Foundation for the Humanities provided an early grant to initiate a study on the role of the arts in the revitalization of Holyoke. The Holyoke Cultural Council, supported by the Massachusetts Cultural Council, provided another grant to defer the printing costs of my Holyoke Community Arts Map in English and Spanish. Generous grants from the Whiting Foundation and the Kahn Liberal Arts Institute, based at Smith College, supported travel to a number of post-industrial cities in New England and the United Kingdom. This travel enabled valuable conversations with many knowledgeable and committed individuals working to identify and develop the creative assets present in their cities. I would like to give special thanks to those who took several hours from their day to share their experiences, their hopes and frustrations: Peter Aucella, Jerry Beck, Brian Cross, Graeme Evans, Fred Faust, Liza Fior, Tom Fleming, Jo Foord, Tony Hughes, Charlie Hunter, Robert McBride, Joe Manning, Phil Niddrie, Justin O'Connor, Graham Riding, Stephen Sheppard, Jeremy Smith, Mic Smith, Ann Wallis, Patryc Wiggins, Erin Williams, Phil Wood and Chris Wyatt.

I owe a very special thanks to the many talented residents of Holyoke, Massachusetts, who are too numerous to mention, yet willingly shared their thoughts, hopes and visions for the city. The Arts Walk Steering Committee and former Holyoke Planning Network provided the base from which to identify many local assets. Those who worked with me directly on the Holyoke Community Arts Map and the arts inventory deserve special thanks. They include, Gisela Castro, Gregory Horvath, Jesse Payne and a number of Hampshire College and Five College students. I would also like to thank friends and colleagues in and around Holyoke who have contributed so much informally over the years to my knowledge of, and appreciation for, the city and all its residents: John Aubin, Esthela Bergeron, Jay Breines, Maria Cartagena, José Colon, Hilda Colon, Nancy Howard, Priscilla Kane-Hellweg, Imre Kepes, Charlie Lotspeich, Gladys Lebrun-Martinez, Betty Medina Lichtenstein, Kathy McKean, Yolanda Robles, Daniel Ross, Don Sanders, David Scher, Yoly Nogue Velez, Sandy Ward, and the late Carlos Vegas, to name but a few.

Over the years, I have benefited from many conversations with colleagues in the larger geography, urban studies and planning communities. Early conversations with Cathy Stanton led to our co-authoring a chapter on cultural workers in New England cities. Members of the Community Economies Project at UMASS, and my dear late colleague in geography, Julie Graham, provided an early sounding board for my research into the alternative forms of creative economy evolving in New England. Smith College's Kahn Institute seminar on Cities and City Life provided a venue from which to explore how smaller cities approach economic revitalization. This initiated a brief collaboration with Italian scholar, Federica Anichini, on the morphology of medieval Italian cities and its symbolic application in Barnsley, England's *Renaissance Plan*. Jane Rendell, at the Bartlett School of Architecture in London, introduced me to two eventual contributors to this volume. Finally, I owe a special thanks to my colleagues at Hampshire College, especially my co-workers in the School of Critical Social Inquiry and in Community Partnerships for Social Change, who have provided generous collegial support over many years.

As with all lengthy research projects, the support of family and friends is essential, and I have had that in spades. I want to thank my London cousins for providing a welcoming home base for my travels, and my UK traveling companion, Anne, who visited many post-industrial cities with me, expressing a healthy skepticism for the replacement of what she would call "real production" with new "creative industries". Closer to home I have appreciated the support of my daughter Molly, her husband Patrick, who has an intimate connection to the post-industrial city of Lawrence, Massachusetts, my son Jonathan, his wife Tiana, and an entire extended family of cousins, friends and neighbors in Western Massachusetts, who endured long conversations about post-industrial cities and the writing process in its final stages. A special thanks to my granddaughter Sophia who always makes me laugh and provides a joyful distraction whenever we are together! Finally, no one has lived more closely with the process of researching and bringing this book to completion than my husband Billy. I am truly grateful for his unwavering support and immense patience in reading and commenting on my countless drafts.

Chapter 1

Introduction: Examining the Creative Economy in Post-Industrial Cities: Alternatives to Blueprinting Soho

Myrna Margulies Breitbart

> We are in the development business, not the arts business ... We find that actors walking around the streets, theater activities in general, give off good vibrations; they make an area more renewable. If cement factories did that, we would be putting them in.
>
> (Frederic Papert, former President of the 42nd Times Square Street Development Corporation)

Much has changed since such commentaries drew the attention of urban policy makers to the economic role of the arts and culture in global cities such as New York. Cultural forms of regeneration and the cultivation of creative industries are now common elements in most urban economic development lexicons. Many pages have been written, and stories told, of how cultural tourism, historic preservation, and the Soho-like transformations of old factory districts attract new business, consumers and a young "creative class". Two decades ago, Sharon Zukin began to take note of the growing relationship between culture, design, and urban redevelopment (1982; 1995). The story has since been retold in several versions. The most common scenario begins with a neighborhood that may contain everything from less populated warehouse districts to low income and working class residential areas, but has seen better days. For a variety of reasons, including a long period of disinvestment, the neighborhood starts to attract the quintessential starving artist in search of cheap living and working space. Once the numbers of such artists rise to a certain level, amenities (e.g. cafes, galleries) move in and the area is secured and considered "desirable" enough to attract a middle class, and perhaps also a large-scale public/private redevelopment scheme. A familiar gentrification process follows in which the arts and artists shift from pioneer to victim. As outside perceptions of the neighborhood change and land values rise, increasing rents lead to the displacement of marginal occupants (including the starving artists), and a generalized cleansing of surrounding public space occurs (e.g. MacLeod 2002; Dutton 2005; Shaw 2006). Depending on the relative location and importance of this neighborhood to downtown and larger capital interests, the process goes on to engender the harsh fortressing measures aptly described by Mike Davis (1990), Neil Smith (1996), Roslyn Deutsche (1996) and many others.

While this familiar scenario does not play out in exactly the same way from place to place, many large cities continue to transform abandoned manufacturing cores into enticing consumer landscapes of visual appeal. Spatial evidence is found in the proliferation of downtown corporate plazas bedecked with public art, festival marketplaces, loft conversions, and riverfront and sidewalk cafes. In spite of cautionary warnings about the uneven development this can engender (Bianchini 1993; Grodach and Loukaitou-Sideris 2007), the momentum to enhance and market urban cultural amenities grows, as one city after another vies for recognition as a premier cultural enclave. As Zukin points out, what is produced and consumed are not traditional manufactures, but rather, art, food, fashion, tourism, and the industries and services that cater to them (Zukin 1995). The processes used to achieve these economic makeovers vary, but nearly always involve the transformation of physical spaces of decline to new uses and functions, and the re-formation of the public's image of these spaces (and the city as a whole) – a process that both results from, and provides justification for, the former. Culture, in Zukin's words, has become the "business of cities" (1995: 2).

The use of the arts and culture to enhance the economy and draw attention to the rising importance of a city is not as new a phenomenon as it would seem. Connections between culture, urban space and capital have a long history that includes the diffusion of ideas and practices from large and growing cities to smaller ones (Hall 1998). As Evans points out, the urban renaissance in continental Europe in the 16th century quickly spread to London and, by the early 17th century, to a number of smaller provincial cities and towns that would try to mimic its cultural developments and attractions (Evans 2001: 51). While based on a long history, the replication of models of urban cultural regeneration and their diffusion from large global cities, such as New York and San Francisco, to more medium-sized and even smaller cities and towns, has accelerated. Attention to the particulars of this largely uncritical dissemination and translation process has not kept pace with its extent, impact, or potential. This is one of many circumstances that motivate this book.

Creative Economies in Post-Industrial Cities follows in the path of Bell and Jayne (2006) and van Heur (2010) in its movement away from a focus on large global cities and towards an examination of how culture-infused economic opportunities are being understood and incorporated into planning, often in very distinct ways, in smaller former industrial cities and working-class neighborhoods. Authors consider to what extent places rooted in an industrial past are enabled to envision a different economic future. They examine whether these visions replicate strategies employed in larger cities or put forth plans that better suit their unique histories and challenges.

Many writers within the vast and growing literature on the creative economy note what one critic calls "eclectic conformity" – a type of cultural homogenization process that masquerades as local distinctiveness (e.g. Holcomb

1993: 142l; Breitbart and Stanton 2006). The case studies in this volume consider whether smaller places outside the global city limelight employ similar cultural-led planning and creative industry development strategies, or ones that look to build the capacity of residents and envision a wholly different economic future. The contributors consider many questions: Where are the visions for change in post-industrial communities derived, and how, if at all, are residents brought in to the planning process? How do efforts to appropriate space help residents resist regeneration strategies that ignore their needs? In what ways are everyday spaces beyond downtown the site of socially engaged creative activity? How does the application of cultural development initiatives in post-industrial cities depart from mainstream approaches to build upon assets or envision a more sustainable future? What new relationships, networks of exchange and economic opportunities are fostered and obstacles encountered? What, too, is the potential of these approaches to effect long-term change?

Many of the authors included in this volume examine the potential role of the built environment in sustainable urban regeneration. They consider to what extent space can be used to disrupt old modes of planning and provide a new context for the incubation of more durable forms of economic and cultural development. There is also a focus on the creative appropriation of space by residents who resist regeneration schemes that ignore local assets or reinforce uneven development. This draws attention to everyday neighborhood spaces and locations other than downtown as potential sites of socially engaged creative work. How, for example, might new spatial formations (e.g. networks and alliances among diverse residents) facilitate the transfer and exchange of knowledge, or provide connections to new skill-building and income-generating opportunities in the growing creative sectors of a sustainable economy? Case studies consider such questions and begin to shift the focus to a critical examination of the potentials inherent in less conventional grassroots initiatives. Examples also focus on a diverse array of creative actors who would not necessarily be recognized as members of a "creative class" by those who research or promote the creative economy.

Sliding Down the Urban Hierarchy

During the heyday of industrial capitalism in the late 18th and early 19th centuries many paternalistic manufacturers designed social structures, and built whole cities, to try to remedy the horrors of factory life prevalent in early manufacturing centers. In New England, the first home of industrial capitalism in the United States, a variety of industries that produced consumer necessities, such as apparel, textiles, paper and shoes, dominated prior to the Second World War. By the early 1970s factory-based employment in these industries fell to one-tenth of the jobs in the region, engendering severe unemployment and out migration (Harrison 1982: 5). Once a "metaphor for social and technological progress", the image of abandoned New England factories and declining cities made their way into the popular imagination (Breitbart and Stanton 2006). Marketing brochures that feature once bleak manufacturing environments as the backbone for cultural and economic rebirth now challenge these images.

The incorporation of culture-based planning into urban regeneration schemes did not happen over night in most post-industrial cities. Indeed, a lot of convincing of planners and politicians was necessary before this path was considered. Much of the "buzz" that led policy makers to consider such options can be traced to the multitude of research reports that were commissioned at the turn of the millennium confirming the positive economic impact that creative industries and arts-related activities had on individual cities and regions. For example, an early study presented by Mt. Auburn Associates to the New England Regional Council presented a wide variety of data to measure the "creative sector". This included cultural organizations, non-profit and commercial enterprises, workers within these enterprises, and individual artist-entrepreneurs (Mt. Auburn Associates 2000). Claiming to support a "new way of looking at the arts and culture", the report examined these activities as it would any industrial or financial sector, assessing its potential to contribute to the region's competitive position. The report also used some rather traditional analytical tools, such as cluster analysis, to group art-related activities geographically, and measure their concentration and impact on one competitive variable, quality of life.

Although the methodology and analysis within this and a multitude of reports that have been written since may be criticized (Markusen and Gadwa 2010), the enthusiasm with which policymakers eventually greeted such findings is key. This can be seen either as a real shift of paradigm or as an extension of an old one. As Robert Goodman pointed out in *The Last Entrepreneurs* (1979), the standard response to deindustrialization had been to enhance a city's competitive advantage by offering traditional tax and other incentives to manufacturers in an effort to lure them back to create what were often very undesirable and precarious jobs. Influenced today by what is now referred to as the "creative economy," the spotlight has shifted from manufacturing to cultural and creative production. Not surprisingly, the economic development packages look quite different, as they seek to retain, develop and attract entrepreneurs and tourists rather than lure a multinational corporation or factory back from Mexico or China.

Charles Landry in the UK and Richard Florida in the U.S. are two high profile figures responsible for promulgating and disseminating the idea of a creative economy and creative city planning. A big turn in urban and economic policy in the U.S. occurred after the publication of Richard Florida's *The Rise of the Creative Class* in 2002. Challenging the conventional wisdom in urban economic development, Florida suggested that companies do not necessarily drive innovation, and that policies geared only towards providing incentives for industries miss the mark. Rather than emphasize manufacturing or service jobs, he points to the importance of attracting creative people and to the increasing dominance of occupations that involve creative work – everything from idea-generating jobs and the arts to advertising, engineering, design and computer graphics.

While there were always select concentrations of culture in places like Greenwich Village, Florida points to the wholesale international sorting of metropolitan areas into categories that either welcome and grow creative enterprise, or remain a bastion of the working class and low level services. He argues that communities need to come to grips with those qualities that attract creative workers because these newcomers will generate and attract new business. Among the qualities he notes are a vibrant cultural life, a sense of place, tolerance for diversity, hip cafes, and historic buildings. In this and subsequent writing, Florida sets out the three criteria for cities aspiring to creative city status: "technology, talent and social tolerance". He goes on to produce many indices ranking cities from best to worst in terms of their success in these three broad areas (2005a, 2005b).

Paralleling the exponential growth of interest in the "creative class" and the creative economy are an ever-expanding number of critiques of Florida's research that come from both the Right and the Left. The range of criticisms move from challenges to Florida's Creativity Rankings of cities according to various indices (Bohemian, Gay, Melting Pot etc.) to more encompassing critiques of the overall theory and its lack of complexity (see, for example, Daly 2004; Shea 2004; Kotkin and Siegel 2004; Malanga 2004; Christophers 2007). One of the most salient critiques centers on social class. It argues that Florida places far too much emphasis on quantitative occupational data generated for large cities, and on a young creative class whose demographic devalues the traditional working class, failing to see, let alone look for, creative assets in unexpected places (Peck 2005; Bell and Jayne 2006). Rather than bemoan the loss of well-paid manufacturing employment, Florida seems more interested in the possibilities that disinvested property hold for new uses such as loft apartments. Other critiques take note of cultural barriers that can be erected between rich and poor, young and old, when new entrepreneurial types move into more ethnic and working class communities (Zukin 2005).[1] Still others suggest that Florida's urban revitalization theories mistake the "side effects" of a once-booming economy for the causes of growth, and that his ideas have been incorporated into neo-liberal political agendas to justify the removal of social safety nets and support for community-based cultural organizations, in favor of an entrepreneurialization of the arts and culture (McRobbie 2003; Maliszewski 2004).

Florida actively addresses these criticisms in paid lectures and through his website, arguing that he is not elitest, and that he believes that every person and town has the capacity to be "creative". Officials in smaller post-industrial cities searching for a foundation upon which to build a new economic base, fear becoming "a Detroit" if they do not, in Peck's words, adopt the city planning

1 After Florida moved in 2007 to the University of Toronto to head up a new research institute, a group called "Creative Class Struggle" formed there to protest his role in building "money-making cities rather than secure livelihoods for real people" (http://creativeclassstruggle.wordpress.com/mission/). The group uses Florida's ideas about the "creative class" to spark larger debates about inequality and neoliberal policies.

mantra, "be creative or die" (Peck 2005: 740). They receive Florida's message that any place can become a creative city with enthusiasm. Whereas large-scale flagship cultural tourist complexes require a massive infusion of capital, the "takeaway" from Florida's work for many local officials seems to be that a beautiful physical setting, and proximity to a healthy natural environment, along with smartly transformed historic structures that reference a gritty past, or even just a liberal political climate to support alternative lifestyles, may be sufficient to attract and keep the creative class. These amenities do not necessarily require massive investment. The fact that many formulaic urban development strategies designed to promote these ideas end up supporting neoliberal planning based on place branding, middle-class consumption, gentrification and competition among individual entrepreneurs and cities, generally goes unnoticed (Harvey 2005).[2]

In spite of the critiques, the proliferation of festival marketplaces, artist live/work spaces and riverfront walks in large cities provide evidence to planners working in smaller post-industrial cities, of culture as the new urban "business". Lowell, Massachusetts provides an early example of culture-led regeneration in New England that is often touted as a success. By the 1920s most of the industry was gone from Lowell and it was considered to be one of the region's most down-and-out cities. Officials have worked feverishly to replace its old abandoned manufacturing core with a more enticing consumer landscape based, in part, on a Heritage Park that capitalizes on the city's unique labor history. The city also created a new Artist Overlay District that seeks to promote the renovation of historic abandoned buildings for commercial use, encouraging artists to live and work in downtown (Breitbart and Stanton 2006).

North Adams, Massachusetts provides a similar example. Abandoned by manufacturing, including its premier manufacturer, Sprague Electric, North Adams was described in the 1980s as the "city with no future", and was next to last in per capita income in the State. The tale of how one of the largest museums of contemporary art, MASS MoCA, ended up in the complex of the abandoned Sprague Electric buildings is told in brief by Sharon Zukin in *The Cultures of Cities* (1995) and more personally in the film *DownsideUp* by Nancy Kelly, a California filmmaker who grew up in North Adams and returned to document the change. The presence of this new museum with massive modern art installations the size of a football field has been accompanied by the attraction of a growing number of high tech dot.com companies and a "new media industrial park". On a field trip with a class several years ago, we visited a few of these companies, including Kleiser-Walczak, the computer animation firm that made "Spiderman" and "The X-Men". Keiser-Walczak was attracted to this remote city many years ago by what was then a $7.00/square foot cost of space. Its North Adams site went

2 David Harvey defines "neoliberalism" as "a theory of political economic practices that propose that human well-being can best be advanced by liberating individual entrepreneurial freedoms and skills within an institutional framework characterized by strong private property rights, free markets, and free trade" (2005: 2).

on to become its main headquarters, with subsidiaries in NYC and Hollywood California, paying about $45/square foot for space. While these developments have brought some change to the town in the form of renovated Victorians, a new coffee house right beside the local barber shop, the renovation of tenements into a high-priced inn, and a few new art galleries, the locally anticipated high paying jobs for residents and tourist spillover effects have yet to materialize. On a positive note, these changes do not seem to have led to significant gentrification either.

Kleiser-Walczak's computer-generated morphs, characters that visually transform before your eyes on the screen, seem an apt metaphor for what entrepreneurs in many former industrial cities envision. This new "innovation economy" is so tied to the, still untapped, potential of technology that its components cannot even be fully articulated by their would-be promoters. Adherents to this development strategy are driven by economic intuition and a fascination with the mysteries and powers of innovation. They admit, however, that they know little about the indefinable chemistry that is believed to result from the placement together in space of artists, computer software and hardware firms, cultural institutions, designers, and product entrepreneurs. Nevertheless, they believe in its potential. The question of "why" is addressed later on in this chapter.

Writing and consulting on creative "place-making" is now an industry unto itself. In the UK, Charles Landry founded the think tank, Comedia, in 1978, way before Florida began thinking about the topic of creative cities. He now provides advice on how cities can transform themselves into thriving creative places around the globe, and rivals Florida in the development of toolkits and consultations (see, for example, 2006; 2008a; 2008b).

The large and growing body of research on the cultural economy includes a number of focus areas that will not be highlighted in this volume (e.g. cultural and heritage tourism; the branding and marketing of cities; as well as studies of creative urban clusters and public/private partnerships).[3] Comprehensive sources already exist to demonstrate how an evolving cultural economy challenges the concept of an "economy" and forces us to rethink economic categorizations by adding culture to the mix (see, for example, Amin and Thrift 2004). Additional research focuses on smaller cities, and the quest of their political and economic decision-makers for a new competitive advantage (e.g. Bell and Jayne 2006; van Heur 2010). More narrowly focused collections examine the strategy of developing cultural "quarters" and innovation districts that evoke new lifestyles in order to draw both cultural producers and consumers. This research also suggests alternative conceptualizations of what is called "vernacular creativity" (e.g. Bell and Jayne 2004; Edensor et al. 2009).

3 A small sampling of this writing includes Bianchini and Parkinson 1993; Scott 2000; Evans 2001; O'Connor 2002; Amin and Thrift 2004; Bianchini and Ghilardi 2007.

There have been notable shifts within empirical research and planning practices. These have gone from from an interest in community arts for their intrinsic value, and for their role in addressing problems caused by structural unemployment and neighborhood decline, to more of what Evans calls an "instrumentalisation of culture" in urban and regional policy (Evans 2010). The shift in emphasis coincides with a change in the language from "cultural planning" to "creative industries". The latter emphasizes the economic value of knowledge-based creative work as a commodity, and as the "basis for competitive advantage," over non-profit art and cultural organizations (García 2004; Evans 2009). According to Foord, the trend now in urban economics is to subsume a greater number of activities under the creative industry and larger knowledge economy umbrella (Foord 2008). Information and communication technologies, media, design, software development, architecture, and other categories, such as publishing, advertising, and even financial services, are now lumped together as part of an ever-bulging class of researchable creative income-generating activities.

The use of many terms such as cultural economy, culture-led regeneration, creative industries, and creative place-making can confuse efforts to define, measure and quantify the impacts of innovation and creative work. According to Foord, use of the term "cultural" rather than "creative" industries prioritized the social aspects of cultural production and the importance of collective, not-for-profit work (Foord 2008: 95). "Creative industries" referred to the "value from copyrighting and distribution of creative content".

> This distinction was critical for those devising creative strategies in developing areas where the creative resource was identified as located in local (indigenous) culture rather than in the production processes of global creative industries (Foord 2008: 95).

Yet many artists and cultural producers object to measuring the creative vitality of a community using official occupational categories that include a large number of professional occupations not ordinarily associated with creativity – e.g. legal and financial services (Foord 2008: 96). They see this categorization as simultaneously too encompassing and too narrow, as it overlooks smaller enterprises and individual practitioners. It also ignores many important forms of creativity and collaborative work that exist among organizations and people at a local level. These interactions give value to a community's existing assets. They also acknowledge local heritage and culture.

Authors in this book explore these factors through case studies that examine whether and how a new creative or culture-led form of regneration has a presence in smaller and medium-sized post-industrial cities and neighborhoods. Some of the examples point to the forces of creative resistance and re-interpretation spawned by efforts to revamp local economies. In the spirit of a "people's geography" this often involves a search for the alternate functions that the arts, artists, and creative industries might play in the lives of marginalized urban dwellers and in a differently conceived set of urban community development approaches. Attention here is directed to acknowledging and building upon the assets of residents who would ordinarily escape the gaze and interest of creative class proponents.

Creating a *Different* Scene

My first glimpse into an alternative conception of a creative base upon which to build a new urban vision came over a decade ago when I joined a small planning group in Holyoke, Massachusetts. We were charged with organizing the first ArtsWalk Festival alongside one of the city's three canals. This small post-industrial city had just released a Master Plan that was a radical departure from the past in its incorporation of an art-based strategy to address economic challenges wrought by deindustrialization. Following the lead of other former industrial cities, the plan conceived of an Arts and Industry District with commercial and residential development for artists alongside the canals and architecturally significant brick mills that once housed Holyoke's famous paper industry. A centerpiece of the project was to be a two-mile-long Canal Walkway.

The group I joined under the auspices of Greater Holyoke Inc. (an independent non-profit organization established for economic revitalization) was comprised of representatives from local community-based organizations, city officials, and owners of former factories who were converting their buildings into artist spaces. The idea of organizing an Arts Walk Festival emerged to fill the lag time between conception of the plan to build a Canal Walkway and its execution. The event was intended to bolster resident enthusiasm for the plan by identifying local talent in the city, and by bringing that talent outside for a day of celebration along a path where the walkway would eventually be built. My job was to begin to compile a database of residents with arts-related talents that included young people, and represented Holyoke's diverse population. Everyone on the list was invited to participate in the event.

Two facts became apparent almost immediately upon the launch of ArtsWalk organizing. First was the extent to which the arts and culture, though hidden from view, were already deeply embedded within some of Holyoke's most economically challenged neighborhoods. Second was the expansive and inclusive definition of the "arts" and "artist" that residents harbored in their identification of local talent. Though Richard Florida's book on the creative class had yet to be published, and few, if any, urbanists were talking about something called a "cultural" or "creative" economy, these early observations of the role of culture in the life of this small city raised many questions that now challenge some of the foundational principles underpinning creative economy-promoting policies.

Fast forward to the present and the intersection of Cabot and Main Street in Holyoke. This is the location of a modest building occupied by Nuestras Raices, a community organization that fosters sustainable economic and cultural development through the cultivation of a large farm, community gardens and job training in the growing green economy. A visit to this site today would likely reveal several meetings underway at the Centro Agricola. In one room, a group of unemployed young men and women aged 17–21 might be gathered for an information session on a new green jobs training program they are about to begin. They might also be learning about a new community-based business enterprise that is designed to promote innovation in the area of sustainable affordable housing and commercial development. In an adjacent greenhouse where a Marine Reef Habitat aquarium business was spawned and moved on, tiny plants are now cultivated for future transplanting, while an adjacent restaurant prepares food. Meanwhile, in the outdoor plaza near a fountain and beneath a large mural, a number of neighborhood residents might be meeting to plan the entertainment for the weekly farmer's market and the yearly Festival del Jibaro that draws visitors from the larger region and features local performers and artistic crafts.

This scene stands in obvious contrast to the kind of "scene" described in much of the literature on the new creative economy, where the main protagonists, the "creative class", are depicted as young big city dwellers with lifestyle preferences that favor trendy new restaurants, mountain biking, and late night clubbing. Clearly there was and is something to be learned about regeneration from smaller cities. Observing these differences motivated my desire to visit a number of other post-industrial centers in New England with the specific goal of uncovering the sites of creative enterprise that are hidden from view in places located down the urban hierarchy. This process, begun 10 years ago with interviews of policymakers, artists and residents throughout New England, is summarized below.

Nowhere to Go But Up: Imagined Futures

In a murder mystery about drug trafficking in Vermont, author Archer Mayer describes the post-industrial town of Bellows Falls as,

> … a troubled community … developmentally stalled since the Great Depression, [with] … a dour and pessimistic self-image out of all proportion to its size … It owes its existence to [the] water's energy, which in the early years gave the upstart, industrially minded settlers an advantage over their more agrarian neighbors. For a succession of gristmills, rag paper plants, and pulp mills, the ceaseless water became literal lifeblood, supplying power …. [this] all this town has to brag about. It is a pantheon of long vanished industrial might … Periodically, the village erupts with face-saving activity. Meetings are held and committees formed to identify and solve the place's underlying problems. But whether it's half-heartedness from within, or the sheer magnitude of the task, these groups never seem to last long, … Another movement was afoot right now, in fact, dedicated to the usual renaissance. It seemed better organized than its predecessors, but no one I knew was placing any bets. A museum of glories

> past, the name Bellows Falls had become solely equated with failure (Mayor 1997: 18–19).

In spite of its less than complementary tone, I found that most people in Bellows Falls who have read the book were amused by their featured prominence. They recognize themselves, their built environment and the negative image that the rest of Vermont has of their community. That in no way prevented Robert McBride, an artist who migrated from New York City to Bellows Falls, from promoting his adopted town as one of the State's leaders in progressive planning and arts-based economic development.

> Time had just kind of forgotten it and what I found amusing when I'd go to State meetings is that you'd be there and suddenly lunch would come and you'd be talking to someone who would ask "Where are you from?", and you'd say "I'm from BF," and really they'd sort of like just get this like frozen grin on their face and sort of like turn away from you. I loved that coming from Manhattan. Wow this is really interesting, you know they don't want to talk to me or they don't think there's anything to talk about. So I love the whole idea of a challenge. There was a challenge here. There was only one way to go and that was up.

Bellows Falls is now one of several poster children in New England for the new creative economy.

With a few exceptions, Master Plans are not generally the places to look for a detailed articulation of arts-based visions of renewal in places like Bellows Falls. Even though one can find references to cultural planning in some of the more recent Master Plans, and some communities have written formal Cultural Strategies that read like Master Plans, nothing reveals more about personal visions for change, and the host of contradictions and competing ideas that co-exist within the same city, than talking to a range of people. While such conversations can make generalization difficult, the inquiry surfaced some fascinating speculation about where these visions come from and why they are so appealing.[4]

In the quest to learn more through interviews and case studies I found that many planners and economic developers were surprised by the economic turns their communities had taken away from traditional manufacturing. They had a hard time predicting or expressing a preference for a future agenda, yet elements of commonality emerged. Most visions of economic and social revival in these small post-industrial cities included an express desire to rid the landscapes of both the symbols and reality of decline and decay. The absence of physical reminders of bleak and protracted periods of de-industrialization was thought to provide the bedrock for a new prideful reconstructed image of place. This could involve the removal of some existing businesses, such as problematic bars and X-rated movie theaters, not to mention inoffensive locally owned businesses that a developer

4 Some of this material is explored in Breitbart and Stanton 2006.

might see as an under valuing of property. In a media-driven society, everyone is aware of the importance of marketing and self-presentation. Explaining her decision to rent a storefront in Easthampton, Massachusetts for free for six months to a new pastry business, property and artisan gallery owner, Mai Stoddard said, "Good business decisions don't always translate immediately into money." She believed that changing the image of her post-industrial city and "putting it on the map as a haven for artisans and craftspeople, and a destination for their customers", is likely to pay off in the future. "Sometimes", Stoddard said, "It's more important to look good than to feel good" (quoted in Brown 2003).

Remaking an image is not *all* that drives these visions for change, however. Most people I spoke with share a real appreciation for their city and their industrial past. They feel strongly that change is possible and believe that their home has some very unique and endearing qualities that should have wide appeal, whether they derive from the physical setting or the social atmosphere. In Lowell, Massachusetts, for example, nearly everyone references Senator Paul Tsongas's and Patrick Mogan's[5] undying beliefs that their city could become *the* premier medium-sized New England city in terms of its cultural offerings and amenities.

Strong beliefs in the unique qualities of their post-industrial cities proliferate among optimistic visionaries. "Easthampton is all about community", Michael Garjian, a resident and the former small-business director of the Valley Community Development Corporation, now director of Vee-Go Energy, said. "It's what makes this a great city. It's a blue-collar city … the sense of community in this town is strong" (Brown, May 2003). Rather than erase the troubled industrial past, Garjian and others want to embrace this past with pride. According to Peter Aucella, affiliated with Lowell National Park in Massachusetts,

> The beauty of [Lowell] is that most people have the basic vision …. historic preservation, reuse of buildings for a variety of uses – residential, commercial, industrial, and trying to be essentially what we are, rather than be someplace else with skyscrapers or whatever …. If you see an old building, you renovate it. That's your character. That's your economic development program.

Mr Aucella was firm in his belief that Lowell needed to use its built environment to draw on its industrial past in order to devise a vision that affirms what it is today.

The post-industrial visions embraced by many planners and economic developers in these smaller cities focus on a renewed physical landscape that includes a revitalized pedestrian-lively downtown, with a refurbished historically-preserved architecture, and noticeably fewer vacant lots and abandoned shops. While this adaptive reuse of buildings is designed to attract artists and arts-related activities, cultural tourism is not the sole intent. Nearly everyone I spoke with

5 Patrick Mogan was a former superintendent of schools in Lowell, Massachusetts, largely responsible, along with Senator Paul Tsongas, for helping to turn Lowell's mill district into a National Park.

heeded the lessons from the past concerning the dangers of depending on a single money-generating enterprise. A large company moving in and then moving out had burned more than one small New England city. Sharing a history of abandonment by single employers, most visions for the future are now built upon a foundation of smaller businesses and creative enterprises. Unwilling to place all their eggs in one basket, diversification is the new mantra for most post-industrial cities. Worcester, Massachusetts does not bank on the arts to pull its economy back. It hopes for the direct economic benefits that can derive from arts and cultural activity, yet acknowledges the indirect benefits that an improved quality of life can have in the quest to attract the technical and medical people necessary to build a growing biotechnology research complex. The same vision of a diversified base is true for Burlington or Bellows Falls, Vermont.

A decade ago, many planners spoke of the willingness of artists to move into substandard buildings and make improvements to the spaces. It was assumed that these improvements would then eventually diffuse to other neighborhoods. Since invigorating new lifestyles must be visible if a city's image is to change, this can mean planting these new artists in a prominent downtown location. According to Aucella,

> Artists have traditionally been pioneers. They're going to move in before your empty nester that is not ready to sell their house in Belvedere, one of the better neighborhoods in the community. They're not all of a sudden going to walk in here and give up their parking spaces and walk up four flights to a loft. It's just not going to happen. So I think that the artists, much as the National Park, have reminded people in Lowell that there is something special going on here. Sometimes it takes people to see something from outside to be reminded of that.

But the powers attributed to artists go further than their monetary investments downtown or contributions to a changed place image. Many small city boosters I spoke with described how new ideas and new activities and spaces for performance and artistic display draw residents and tourists downtown to support local business. According to the vision, artists can also attract outside entrepreneurs – creative entrepreneurial people who invent new economic ventures. The creative synergy that develops when people of diverse talent live in close proximity to one another is presumed to take off when public and private entities begin to work together. In New England, North Adams, Massachusetts is often used as an example. The State and the city worked closely with Williams College alums, internet start-ups, and venture capitalists to construct a plan for the transformation of the old Sprague Electric Company into a contemporary art museum. Efforts to copy this example of a public/private partnership abound, and many planners and economic developers who work in similar cities now rely upon individuals who can act as partner intermediaries. One eye prowls outside the city for prospective new enterprises, while the other squarely focuses on development opportunities and property changes within the city.

But how do artists and cultural planners in small post-industrial cities vision change in their old or adopted cities? Do they share the visions described above or depart from them? If they are not long-term residents of the city, what attracted them and what will keep them there? From my earlier survey, it seems that what attracts creative people in the first place is also what they seek to build upon in the future. Artists can be attracted by the general physical setting and the grittiness, as well as by specific elements of the built environment from an industrial past (e.g. old mills, derelict trains and their abandoned stations, historic architecture, and so forth) – the very things that planners think negatively impact their city's image and are deterrents to renewal. Describing his attraction to Bellows Falls in the early 1980s when he came up from McDougall Street in New York to attend a dinner party, Robert McBride said,

> The great thing that initially attracted me to BF was … the train and those train bridges and the mountain and the dense little community and even though it was worn, it wasn't abject poverty depressing …. you could just feel all of the potential. I mean you know BF is like Rome to me. Everywhere you look there is a collage of architecture, sky and something else. And we weren't a big enough town to have all of that urban renewal crap to have someone come in and say well we're going to level these 12 blocks … So it was kind of almost like a place that just needed to be really dusted off and cranked up again.

Jerry Beck, an artist who moved from Boston to Lowell, described a similar attraction to the physical plant as well as the welcoming atmosphere.

> I've always loved Lowell, I mean you can see the beauty of it and the canals, and I love industrial spaces and the river and, you know, the light off the building, and the people were different than the people from Boston, definitely. It seemed like people were much more friendly here and it's different, there's a different sensibility. They were very welcoming here.

As Charlie Hunter, an artist and entertainment promoter from Bellows Falls, explained, artists either want "cute farm houses in the country" or Tribeca-like old mill towns that are somewhat gritty and "look like a place". Artists, he said, "want to live in places that look like places" and are affordable. According to Beck, who is the former director of Lowell's Revolving Museum, the appeal of raw spaces also comes from the particular way that creative people experience their environments. They derive pleasure from envisioning ways that space can be transformed, and even though the reasons behind that joy can differ from planners and developers, the expression of this appeal can sometimes make them sound a lot like them. "Artists", says Beck,

> … are drawn to the quality of that space for a reason. I don't think there's an artist I know that wouldn't mind just looking at spaces just to fantasize. I still

> do that. If I had more money now, I'd probably be buying buildings and being a developer. And I'm sure many artists would love that idea because they're in love with the spaces and the building and the aesthetic and the industrial revolution. And the idea of transforming it is an artwork and this is like a diamond.

Besides being attracted to the idea of turning old industrial spaces into work and living spaces, artist's visions include the possibility of adding more visual stimulation to the public landscape in the form of both temporary and permanent public spaces and public art.

At the same time, other artists I spoke with said that they did not want to see hundreds of new cookie cutter galleries and craft shops displace local pharmacies, bookstores, hardware supply shops and small groceries, though they would not mind a few more clothing shops and cafes. It's less the *type* of business or enterprise, some artists said, than the attitudes of the proprietors. For example, Charlie Hunter was put off by "do-gooders" who come to Bellows Falls thinking they know what is best for the city.

> You know like "Harvey should have a food coop so that …" and people want to have a farmer's market so that … mothers will buy healthy foods. It's not going to happen. It's not going to happen in five years anyway … And the people who so approve of what I am doing, um, because it is going to improve the musical taste of, you know, like the people who are house painters. Well the house painters like to listen to classical rock and it is contemptuous to say, "Oh, you shouldn't be listening to that. You should be listening to, you know, my folk music." I hate that …. it's so patronizing … I like almost everybody in town.

The perceived social diversity of old industrial cities is something that attracts many artists to smaller industrial cities and is incorporated into their visions of the future. This is true even if it means living amongst people with whom you share little in common. Sterility is repugnant to many new residents, and, according to McBride, a little tolerance can go a long way.

> you know, it's like talking about families. I mean you got the horrible uncles, the horrible aunts. I mean there is sort of a DNA. If there's life there's life. I mean I don't know if I could survive in an outright horrible suburb. Yeah I might not be able to.

Hunter adds that, "the best of all possible worlds would be to be a good functioning downtown, you know, just a good functioning town with a range of people from poor to well off."

The embrace of cultural and social diversity in artists' visions for revitalized smaller cities stems from more than an interest in avoiding sterility and homogeneity. Since a number of older industrial cities have large immigrant populations, many see the arts playing a role in unifying people that are increasingly drawn from different backgrounds. "We are despaired individuals", Daniel Malloy, the former Mayor of Stamford, Connecticut, now Governor, said at a cultural economy conference. "We don't concentrate on what brings us together; we focus on what separates us .... The arts can [help us] shape a common culture." In a similar vein, Jerry Beck, in Lowell, expressed the desire to utilize his talent to work with children and youth in empowering ways that awaken their latent creativity and inspire hope. He sought to generate a new paradigm to supplant the individual artist's inwardly directed work with more collaborative projects. The "Revolving Museum", an interactive, participatory, constantly changing public art program, now works directly with schools and after-school programs to facilitate such work. The broad intention behind this vision of the intersection of art and urban development is to counter the isolation of artists and integrate them into the community at large, while also incorporating creative activity into everyday life and space. Part of Beck's paradigm also includes an alternative economic development model that incorporates art making into every city department, and seeks to alter the curricula of trade schools to enable young people to learn their skills through apprenticeships with the city that include making a contribution to improving the built environment. The clear focus of all of these artist-driven visions for urban change in smaller post-industrial settings is on identifying potential assets, including relevant history, *from the inside*, while also seeking to attract new creative talent *from the outside*.

Have Vision, Will Travel

One might ask where these visions for culture-driven revitalization in small and medium-sized industrial cities come from? How do places that have lived with notoriously negative images, anachronistic economies, and numerous sites of industrial decay, come to believe that at least a part of their economic recovery depends upon something as elusive (or material) as culture, art and other forms of creative enterprise?

It turns out, not surprisingly, that the desired futures that people construct depend heavily on past individual and collective environmental autobiographies (Breitbart and Stanton 2006). All of the good and bad living situations that planners, developers and artists have either observed or experienced directly help to construct a sense of current possibility. For economic developers, realtors and planners, the failure of past revitalization strategies is a key factor motivating them to take the city in a dramatically new direction. In New England, all of the older former industrial Gateway cities experienced deindustrialization and a range of failed attempts to lure industry back through the construction of industrial parks, the replacement of manufacturing with centrally located suburban-like retail shopping outlets, and similar projects. The outspoken (former) Mayor Barrett of North Adams recounted his surprise when first approached with the idea of transforming a large complex of abandoned factory buildings into a museum of contemporary art. While he proudly admitted that he would never "cross the street to see the exhibits", he nevertheless acknowledged his willingness to go along with the idea for the lack of other alternatives and a 14% unemployment in the town. But how, he wondered, could he sell this idea of art as economic development to his working class constituency? That was the challenge. Barrett saw art as the "only game in town", and as such, it had appeal.

Many of the artists I spoke with have lived in a gritty warehouse district of a large city and unwittingly become part of an irreversible gentrification process. Living communally amongst other artists can be an invigorating experience in which even the frequent moves contribute positively to the creative process. But, as artists pointed out, moving can wear thin over time, especially when your art consists of large sculptures that are difficult to transport. Others simply become tired of living the life of a nomad. Some artists become landlords themselves along the way in order to survive, as Jerry Beck did when he rented a very large warehouse along the Boston waterfront and then sublet it to other artists. His vision of artists owning their own spaces derive from both his own displacement experiences and the difficulties he subsequently encountered when he tried to acquire some stability on the backs of other artists.

Artists and cultural planners readily supply examples of the places they do *not* want to become. They seem to share a uniform antipathy to suburban-like environments, fast food, and dead downtowns. While some express an attraction to gritty industrial architecture and interesting public spaces, many of those living in smaller post-industrial communities also seem to appreciate the everyday-ness of their environments. Contrasting Bellows Falls to more trendy small cities, Charlie Hunter suggests that,

> Northampton is just turbocharged. What's a really useless [place full of artists]? New Hope, [Pennsylvania]? Some little completely artsy town where if like the supply routes were cut off everyone would starve within once the food was gone from restaurants.

When asked if he thought Bellows Falls might one day "take off" like other arts towns, his response was a definitive "I hope not!"

> I don't want it to. If I can put myself among the movers and shakers, that none of us want it to take off … I think there's a nice meeting of the minds of all of these different communities, of the local business people and the person in the local bank branch, I mean he grew up in town. And Bill at the hardware store. I mean we're all on committees together so we all want it to be basically a town in which arts are a component as opposed to an arts town. ….

When Hunter gets together with other residents at the local diner and engages in discussion about what is "missing" from town they focus on basic shopping needs rather than cultural events or the need for more galleries.

> I mean the thing I like about downtown Bellows Falls is that when we sit around and help our great aunt brainstorm about what we need downtown, I mean it would be great to have a women's clothing store. You know. One of the disappointments was the Newbury//Millbury building that I bought with two other people and then realized that … we were buying commercial real estate and things I don't know anything about. I mean it would have been great, I would have loved it if a woman's clothing store had gone in. Instead, it's going to be a Pretzel Company.

As this comment illustrates, many artists are aware of the role they play in often unwittingly changing the place they were attracted to in the first place. Half jokingly, Hunter expressed this well.

> You always ruin whatever it is that you love anyway. No. You make a nice nest, the other birds are going to find out. Because there are a lot of people without a lot of imagination, but with money, and in America everyone is so sad and wants a place that means something, they're just going to descend like locusts.

When I asked him what, if anything, would ever lead him to consider moving away from his adopted home of Bellows Falls, his response was unhesitating and centered on his antipathy toward homogeneity.

> Well if horrible yuppies came in then I would realize that it had succeeded and that my vision and worst nightmare had come together in one horrible moment. Um … I think I have two arguments in my head. One is that it is inevitable that Bellows Falls (BF) is going to [gentrify] …. I mean the best of all possible worlds would be to be a good functioning downtown, you know, just a good functioning town with a range of people from poor to well off. The worst that would happen is if it homogenized in one direction or another.

Hunter has already lived in places that have gone through artsy transformations and is thus aware of how the appeal of smaller post-industrial towns can become a double-edged sword.

Many of the visions expressed by developers for a new cultural economy in other post-industrial cities are less nuanced and uncritical, as they reference examples from larger cities. In Lowell, Massachusetts, developer Fred Faust was quick to admit the direct influence on his work of examples of arts-driven revitalization from other cities.

> The arts district really came from looking at what was going on in Providence, [Rhode Island]. The original arts development that we did when I was on the Preservation Commission, which is at Market Mills Artists Working Space, came about because when I was living in Alexandria, [Virginia] I saw the Torpedo factory down there, and we were trying to create something similar here. And certainly we've gone to a lot of other places. The Preservation Commission used

> to make field trips. We went to Savannah and San Antonio and other places like Portsmouth, New Hampshire that have had their successes and that's stimulated ideas and different ways to do things … No one should be that proud that they can't try and take other formulas for success.

Mayor Barrett of North Adams mentioned Soho as a positive example of change, while Robert McBride referenced Soho and Greenwich Village in New York City, as well as the Bay Area of San Francisco. However, these examples were not universally seen as places to emulate. On occasion, such places were mentioned as sites from which to learn how *not* to use the arts as a vehicle for transformation. Just as Charlie Hunter drew on his observations of Tribeca in New York, and a number of booming "artsy towns" such as Hudson, New York and Brattleboro, Vermont, in order to explain what he would like to *avoid* in Bellows Falls, Jerry Beck's travels across the country gave him reason for pause.

> Look what happened to every arts community … especially with the Internet boom and icoms … I had friends in San Francisco. I used to look up to their alternative program there and the spaces, and I was like blown away …. when I found that that whole community was devastated, and what was happening in Boston and New York and is continually happening. It's very clear. If you didn't own the building you're gone. It's as simple as it gets. It's the old cliché and you know it as well as anybody. So artists move in to these forgotten spaces and gentrification just follows.

The most common places from which people draw specific ideas are closer to home – other small cities that have industrial pasts similar to theirs. Some openly admit to importing ideas about culture-driven development directly from their neighbors. Mayor Barrett, for example, said he did not need "high-priced consultants" when he could "steal" ideas from other Massachusetts cities.[6] Cultural intermediaries,

6 Both Peter Aucella and Fred Faust in Lowell acknowledged Providence, Rhode Island as the inspiration for their Arts Overlay District and live/work zoning ordinance, even as they proclaimed their city to be the first in the state to develop a true cultural plan.

> "Four or five years ago people started looking at Providence to see what they were doing. …. Lots of new things coming here, so someone said, 'We have a plan for development of the city. Why don't we have a plan for cultural development?' So a business group in city leadership went to the Massachusetts Cultural Council and said they wanted to do a cultural plan, and the Council was beside itself. Lowell was first city to be funded by MCC to do a cultural plan. This was, I want to say, 1984 or 85. I think ten years later, there was an update of the plan and they documented everything that had happened. Mind-blowing amounts of stuff. Just amazing. That was all before city administration went to Providence and saw they had artist living spaces, got ideas like we could add excitement to the downtown. Went down, came back and basically cannibalized the Providence live-work city zoning ordinance."

regional conferences, and local and state arts councils also play an important role in disseminating ideas about culture-led regeneration to post-industrial cities, as do research reports that measure the economic impacts of the creative economy.[7] Conferences that unveil the findings from these reports open a communication corridor between cultural advocates and local policymakers. Examples of successful projects diffuse rapidly across the region and gradually move into the planning lexicons of city planners. This accounts for the low-key competition that allows Lowell to claim to be the first small New England city to design a cultural strategy, and Worcester to claim to be the first to create a new position of Cultural Development Officer.

Introducing the Case Studies

The case studies that follow focus to some extent on whether and how post-industrial cities and neighborhoods incorporate the same ideas about culture-led regeneration and creative industries into their planning as larger cities, or modify these templates to better suit their needs. All of the authors maintain a critical edge and their work uncovers some examples that depart from mainstream creative economy practice. Such critiques of mainstream approaches require a focus on the structural barriers, and unequal distributions of power, that make the search for viable urban development alternatives especially difficult for smaller cities, and risk derailing even creative, well-thought out grassroots initiatives.

While acknowledging these obstacles, there is a shared desire among authors to move beyond critique and to focus on how the growing economy surrounding culture, the arts, design, media and technology can be harnessed and transformed to best benefit post-industrial cities and improve the quality of life for residents. How, for example, might particular forms of citizen-involved planning generate new and more meaningful economic opportunities for people who are *not* currently

Holyoke's former Director of Planning traveled in the 1990s to New Bedford, Worcester and Providence before promoting an Arts/Industry District in his downtown, while New Bedford's Mayor admitted that the idea of AHA! (Arts, History and Architecture) came directly from Providence, Rhode Island's gallery night. Not to be upstaged by its neighbors, Worcester, Massachusetts claimed to originate ideas that Lowell and Providence had never even conceived of – arts-based development that goes beyond the downtown to incorporate neighborhoods – while Providence claimed that its Performing Arts Center was the inspiration for Worcester's proposed Center for the Performing Arts. And so it goes – a lot of backward and forward sharing of models that are themselves in a constant state of flux, looking for the next good idea to import.

7 The "Creative Economy Initiative: A Blueprint for Investment in New England's Creative Economy", released in June 2000, became the focus of widespread discussion. This report was commissioned to provide dramatic and convincing quantitative evidence to support the wide impact of culture and the arts on the region's economy, and to entice private business people and politicians to think more seriously about the arts.

defined as members of the "creative class" and yet value the aesthetic and material experience of the city? In response to this question, some contributors examine alternative definitions of "creative" in the context of bottom-up initiatives that emerge from within everyday neighborhood spaces. These activities are often ignored in the burgeoning literature on cultural economic planning, and even in radical critiques. Many of the individuals and activities highlighted in this book remain invisible to on-the-ground creative economy strategists and local planners and politicians. Greater awareness of the obstacles this invisibility creates for those working to provide new economic opportunities to a more racially and class diverse population is crucial.

Most of the cities featured in this book were key centers of innovation in their regions during the early days of industrialization and into the 20th century. While they are typically dismissed now as potential sites of creativity or future economic viability, the case studies provide a window into the struggles and occasional triumphs that accompany efforts to reverse economic decline. In general, these examples do not dwell on policies that are designed primarily to import a new young "creative class" or encourage cultural tourism, though many local planners in these post-industrial cities would hardly object to this goal. Case studies focus more on efforts to invite the wider participation of residents, and to cultivate new economic and social relationships. Some examples go further than others in promoting inclusive and collaborative experiments in creative and culture-based economic development. In general, however, creativity and talent are defined broadly, and value is ascribed to both the economic and non-monetary outcomes of cultural and creative activities. Authors also discuss the obstacles that affect implementation, and they consider wide-ranging definitions and likelihoods of success.

The chapters in *Creative Economies in Post-Industrial Cities: Manufacturing a (Different) Scene* divide imperfectly into two sections that address similar questions yet vary in emphasis. The empirical case studies in Part I, *Exploring alternatives to the "cultural buzz"*, document how policies and practices related to the arts, culture and creative knowledge-based industries are evolving in a number of small- to medium-sized post-industrial communities in the U.S., the UK and Canada. These are often part of more comprehensive approaches to regeneration. The aim of exploring the intersection between a cultural and sustainable economy necessarily provokes questions that get at the heart of how urban regeneration is approached and how neighborhood needs and assets are acknowledged and addressed. Together the chapters provide an emerging picture of how mainstream creative economy practices are transformed and applied to tackle some of the social and economic challenges faced by residents. No claim is made regarding the comprehensiveness of the examples or their ability to be generalized. Emphasis is placed instead on using the case studies to raise new questions and re-direct future research. Though some recurring themes are explored in the Afterword, examples vary from one post-industrial setting to another almost as much as they vary together from what transpires in many larger urban settings.

Part I begins with Providence, Rhode Island, an early industrial powerhouse in New England. In this chapter, Salkind describes how various cultural institutions, including non-profits and preservation coalitions, have played a role in shaping the cultural economy of the city and public policies to promote it. Taking us chronologically through a series of efforts to revitalize the downtown, Salkind documents the important and conflicted role of historic preservation in promoting both for-profit and non-profit development. He examines changes to the built environment in downtown, including how they leverage cultural resources to create a new enticing artsy "scene". We see the occasional convergence of roles played by non-profits, city agencies, and artists, who are themselves real estate *developers* and co-constructors of public policy. Noteworthy politicians such as Mayor "Buddy" Cianci, later convicted of a felony, but who many credit with sparking the city's revival as a cultural destination, are introduced as key players in the Providence Renaissance. To illustrate the contradictory elements in this evolving creative economy, Salkind explores the underground, anti-capitalist art practices that contributed to the development of Providence's post-industrial neighborhood economies in relation to the desire of non-artist residents for more traditional forms of real estate development. Salkind also contrasts policies and practices promoted by the city with the creative enterprises launched by more ad hoc and informal social networks. This allows him to raise important questions about the relationship of property ownership, non-profits and sustainable cultural development.

The focus of the second case study in Chapter 3 is Holyoke, Massachusetts, the first planned industrial city in the United States. Here Breitbart tells a story of two parallel, on-going approaches to urban regeneration, both of which incorporate cultural planning and seek to promote creative entrepreneurship in some fashion. One strand follows more mainstream "big city" practices. Special districts for the arts and innovation are designated, and there is a growing focus directed at technology-based and artist-driven development to improve the tax base and attract new economic activity and middle class residents. The other is a more grassroots approach that seeks to utilize the city's established ethnic diversity to draw attention to, and build upon, the talents and aspirations of existing residents, a near majority of whom are of Latino descent. Holyoke is described as being on the verge of major change, having just elected a young Mayor who is a strong proponent of the creative economy and entrepreneurial growth. The chapter raises questions concerning how the serious educational, economic, and general quality of life issues currently affecting Holyoke's existing residents will be addressed within or alongside this new agenda as it evolves.

In Chapter 4, Bain and McLean contrast the long history of grassroots arts mobilization in artist-run centers in the post-industrial cities of Thunder Bay and Peterborough, Ontario, to the approaches to cultural planning espoused by current gurus such as Richard Florida. The authors argue that eclectic creative practices and cultural community building through informal social networks are under appreciated resources that have the potential to draw more people into creative practice, and help post-industrial cities improve the economy and the local quality of life. Bain and McLean examine how these more informal social networks foster local creativity and challenge top-down models of urban development, including pre-fabricated toolkits that promote cultural tourism and city branding. They describe the appeal of smaller post-industrial cities to many cultural workers who appreciate apprenticeship models of learning, and wish to escape the constraints of disciplinary boundaries and high barriers to cultural sector participation. Much of this chapter focuses on how less structured and more experimental artistic practices and struggles over postering as a method of community outreach, contrast to more mainstream and, they would argue, less democratic and inclusive cultural development models.

Chapter 5 moves to the North of England and examines changes in the collective production of art through different political eras and ideological shifts. Starting with the 1980s and moving to the present, Gee describes a chronology of different approaches to economic development on the part of labor, liberal and conservative governments that all in some way recognize the growing importance of the arts and a service economy. He examines, in particular, how more conservative and neoliberal approaches incorporate business models into their top-down management of public funding for the arts, and how local arts associations in post-industrial communities react to such changes by attempting to resist commodification through collective organizing and efforts to draw upon the region's economic and social history in the content of their work. Gee describes the critical content of much of the art as a form of regional opposition to an evolving national agenda that is increasingly seen to promote individualistic entrepreneurial behavior and the privatization of public space. This national agenda is contrasted to some of the more collaborative exchanges and social change-focused artistic work found in many post-industrial communities in the North.

Chapter 6 focuses on Barnsley, England, once the center of the mining industry and the bitter miner's strike of 1984. Like all of the post-industrial cities in this book, Barnsley suffers the effects of job loss and faces massive social and economic challenges. These are not helped by negative media images that inhabit even the minds of many residents. Breitbart describes the role played by a regional agency, Yorkshire Forward, in trying to reverse decline through the promotion of a planning process designed to spark creative ideas and aspirations that enable the community to invent a new and more sustainable economic future. The process of working with well-known architect Will Alsop provokes questions about just how far a town rooted in a very different economic past, and facing such overwhelming health and educational challenges, can convert itself into a 21st-century market and incubator of creative industry. In the face of the current recession, Breitbart examines how changes to the built environment, attention to quality of life issues, and a continued focus on the cultivation of internal assets, undergird even modest success.

The focus of *Part II: Moving beyond Neoliberal Methodologies in the Study and Practice of Creative Economy Planning* is the question of how a city that aspires to become more economically sustainable can engage planning *as a creative process* that enables a broader spectrum of the population to develop its productive potential. This section draws on theoretical perspectives such as participatory design, institutional ethnography and asset-based planning to expand current approaches to urban cultural regeneration research and promote more experimental methods for involving residents in visioning and enacting change. Several chapters challenge highly constrained top-down methodologies currently used to determine what will be regenerated in a city, and how and where this will be accomplished. While these chapters touch upon many of the topics addressed in Part I, they highlight new approaches to both studying the creative economy in smaller places, and to examining or promoting local impacts. Ideas and practices that emerge from these cases suggest the potential of an alternative definition, and crafting, of a local creative economy. These include more inclusive spatial visioning processes that have the potential to cultivate new cultural and economic opportunities for a wider spectrum of creative actors. Taken together with the ideas that emerge from the examples in Part I, the approaches to planning and research in Part II seek to place residents squarely in the planning process and provide a foundation for more relevant and sustainable creative economic practices.

In Chapter 7, Osorio provides a glimpse into the evolving creative economy of Easthampton, a small mill town in western Massachusetts, once noted for its buttons, elastic, home products and brushes. Utilizing social network analysis and "institutional ethnography", a sociological method that examines the context within which everyday social relations evolve, Osorio seeks to complicate and challenge the ideas put forth by Richard Florida about the composition and mindset of a "creative class". Tracing the evolution of two different factory spaces in the city that have attracted artists and a variety of creative activities, he argues that the creative class is not a reductive index of particular attributes, but rather, a heterogeneous mix of people with divergent goals, expectations and perspectives that is continuously evolving. Osorio shows how creative class differences, such as seeing art as a business versus seeing it as social commentary, are manifest in very different and changing physical venues. Most importantly, he makes the case that the creation of a local creative class is itself a co-constructive activity, as much dependent on what is going on in the town as what transpires among creative individuals. The chapter examines how artists are influenced by this unique environment and, in turn, try to exert influence on it. Osorio also describes how individuals with talent extend their skills through collaborative work and shared space that is enhanced by the small size of the city.

In Chapter 8, Fitzpatrick employs Bakhtin's socio-linguistic concept of "dialogic communication" to theorize about creativity as a contested concept and practice. The case study examines the responses of different interest groups in Liverpool, England to the community involvement dimension of the European Capital of Culture, a year-long event hosted by the city in 2008. As a methodology, dialogism allows Fitzpatrick to explore the impacts of this major event among different constituencies that constitute an already existing public sphere. Through in-depth conversations with many cultural players, she unravels and brings to light the multiple forms of cultural expression that already existed within the cultural fabric of Liverpool, and yet were ignored in this event-led regeneration process. Debates about participation, uneven development and cultural policy provoked by the Capital of Culture events are explored through an analysis of the language of event promoters. Here we see how certain assumptions were made about Liverpool as a post-industrial city in decline, and how this conditioned who was given a voice in the events accompanying the Capital of Culture designation. Fitzpatrick draws many lessons from this research, including the observation that existing "spaces of creativity" are important multi-faceted sites of learning and experimentation that deserve greater support.

In contrast to the top-down cultural planning that Fitzpatrick describes in Liverpool, Chapter 9 focuses on Newham, an industrial suburb of London, where an unlikely cohort of elderly residents decide to use performance art and visual media to challenge their own marginalization in the neighborhood regeneration process. Newham is known for its high levels of deprivation, and was the target of many regeneration schemes in the past. Handler describes some of the ways that such schemes privilege youth-centric planning and foster stereotypes of elderly residents as passive victims and uninterested bystanders. She argues that while some older residents once worked in the defunct factories, existing plans assume them to be disconnected from, and irrelevant to, its future. In response, Handler describes how a creative mapping exercise produces a space for elderly residents to reflect critically on their loss of access to public space. She then explores a fascinating array of creative interventions and direct actions that are initiated by older residents attending a weekly dance class, in order to reclaim space and insert their voices into the regeneration process. These creative occupations of space, and the design of artifacts to bring more comfort to city space, unsettle stereotypes and reveal the hidden spatial politics of creativity-based regeneration schemes. These conscious and creative interventions, while reminiscent of the more youth-focused resistance of the early Situationists, or even the recent Occupy movement, not only provoke dialogue, they suggest how we might use creative means to think more broadly and inclusively about neighborhood renewal.

Chapter 10 details the philosophy and creative design practices of muf, a feminist architecture/art firm based in London. The examples described here in brief sound bites by Breitbart and Fior, emphasize the importance of coming to know a place through a "close looking" and detailing of existing assets. This process always includes stakeholders, and the outcomes often disrupt already existing master plans, substituting alternative interventions designed to expand resident access to public space and improve the quality of life in communities undergoing regeneration. Many of muf's interventions are temporary in nature and susceptible to change over time. Since the chapter originated as a conversation, and emerged from a series of interviews and communications over time, key themes form the basis for a brief exploration of the way muf works with post-industrial communities. We see how collaborations with artists and landscape designers are arranged and expand our understanding of more sustainable planning and design processes.

In the final chapter, Breitbart summarizes how post-industrial cities use different forms of cultural production and creative economic practice to develop new economic opportunities and an improved quality of life for residents. The chapter critiques the uncritical adoption of creative economy practices that promote spatial inequities, and directs attention instead toward the creative process involved in cultivating citizen participation and the use of local knowledge in planning. Bianchini's concept of "sustainable creativity" is employed as a framework for exploring more socially just policies and practices that bring sustainable forms of urban regeneration together with the goal of social inclusion. This, it is argued, involves an effort to include many diverse publics in the planning process, and to identify, place value on, and increase the visibility of broadly defined cultural assets. It also includes efforts to build the economic and income-generating capacity of residents through the provision of access to training in growing creative and productive sectors of the economy. Examples of innovative educational, cultural, workforce, and neighborhood development strategies are described to support this approach.

References

Amin, A. and Thrift, N. (2004), *The Cultural Economy Reader* (Oxford: Blackwell).

Atkinson, R. and Bridge, G. (eds) (2005), *Gentrification in a Global Context: The New Urban Colonialism* (London: Routledge).

Bell, D. and Jayne, M. (eds) (2004), *City of Quarters* (London: Ashgate).

Bell, D. and Jayne, M. (eds) (2006), *Small Cities: Urban Experience beyond the Metropolis* (Abingdon: Routledge).

Bianchini, F. (1993), "Remaking European cities: the role of cultural policies", in F. Bianchini and M. Parkinson (eds), *Cultural Policy and Urban Regeneration: The Western European Experience* (Manchester: Manchester University Press).

Bianchini, F. (1999), "Cultural planning for urban sustainability", in L. Nystrom and C. Fudge (eds), *Culture and Cities: Cultural Processes and Urban Sustainability* (Stockholm: The Swedish Urban Development Council).

Bianchini, F. and Ghilardi, L. (2007), "Thinking culturally about place", *Place Branding and Public Diplomacy* 3:4, 1–9.

Bianchini, F. and Parkinson, M. (eds) (1993), *Cultural Policy and Urban Regeneration: The Western European Experience* (Manchester: Manchester University Press).

Breitbart, M. and Stanton, C. (2007), "Touring templates: cultural workers and regeneration in small New England cities", in M. Smith (ed.), *Tourism, Culture and Regeneration* (Oxfordshire: CABI).

Brown, J. (2003), "Remaking Easthampton: this former mill town is creating a new image", *BusinessWest*, May.

Christophers, B. (2007), "Enframing creativity: power, geographical knowledges and the media economy", *Transactions of the Institute of British Geographers* 32, 235–47.
Daly, A. (2004), "Richard Florida's high-class glasses", October. Available at www.artfactories.net/IMG/pdf/Richard_Florida_by_Ann_Daly.pdf [accessed 10 April 2013].
Davis, M. (1990), *City of Quartz: Excavating the Future in Los Angeles* (London: Verso).
Deutsche, R. (1996), *Evictions* (Cambridge: MIT Press).
Dutton, P. (2005), "Outside the metropole: gentrification in provincial cities or provincial gentrification", in R. Atkinson and G. Bridge (eds), *Gentrification in a Global Context: The New Urban Colonialism* (London: Routledge).
Edensor, T., Leslie, D., Millington, S. and Rantisi, N. (eds) (2009), *Spaces of Vernacular Creativity: Rethinking the Cultural Economy* (London: Routledge).
Evans, G. (2001), *Cultural Planning: An Urban Renaissance?* (London: Routledge).
Evans, G. (2009), "Creative cities, creative spaces and urban policy", *Urban Studies* 46:5/6, 1003–40.
Evans, G. (2010), "Creative spaces and the art of urban living", in T. Edensor, D. Leslie, S. Millington and N. Rantisi (eds), *Spaces of Vernacular Creativity: Rethinking the Cultural Economy* (London: Routledge).
Florida, R. (2002), *The Rise of the Creative Class* (NY: Basic Books).
Florida, R. (2004), "The great creative class debate: revenge of the squelchers", *Next American City*, July. Available at http://programmingz.com/b/4566025/Revenge-Of-The-Squelchers-Creative-Class-Group/#.UWYhR89CcTA [accessed 10 April 2013].
Florida, R. (2005a), *The Rise of the Creative Class* (New York: Routledge).
Florida, R. (2005b), *The Flight of the Creative Class* (New York: Harper Business).
Foord, J. (2008), "Strategies for creative industries: an international review", *Creative Industries Journal* 1:2, 91–113.
García, B. (2004), "Cultural policy and urban regeneration in Western European cities: lessons for experience, prospects for the future", *Local Economy* 19:4, 312–26.
Goodman, R. (1979), *The Last Entrepreneurs* (New York: Simon & Schuster).
Grodach, C. and Loukaitou-Sideris, A. (2007), "Cultural development strategies and urban revitalization", *International Journal of Cultural Policy* 13:4, 349–70.
Hall, P. (1998), *Cities in Civilization* (New York: Pantheon).
Harrison, B. (1982), "Rationalization, restructuring and industrial re-organization in older regions: the economic transformation of New England", Working Paper #72, Joint Center for Urban Studies of MIT and Harvard University.
Harvey, D. (2005), *A Brief History of Neoliberalism* (Oxford: Oxford University Press).

Holcomb, B. (1993), "Revisioning place: de- and re-constructing the image of the industrial city", in G. Kearns and C. Philo (eds), *Selling Places* (London: Pergamon).
Johansson, T. and Sernhede, O. (eds) (2002), *Lifestyle, Desire and Politics: Contemporary Identities* (Gothenberg: Centre for Cultural Studies).
Kearns, G. and Philo, C. (eds) (1993), *Selling Places* (London: Pergamon).
Kotkin, J. and Siegel, F. (2004), "Too much froth: the latte quotient is a bad strategy for building middle-class cities", *DLC Blueprint Magazine*, January. Available at halliejones.com/Resources/TooMuchFroth.pdf [accessed 10 April 2013].
Landry, C. (2006), *The Art of City Making* (London: Routledge).
Landry, C. (2008a), *A Toolkit for Urban Innovators* (London: Routledge).
Landry, C. and Wood, P. (2008b), *The Intercultural City*. (London: Routledge).
Landry, C. and Bianchini, F. (1995), *The Creative City* (London: DEMOS).
Lin, J. and Mele, C. (eds) *The Urban Sociology Reader* (London: Routledge).
MacLeod, G. (2002), "From urban entrepreneurship to a 'ravanchist city'? On the spatial injustices of Glasgow's renaissance", *Antipode* 34, 602–24.
Malanga, S. (2004), "The curse of the creative class", *City Journal*, Winter. Available at http://www.city-journal.org/html/14_1_the_curse.html [accessed 10 April 2013].
Maliszewski, P. (2004), "Flexibility and its discontents", *The Baffler* 16, 69–79.
Markusen, A. and Gadwa, A. (2010), "Arts and culture in urban or regional planning: a review and research agenda", *Journal of Planning Education and Research* 29:3, 379–91.
Mayor, A. (1997), *Bellows Falls* (Portland, Oregon: Gere Donovan Press).
McRobbie, A. (2003), "Everyone is creative: artists as pioneers of the new economy?" Available at http://www.k3000.ch/becreative/texts/text_5.html [accessed 10 April 2013].
Mt. Auburn Associates (2000), *The Creative Economy Initiative: The Role of the Arts and Culture in New England's Economic Competitiveness* (Boston: New England Council).
New England Council. (2000), *Creative Economy Initiative: The Role of Arts and Culture in New England's Economic Competitiveness* (Boston, MA).
Nystrom, L. and Fudge, C. (eds) (1999), *Culture and Cities: Cultural Processes and Urban Sustainability* (Stockholm: The Swedish Urban Development Council).
O'Connor, J. (2002), "Public and private in the cultural industries", in T. Johansson and O. Sernhede (eds), *Lifestyle, Desire and Politics: Contemporary Identities* (Gothenberg: Centre for Cultural Studies).
Peck, J. (2005), "Struggling with the creative class", *International Journal of Urban and Regional Research* 29, 740–70.
Scott, A.J. (2000), *The Cultural Economy of Cities* (London: Sage).
Shaw, W. (2006), "Sydney's SoHo syndrome: loft living in the urbane city", *Cultural Geographies* 13, 182–206.

Shea, C. (2004), "The road to riches?", *Boston Globe*, February 29. Available at http://www.boston.com/news/globe/ideas/articles/2004/02/29/the_road_to_riches/ [accessed 10 September 2010].
Smith, M. (ed.) (2007), *Tourism, Culture and Regeneration* (Oxfordshire: CABI).
Smith, N. (1996), *The New Urban Frontier: Gentrification and the Revanchist City* (NY: Routledge).
van Heur, B. (2010), "Small cities and the geographical bias of creative industries research and policy", *Journal of Policy Research in Tourism, Leisure & Events* 2:2, 189–92.
Zukin, S. (1982), *Loft Living: Capital and Culture in Urban Change* (Baltimore: Johns Hopkins).
Zukin, S. (1995), *The Cultures of Cities* (Cambridge: Blackwell).
Zukin, S. (2005), "Whose culture? Whose city?", in J. Lin and C. Mele (eds), *The Urban Sociology Reader* (London: Routledge), 282–9.

PART I
Exploring Alternatives to the Cultural "Buzz"

Chapter 2
Scale, Sociality and Serendipity in Providence, Rhode Island's Post-Industrial Renaissance

Micah Salkind

Introduction

When culture workers in Providence, Rhode Island celebrate their participation in a progressive political legacy, they often invoke the City's founder and favored counter-cultural ancestor, Roger Williams, a dissident rogue barred from the Massachusetts Bay Colony in 1636. As former mayor David Cicilline put it in his 2003 inaugural address, the City's founders provide a cultural compass for the City: "Those of us who live and work in Providence feel a grateful connection to our seventeenth century founding fathers and mothers … [we] owe them a debt for their visionary contribution to the principles of religious tolerance, free speech, and the right to self-determination" (Cicilline 2003a).

Providence's economy has flourished in part because its leaders, beginning with Williams, have dared to do things differently. Whether they were bucking Puritanical fervor, countering the faulty logic of urban renewal, or organizing on behalf of artists creating work in forgotten industrial spaces, the city's bureaucrats, arts and preservation professionals and business leaders have worked tirelessly, and at times with cross-purposes, to shape the contemporary cultural consciousness of their city. Some hope that arts and culture, including live music, visual art and crafts, theatre and performance art, will anchor the city's burgeoning tourism industry and foster appreciation for its iconic architecture. Some hope that the "quality of life" hailed by a visible cultural economy is allowing the city to lure and retain an entrepreneurial and innovative labor force (Creative Providence 2009: 7).

Cultural institutions in Providence often do research and development for local policymakers, shaping the work of the city's Department of Art, Culture + Tourism (Ketten 2011). This chapter looks closely at the ways that creative laborers working with non-profit organizations and preservation coalitions have forged symbiotic relationships with city government and each other, strategically embracing the often reductive logic of creative city building to foster citywide civic engagement and more equitable access to art and culture.

While founders and directors of nonprofits often appear as the protagonists of this story, their work must be understood as being representative of complex collectivities and interpersonal networks. Although much of Providence's post-industrial history revolves around top-down economic development in the city's core business district, underground, non-commercial and anti-capitalist art practices have been essential to the development of the City's post-industrial economy. The following case studies demonstrate that Providence is a city of diverse "neighborhood-based creative economies", as opposed to a cultural-economic hub with marginal spokes (Stern and Seifert 2008: 6).

Paying close attention to the ways that local assets such as available space in historic properties, human and capital resources, social networks and geographic density have played into Providence's regeneration, this chapter illuminates a complex post-industrial cultural context, one in which artist activists and policy elites publicly expose the contradictions of creative city building by selectively appropriating its powerful appeal to meet their converging and diverging social needs.

An Industrial Economy, Infrastructure, and Labor Force

In the 18th century, wealthy Rhode Island families were heavily dependent on the slave trade as a major source of capital. Sixty percent of slaving voyages departing from North America sailed from Rhode Island, a staggering number even without taking into account the fact that many continued to depart illegally after 1787, when state law prohibited the trade (Brown University 2006: 10).

In 1789, reacting against what he believed to be his brother John's immoral ongoing investment in the slave trade, abolitionist and Quaker activist Moses Brown founded a textile-manufacturing firm in partnership with his son-in-law, William Almy. By 1790, with the mechanical expertise of Samuel Slater, Brown and Almy built the first textile mill on the Blackstone River powered by a sophisticated version of the British Arkwright system. The industrial textile mill, which produced inexpensive slave clothing from southern cotton, may not have unhitched Rhode Island's economy from slavery, but it sparked nearly a century and a half of economic growth in the state (Woodward 1981: 9).

Providence's topography, with its many waterways and tributaries, made it a perfect site to incubate America's new industrial economy. Just 35 miles north from open ocean, the city was already an important port in the region, which meant that even prior to the growth of rail transport, which would eventually give Providence increased access to markets 40 miles northeast in Boston and 157 miles southwest in New York, local manufacturers had access to a vast shipping network. Between 1790 and 1860 nearly 300 textile mills opened in Rhode Island, populating the heart of the Capitol City along the Moshassuck River, and later along the Woonasquatucket and West Rivers. These massive red brick buildings, and the hard manual labor that would take place inside of them, have become integral to Providence's post-industrial story, just as they were to its industrial one.

By the 1820s the immense profitability of Providence's industrial economy had spun off ancillary growth in machine parts and steam-engine manufacturing, industries that helped make textile-manufacturing processes even more efficient. But this immense growth was tempered by competition from British manufacturers. In 1789, 79 Providence artisans founded the Providence Association of Mechanics and Manufacturers (PAMM), whose stated aims were to encourage the growth of local industry and enforce manufacturing standards. Desperate for a piece of the pie in an increasingly competitive marketplace dominated by British-made goods, PAMM activists organized to support import tariffs and helped lobby Rhode Island officials to ratify the new U.S. constitution (Fink 1981: 4). Just as the built environment of the industrial city would come to characterize its post-industrial identity, the personality of its industrial labor force foreshadowed the rebellious disposition of artists who would live and work in the mills at the end of the 20th century.

As Providence became wealthier and more populous in the second half of the 19th century, an outsized cultural economy began to take shape in its downtown neighborhood. By 1915 there were 13 theatres in downtown Providence alone, in addition to several on Federal Hill and in Olneyville (Brett 1976). By the 1920s Providence's industrial machine, and the leisure economy it supported, began to falter as employers left the state for the South. Between 1929 and 1931 employment in Providence declined by 40% in the textile industry, 47% in the jewelry industry and 38% in the base-metal industry (Fink 1981: 30).

Like other cities along the Eastern seaboard, 1950s Providence experienced intense urban decay as a result of accelerated white flight and deindustrialization. What few remaining textile mills had not already left the state shut their doors, and the worsted goods trade suffered heavy losses to the growing synthetics industry. Between 1950 and 1960 the population of Providence declined 16.6%, the second largest drop experienced by any American city at the time. By 1970 it had dropped by another 13.7% (Leazes and Motte 2004: xxiv).

The local arm of the Federal Urban Renewal Program, The Providence Redevelopment Authority (PRA), attempted to stanch economic decline in the City with little success, in part because there was almost no buy-in from city or state leaders. The demolition of industrial infrastructure and neighborhoods that did take place, primarily around the construction of the I-195 highway in the Jewelry District and in the northern part of the city, stimulated little economic growth, displaced poor residents and cost Providence acres of historic architecture (Fink 1981: 32). Providence's transition from an industrial city to a post-industrial one began in the 1950s when East Side elites, many the descendents of the City's first families, organized in response to the failed urban renewal projects of the PRA and the expansion of Brown University.

Building an Historic Preservation Coalition on College Hill

John Nicholas Brown, a descendent of the University's namesake John and industrial pioneer Moses, called the first meeting of the Providence Preservation Society (PPS) in February of 1956. Under the leadership of Brown and art historian Antoinette Downing, PPS partnered with the City of Providence to solicit $50,000 from the U.S. Department of Housing and Urban Development (HUD). The funds were used to research and publish *College Hill: A Demonstration Study of Historic Area Renewal*. The study, popularly known as the "College Hill plan", became a blueprint for preservation-based residential development in the City (Greenfield 2004: 164).

The College Hill plan recommended that a combination of public and private financing be used to speed the gentrification of North Benefit Street, a section of housing occupied mostly by poor Cape Verdean and African American home owners and renters. The PPS boosters used the plan to build support for the creation of several for-profit neighborhood development corporations designed specifically to raise the overall value of property along the street. PPS members oversaw the demolition of decrepit structures, the importation of properly historic one- and two-family properties from other areas of the city, and the rehabilitation of exteriors, re-selling new and old properties on the street at newly inflated rates.

Many former residents have called the gentrification of Benefit Street a nail in the coffin for the area's Black community. A good deal of public space on College Hill, and its physical connection to what had once been a working waterfront in Fox Point, had already been destroyed to build I-195. Losing access to affordable housing on the edge of Brown campus drove what had once been a small, vibrant neighborhood of African Americans and Cape Verdeans to disburse across the City's outlying metropolitan area. Clifford Monteiro, past president of the Rhode Island NAACP and a former resident of 5 Benefit Street, recalls the pressure put on his family by developer and PPS "pioneer" Roger Brassard:

> Redevelopment panicked black people. It was about fear. They scared my mother, a good, religious person. Roger Brassard was the friendliest folk you ever met. He told my mother that redevelopment has certain standards, was going to make many improvements that we couldn't afford. He said our house wasn't worth $5,600, but he's going to give us the money anyway. My mother bought the line (Rickman 1999: 11).

By focusing on the aesthetic conditions of the built environment and inflated real estate markets at the expense of a marginalized community, the College Hill preservation movement presaged many of the issues that would later arise in the debates over development in the city's industrial Valley and Olneyville neighborhoods. Most importantly, preservation on College Hill helped to foment a coalition of upwardly mobile homeowners, bureaucrats, and urban elites that would be a fixture in the City's future pro-preservation activism.

Increased property values along Benefit Street promised higher tax levies for the city, proving the value of historic preservation as an effective economic development tool.[1] Additionally, the newly polished street was advertised as a tourist destination. The Rhode Island Historical Society leads popular walking tours of Benefit Street throughout the summer months; tours of the neighborhood are now offered as self-guided audio podcasts as well (Rhode Island Historical Society).

By placing historic preservation at the heart of development discourse in Providence, the Benefit Street coalition helped create a pro-growth framework leveraged on cultural assets. Beginning in the late 1970s, Mayor Vincent "Buddy" Cianci, with the support of Senator Claiborne Pell, Governor Lincoln Almond and a shifting network of pro-development preservationists, would capitalize on this emergent framework and drastically reshape Providence's core business district under the banner of The Providence Renaissance.

Branding the Renaissance through Infrastructure and Art

The Providence Renaissance was comprised of eight separate but related construction projects realized between 1981 and 2000 at a cost of about $1.2 billion. Some projects were federally funded, some were funded by the state, and others were privately funded or funded through mixed private and public investment. The $169 million, federally funded Capital Center project, which incorporated the railroad and river relocations; the construction of the new Providence Station rail depot; the construction of a new highway interchange with Interstate 95; and the construction of Waterplace Park, was the Renaissance's economic lynchpin (Leazes and Motte 2004: 73).

The most visually striking, and perhaps the most popular, aspect of Capital Center was the transformation of the city's urban waterways. Architect Bill Warner's *Memorial Boulevard/River Relocation Plan* called for moving the confluence of the Moshassuck and Woonasquatucket Rivers 100 feet to the east and replacing the gigantic expanse of the Crawford Street Bridge covering the majority of the rivers with a series of smaller "historically themed vehicular and pedestrian bridges". The newly constructed bridges, as well as a federally funded park and amphitheatre (Waterplace Park), opened new public space to enjoy vistas of both the river and the historic downtown streetscape.

---

1 Briann Greenfield, a historian of preservation, says that the College Hill plan worked, at least for the preservation community and middle-class homeowners, because it was more inclusive than previous preservation projects. Rather than preserving one property that may not even be inhabitable, developers targeted the entire neighborhood, creating a sense of ownership and accountability to the built environment that could never have been maintained through advocacy around isolated, uninhabitable historic properties (Greenfield 2004: 166).

According to political scientists Francis Leazes and Mark Motte, Cianci's Renaissance project was both a major overhaul of infrastructure, and a major revamping of the Providence brand. To stabilize his big-tent pro-development coalition, and maintain popular support for his costly and disruptive construction projects, the Mayor needed to convince city residents that they were buying into something that hailed them as unique, creative consumers and culture producers – and he needed them to want to be in the city center. An art installation on the Providence Rivers accomplished both tasks (Leazes and Motte 2004: 37, 123).

Incorporated in 1996 by Barnaby Evans, WaterFire is a site-specific sculptural installation – a network of bonfires on metal braziers set like a sparkling necklace across the City's newly recovered downtown rivers. The annual lighting of WaterFire, which occurs on many weekend nights each summer and early Fall, has become a significant tourist attraction and an oft-cited example of the Providence Renaissance's economic and cultural multiplier effects. WaterFire's enormous popularity can be largely attributed to the way it creates a serene, contemplative space in the city center where a huge number of local residents and tourists can convene. The installation also activates Providence's urban built environment by shedding flickering light on the City's uncovered rivers, making them a central part of a local ritual/spectacle. WaterFire at once memorializes Providence's industrial past and hails the City as a site of industry and creativity in the present.

Barnaby Evans' ingenuity and perseverance transformed WaterFire from a one-off installation held during 1994's New Year's Eve First Night celebration into an event that has been successfully restaged year after year with inexhaustible success. Evans is not an isolated genius, however. He is a cultural descendent of the PAMM activists who collectively rallied on behalf of industrial markets, and the College Hill preservationists who banded together to recognize the under-appreciated aesthetic value of their city. According to Lynne McCormack, Director of Providence's Department of Art, Culture + Tourism, Evans marshaled public and private financing to support his own artistic work, but also that of other artists. When Evans and a cohort of Providence artists were evicted from their studios in The Foundry Mill in the late 1980s, Evans took the lead coordinating the financing and political support needed to develop The Regent Avenue Studios, which now houses the offices of WaterFire as well as artist live/work space for many of those who were displaced (McCormack 2011).

Branding the Renaissance for Artists

While Mayor Cianci marshalled the historically-inflected redevelopment of downtown Providence's urban landscape, which eventually incorporated new hotels, a convention center, a revamped historic vaudeville theatre and an enormous (and enormously controversial) urban mall, he used his cozy relationship with *The Providence Journal* to exhort artists to invest in the city. Cianci realized that artists could add value to downtown development and that luring them would require little or no additional capital investment. Sebastian Ruth, founder and Artistic Director of Community MusicWorks (CMW), recalls seeing a photograph of the mayor in front of the newly refurbished Providence Performing Arts Center (PPAC), arms outstretched, with a caption that said something to the effect of: "This is the Renaissance City … come here and create" (Ruth 2011).

In 1997, Ruth and his classmate Tyler Denmead, with the support of mentors and funding from Brown's Swearer Center for Public Service, founded two youth arts organizations that have become nationally recognized models in the field. CMW offers instruction in violin, viola and cello to more than 100 children living in the West End, South Side and Olneyville neighborhoods of Providence. New Urban Arts fosters youth development through arts-based mentorship for public high school students. The commitments that Ruth and Denmead made to stay and develop institutions in the city were predicated on a shared belief that their efforts would be valued as part of the larger artistic Renaissance. They also knew that Cianci was focused on bringing in and retaining outside talent, not nurturing the City's existing population. Aware that there were very few options for artistic training available to public school youth in the city, Ruth and Denmead simultaneously took advantage of the possibilities created by, and implicitly critiqued the limitations of, the Providence Renaissance (Hocking 2010; Ruth 2011).

Service-learning programs based at Providence colleges and universities incubated projects like NUA and CMW that sought to address long-term inequity in public funding for the arts. Cianci's public pronouncement and infrastructural improvements also helped entice entrepreneurial and civically minded students to stay in the city, but these factors alone cannot explain how student projects gained legitimacy and stability. The third key element contributing to the success of students like Ruth and Denmead, who wanted to start non-profit organizations, was that they could find, and gain access to, all the relevant institutional gatekeepers in the City's cultural economy within a radius of just a few miles of their colleges and universities. Providence's geographic and cultural scale made it possible for students to transform service projects into legitimate, local institutions.

The growing popular consensus that Providence could be a place to stay, especially for mostly upper-middle-class Brown and RISD graduates, calls to mind the demonstrated investment of the city's old guard and policy leaders. Unlike their counterparts in Philadelphia and Hartford, many Providence elites stayed put during the 1950s and 1960s and sought new ways to use their cultural capital to shape their city in an image that would both suit their aesthetic dispositions and secure their property investments. The creation of new, but historically-inflected, infrastructure that appeared to benefit city residents, combined with rhetorical invocations of Providence's history as a capital of arts and culture, created a powerful mythological dyad that inspired more creative, industrious people who, whether they saw it as part of their work or not, helped realize Mayor Cianci's Renaissance vision.

Artists and cultural workers who put down roots in Providence during the 1980s and 1990s joined another robust cohort of cultural entrepreneurs, who committed to making work in the City as early as the 1960s and 1970s.[2] While this foundational cohort more than merits a study of its own, one figure from the early days stands out – Umberto "Bert" Crenca, artistic director of AS220. Crenca helped to lay the groundwork for much of the cultural development that shaped the Providence imaginary in the 1990s and early decades of the 21st century.

A Pilot Light in Sync With a Pro-Growth Agenda: The AS220 Collective

In 1983, new development and civic boosterism in Providence did not mean that arts and culture were understood to be economic growth areas deserving of special status. During this time, Crenca began to work with a collective of local artists who were vexed by Providence's lack of access and appreciation for those who were not part of the RISD machine, or who did not see themselves as part of Cianci's Renaissance vision for a Florentine artisanal economy where artists would work in open studios along quaint, historically-themed streets.

The AS220 collective, as it has come to be known, penned a manifesto, produced an un-juried show, and set up shop above the Providence Performing Arts Center (PPAC) at 220 Weybosset Street (Cook 2010). Crenca recalls an incident from the early 1990s, before AS220 was incorporated as a nonprofit and comfortably housed in a legal space it owned, but after it had moved from the PPAC building to Richmond Street a few blocks away:

> When the City came knocking on the door and we got a letter that said "we know you are operating a public venue and you need to see us before we come see you." It was almost phrased exactly like that: both a threat and an invitation! I went in and, very strategically, I had a flannel shirt on and jeans. You know, very working class first-generation Italian, which I am! I wasn't putting on any airs. I threw my arms on the counter at the bureau of licensing and said, "Close us down! I mean there is nothing else happening. We are like a pilot light for the city. If you want to put it out, put it out!" (Crenca 2011).

Crenca's performance endeared him to Cianci and AS220 survived its official inspection on Richmond Street, but not without some artistic modifications to its plumbing infrastructure. Crenca and Susan Clausen, a co-founder and current property manager, brush-painted PVC piping to look like copper so they could keep the doors open to the public. Since AS220 operated in a legal grey area

2 As Lynne McCormack points out, an entire generation of artists like photographer Salvatore "Sal" Mancini, oil painter Paula Martiesian, gallerist Bérge Zobian, storyteller Len Cabral, and, McCormack's mentor and colleague at the Department of Cultural Affairs, Bob Rizzo, made Providence their home long before the oft-heralded Fort Thunder collective made waves in national media outlets during the 1990s (McCormack 2011).

during its infancy, Cianci's patronage helped keep the organization insulated from legal challenges to its survival (Crenca 2011).

While AS220's collective was gestating on the outskirts of a neighborhood adjacent to downtown called the Jewelry District, Cianci's Renaissance coalition was fretting over how to maintain popular support for downtown infrastructural improvements. Aside from the successes of PPAC and The Trinity Repertory Company, and the survival of a smattering of clubs and bars that presented live music, there was little in the way of visible arts and culture in Providence's downtown. If the city was crawling with artists, as Cianci was fond of saying, then where were they?

In the Fall of 1992, a group of stakeholders convened by the Providence Foundation, a non-profit branch of the Greater Providence Chamber of Commerce, brought nationally renowned New Urbanists Andrés Duany and Elizabeth Plater-Zyberk to town. This was partly in response to the paucity of visible art making in the city center, and partly in response to concomitant fears that the Renaissance Development hadn't born fruit in terms of rising property values and increased occupancy. Renaissance advocates hoped that outside expert opinion could help jump start stalled private investment and give them new ideas for catalyzing the growth of downtown artist communities.

Figure 2.1 AS220'a first rented space on Richmond Street
Source: Courtesy of AS220.

The Providence Foundation published *Downcity Providence: Master Plan for a Special Time* after several days of semi-public charettes at Brown. Duany and Plater-Zyberk told the Providence Foundation that the City's economic health depended on whether artists could be brought downtown, though preservation and a unique sense of place were also deemed critical to cultural development (Duany 1992). Following the convening at Brown, the Mayor brought a second group together under the leadership of his Chief of Administration to come up with a practical strategy based on Duany and Plater-Zyberk's findings. This Arts and Entertainment District Task Force (AEDTF) included Crenca as well as leaders from Trinity Rep, Johnson and Wales University and the Chamber of Commerce. Notably the AEDTF did not include Lynne McCormack or Bob Rizzo, who as the producers of the Convergence Arts Festival and staff of the Parks Department's Office of Cultural Affairs, had valuable insight into how artists lived and worked in the City (Schupbach 2003: 51).

The AEDTF recommended that a central part of the downtown neighborhood be designated as the Downcity Overlay District and governed by a new Mayor-appointed Downcity Design Review Committee. "Downcity" was a vaudeville-era name for the neighborhood that the consultants and the Taskforce seemed to think fit snugly with the Renaissance image of rebirth. This new committee, created very much from the top down, with little input from unaffiliated artists in the city, regulated new construction and rehabilitation in the downtown, mediating between the City, developers, preservationists and the downtown business community, for the duration of Cianci's tenure as Mayor (Werth 2010: 43). The AEDTF spent the next three years lobbying the state legislature to pass three new laws to give substance to the nascent arts district. These laws incentivized artistic production and residential development through a state tax exemption for artists making their living in the neighborhood, exempted sales tax on the sale of artistic work there, and allowed for a municipality to discount the tax burden on the entire assessed value of a residential building, whether it was housing artists or not. It is difficult to say whether tax incentives and re-districting brought new artists to the city center (Schupbach 2003: 53–4). It is clear, however, that the incentive for developing residential property helped expand the availability of housing stock in Downcity. By 2010, census data showed Downcity's population to have increased by 65% from 2000 (Providence Plan 2011).

In 1993, about a year after Crenca sat on the AEDTF, AS220 moved several blocks closer to the heart of the Downcity district. The organization was allowed to rent 115 Empire Street for $1 a month from the City until it could pull the financing to redevelop the space together (Deller 2010). AS220 called in favors with well-connected bureaucrats and bankers left and right, selling their project as a force for stabilization on a street known primarily as a haunt for sex workers. The organization was eventually able to complement the $300,000 it had raised from a potpourri of fundraisers with a crucial $200,000 gap-funding loan from the City, and another $528,000 of loans from a consortium of banks mobilized by The Providence Foundation (Cook 2010).

In 1992, the best way to get material support from the city was to have Cianci on your side, and this meant buying into the pro-development logic of the Renaissance project (Leazes and Motte 2004: 65–6). Crenca's personal relationship with Cianci undoubtedly helped AS220 garner the financial support in needed to purchase 115 Empire, but it was the organization's mission to bring artists into the city center, a goal perfectly in sync with the Mayor's new plan for Downcity, that assured AS220's success.

From AS220's beginnings, long before it became an organization with a $2.5 million budget, its financial model relied heavily on the sweat equity of its founding collective and resident artists. Their labor supported low-cost programming while overhead was paid with rent from artist studios. The collective rarely came up with all the money needed to run operations in the early days, but its earned-revenue model insulated it from the influence of capricious grant-makers. Eventually, with property of its own, AS220 opened a bar and restaurant, added artist studios, and began to operate with numerous efficiencies, all the while leveraging its equity to obtain new loans and finance further expansion.

Today AS220 is one of Providence's most visible nonprofit arts institutions. Its staff runs several un-juried galleries and public programs at least six nights a week. AS220 Youth Arts, formerly Broad Street Studios, offers instruction in music production, dance, photography, visual arts, personal fabrication and literary arts to teenagers from Providence as well as to young people serving time in the city's juvenile justice facility. Premiere Baptista, a 17-year-old involved in AS220 Youth Arts, says the organization put her on her feet: "I wanted to start somewhere with my music. I didn't know where I should go. If you come to AS220, they help you every step of the way" (Cook 2010).

While AS220's immense importance for artists in the city cannot be overstated, its most important impact on Providence may turn out to be its work as a developer. Today, after nearly thirty years of continuous operation, AS220 owns two newly refurbished, historic buildings in addition to their main organizational headquarters at 115 Empire Street: the Dreyfus Hotel at 95 Matthewson Street and the Mercantile Block Building next door on Washington Street. The organization rents and leases units in these properties mostly at below market rate, ensuring that artists and others who identify as part of the City's "creative community" can live and work in the city center. To add to the diversity and livability of the neighborhood, AS220 also solicits commercial tenants, like barbershops, bars, locksmiths and restaurants.[3]

3 Financing the redevelopment of The Mercantile and The Dreyfus could never have happened had a coalition of developers and preservationists, lead by Arnold "Buff" Chace of Cornish Associates, not successfully lobbied the state legislature to pass a landmark historic tax credit in 2001 (Marsh 2011). The credit, which was also crucial to the financial calculus behind The Steel Yard, an industrial arts organization founded in 2001, made it possible to write off 30% of qualified rehabilitation costs of historic properties. When combined with 20% federal credit and other local incentives, the tax credit made projects

AS220 occupies a liminal space between a pro-growth group of developers and property owners, and the artists who, in different circumstances, they would be likely to displace. As founding board member Geoff Adams says, owning property was partly about insulating the artists involved in AS220 from the effects of gentrification caused by other high-end downtown development:

> The pattern that was well known to everybody was that artists come in and break new ground in an urban neighborhood, and then they get quickly priced out by developers. So the artists make it a place to be, and then the developers make it a place to be for people who are priced way above what artists can afford. AS220 didn't want that to happen. So the idea to buy a building would be one of creating permanence in the city (Cook 2010).

Owning property is also about giving the arts a visible presence in the city's central business district, a process that factors both into the ways the City brands itself as an arts and culture destination and the ways that developers have been able to sell the neighborhood to well-heeled tenants.

AS220 benefits from its close proximity to the city hall and the seat of government, and from adopting a mission that is often in lock step with that of the Providence Planning Department. These conditions make it easy for the city planner to help Crenca secure HUD money and other resources, as he did for the Dreyfus project and a recent public mural painted by the artist Shepard Fairey. AS220 and the Planning Department are both working to cultivate, as city planner Thomas Deller puts it, "whole neighborhoods" where retail, housing, jobs, and transportation all contribute to neighborhood vitality (Deller 2010).

Crenca, whose work as an arts manager emerged from an anti-commercial impulse, embodies the liminality characterizing his organization. While he knows he has performed an economic miracle with AS220, predicated as much on serendipity as on strategy and hard work, he also realizes that his status in the city has changed; no longer an outsider, he has a permanent place at the policy table. Despite this, Crenca is still philosophically committed to AS220's foundational egalitarian and anti-commercial principles, such as a flat pay structure and a ban on music licensed by the American Society of Composers and Publishers (ASCAP) and Broadcast Music Inc. (BMI), both of which manage commercial music rights for recording artists (Crenca 2010).

Ultimately it is inconsequential whether spaces for artists that work by advancing commercial real estate interests raise property values in neighborhoods where there is no one to displace. Rather than wring their hands about what might happen if they aid and abet gentrification of a non-residential neighborhood, AS220 has attempted to make its facilities into spaces that foster racial comity and create equity among differently privileged populations. Since moving to Empire Street,

---

like AS220's redevelopment of The Dreyfus and The Mercantile Block feasible (Goldstein-Plesser 2010: 25).

the organization has created alliances with, and incubated, other neighborhood-based cultural organizations like Youth in Action, College Visions and the now defunct Providence Black Repertory Company.

Preservationist Lucie Searle is unflinching in her belief that AS220 is a boon to the city and its artists. During an April 2011 forum in which non-profit leaders reflected on the ways that Providence's creative economy functions, she called AS220's work as a developer vital to the city's Renaissance project: "The most important thing [AS220] did, and we continue to do … [is] bring stability, and it creates community, which makes other people want to be [downtown], it makes people feel safe, it makes people want to live there, want to work there" (Ketten 2011).

Expanding the Preservation Coalition in Olneyville

While AS220 helped transform Downcity into a more livable area during the 1990s, many Providence artists lived and worked on the industrial outskirts of the city center. In 1995 a group of idealistic young RISD and Brown graduates and near-graduates began to produce rock shows, haunted houses, boxing matches, and other zany performances in Olneyville's Valley Worsted Mill building in Eagle Square, just over a mile down the Woonasquatucket River from Waterplace Park. This playground for protracted adolescence became known as Fort Thunder.

Since its much-lamented demise in 2001, Fort Thunder has been invoked as the paragon of innovative, collective and non-commercial arts practices associated with an iconic "Providence" sound and visual aesthetic. While this particular Providence aesthetic is characterized by an impossibly loud, punk music hybrid called noise rock and silk-screened rock show posters, the Fort's most iconic artistic legacies are its fantastically decorated live-work artist studios.

Making performances and messy but aesthetically appealing live-work spaces was nothing new in 1990s Olneyville. Artists like those affiliated with Alias Stage had been working out of Atlantic Mills on Manton Avenue since the 1980s, producing theater and other non-commercial performance art (McCormack 2011). What was new about Fort Thunder was that the artists living and working there were connected to RISD, and this made their mess worthy of national attention. Publications like the *New York Times* and *Artforum* began to laud the beautiful detritus of what came to be called the "Fort Thunder Collective" (Tannenbaum 2006: 21). As more people came to know about the Fort, its illegal status became more difficult for the City's licensing bureau to ignore. Much like AS220 in its early days, and Providence's industrial architecture during the height of urban redevelopment, the Fort survived as long as it did because of benign neglect; it would likely never have become such a seminal institution had it been incubated under the Renaissance spotlight.

Figure 2.2 Eagle Square protest
Source: Courtesy of Stephen Mattos.

In 1999, representatives from a New York-based commercial property developer called Feldco began taking pictures around Eagle Square. Feldco intended to develop a suburban-style strip mall on the site. The company claimed that to rehabilitate the historic mills in Eagle Square would be practically impossible, and that the Providence Commons could be realized only by razing the entire parcel of land, destroying 26 historic mill buildings, including Fort Thunder, and displacing nearly 100 artists (Marsh 2011).

At a public hearing on the Feldco project in November of 2000, Brian Chippendale, today one of the Fort's most well-known alumni, invoked the Providence Renaissance's most notorious success story: "There's 100 potential Barnaby Evanses sitting around [Fort Thunder]. That's a lot of potential to lose" (Smith 2000). By comparing himself and other artists in Olneyville to Evans, the creator of WaterFire, Chippendale was calling Cianci out on neighborhood development policies that clashed with the priorities upheld in the city center. If Providence's leaders were really serious about the Renaissance project, and its commitment to artists, Cianci would have to take Olneyville seriously.

Around the same time, a group of gentrification watch dogs mobilized out of AS220 organized as Providence Artists United (PAU) to advocate on behalf of The Safari Lounge, a bar that faced eviction from a building in Downcity earlier that year. The Safari was a beer and shots joint where Providence rock bands played in the late 1980s, the Ilarraza family's only source of income, and the type of venue essential to maintaining class-diverse artist communities in the city center. PAU placed an ad with the following text in the January 27, 2000 issue of *The Providence Journal*:

> The Safari Lounge is a rare and almost miraculous embodiment of the publicly stated goals of the so-called "Renaissance" of downtown Providence. We can't help but see the fate of the Safari as a test of these goals. Is the Renaissance really giving an opportunity to artists … Or is it wiping away businesses that grew up independently … to replace them with a planned economy that may not even be viable? (Donnis 2000)

In March, *The Providence Journal*, still irresolutely pro-development and politically aligned with Cianci, printed an editorial portraying the failed eviction of the Safari as a "technicality". *The Journal* also framed the debates that PAU claimed to be about gentrification in terms of passive property value increases ("rents will rise") rather than class bias (Providence Journal Staff 2000). The PAU responded with an op-ed of its own, restating its case for the illegality of the Safari eviction and calling the op-ed's recourse to the wisdom of The Safari's owner a celebration of values reminiscent of those cherished by European feudal elites (Kuehl 2000).

The contest over Eagle Square was a perfect place for the PAU to continue questioning the logic of Renaissance investment while expanding their critique of gentrification and displacement. Ironically, in joining the environmentalists and enlightened developers of The Fort Thunder coalition, PAU was also aligning itself with the city's archetypical gentrifiers, the middle-class preservationists from the Providence Preservation Society.

PAU and Fort Thunder artists knew that Eagle Square was only a viable space as long as it was under the radar; after garnering so much national attention, it was destined for either the wrecking ball or redevelopment. By expending their cultural capital to save Eagle Square, artists hoped they could at least help preserve the historic mill buildings that had been the site of so much creative ferment (Smith 2000). Sadly, by supporting the preservation of industrial space in Eagle Square, they helped create conditions that catalyzed a real estate buying spree in Olneyville, jeopardizing the availability of the low-rent, low-visibility industrial space they continued to inhabit. Their pro-preservation stance put Olneyville artists squarely in opposition to Olneyville Councilwoman Josephine DiRuzzo, and Olneyville property owners, who, according to *Providence Journal* writer Greg Smith, saw development of any kind in their neighborhood, even suburban-style strip mall development, as a "much needed shot in the arm". During the famed November 2000 public hearing at City Hall where the Fort Thunder coalition rallied to "Save Eagle Square," an Olneyville property owner appealed to a sense of misplaced entitlement on the artists' part: "When these kids leave, we'll still be here. To us this is a present from God" (Smith 2000).

Residents and supporters of Eagle Square felt connected to the embodiment of history and aesthetic iconicity represented by industrial mills much as College

Hill preservationists did with their Benefit Street Colonials.[4] Olneyville property owners hadn't anticipated that many artists would stay, and despite repeated evictions over the next decade, invest in their neighborhood with both their sweat equity and cultural capital, like the East Side elites of the 1950s.[5]

A small group of privileged artists were able to purchase properties in the neighborhood, like the Armington and Sims Engine Company Building, and later the neighboring Providence Steel & Iron property (future home of the Steel Yard) and Eastern Butcher Block. Many, however, acted as though they were there because they could not afford to rent such voluminous studio space otherwise. They were also, as legal scholar Matt Jerzyk points out, "risk-oblivious" to the hazards endemic to dwelling in decrepit industrial spaces (Jerzyk 2009: 415).

The buzz created by pro-preservation activism around Eagle Square eventually died down when Feldco was given a final ultimatum from Mayor Cianci and relented to demands that it preserve 25% of the structural integrity of the site and accommodate a planned bike path and access to the historic Woonasquatucket River (Goldstein-Plesser 55–6). Baltimore-based developer Bill Struever, later reviled by local artists for developing nearly $350 million of industrial property in Olneyville, helped create the necessary leverage with Cianci by offering an alternative pro-preservation vision for Eagle Square (Marsh 2011).

The fight over Eagle Square catalyzed two important preservation-driven policy developments on the local and state level: the establishment of an updated, non-contiguous Industrial Sites and Commercial Buildings designation (ICBD) in Providence, which restricted alteration of character and destruction of historic industrial properties, and the aforementioned statewide 30% tax credit for historic preservation, a critical component of the funding implemented by AS220 during its expansion into The Dreyfus and Mercantile Block buildings almost 10 years later (Providence Preservation Society 2002). These pro-preservation policies were central to both artist-driven, slow-growth development and an explosion of development driven by outside capital. The story of The Monohasset Mill Project and its non-profit neighbor, The Steel Yard, demonstrate the success and continued challenges that developers face as they seek to make Olneyville's mills into safe,

4 René Morales, Assistant Curator at the Miami Art Museum and a former audience member at Fort Thunder, reminisces in the catalogue for the RISD Museum's Wunderground exhibition: "I loved those crumbling factories filled with rusty artifacts from the Industrial Revolution … It was electrifying to hear those cavernous spaces shot through with noise rock" (Tannenbaum 2006: 109).

5 Despite the surface-level similarities between the pro-preservation coalition mobilized around Eagle Square, and the College Hill coalition of the 1950s and 1960s, there are many points at which the two moments in Providence's preservation history diverge. Benefit Street was already a residential neighborhood, and the College Hill coalition was a much smaller, wealthier group of stakeholders than the Fort Thunder group; most of them were not residents of the street they hoped to transform.

legal, vibrant spaces where artist communities are able to thrive in and engage broader publics.

Artists as Developers in Olneyville

The Steel Yard is an industrial arts center sitting on the site of the former Providence Steel and Iron Company, two acres of remediated brownfield in the heart of Providence's industrial Valley. The organization was founded just down the block from Eagle Square in 2001 by Nick Bauta, a RISD alumnus, sculptor and descendent of the Canadian food magnate W. Garfield Weston, and Clay Rockefeller, a Brown alumnus, ceramicist and the great-great-grandson of John D. Rockefeller. It offers courses in welding, blacksmithing, ceramics, jewelry, glass casting and the foundry arts and is funded by contributed income and earned income from rentals and public projects.

The Monohasset Mill Project (MMP), The Steel Yard's for-profit neighbor, was developed next door in the former Armington and Sims Engine Company Building by Rockefeller and a group of longtime Armington residents and RISD alumni. Initially, MMP developers saw their project only in terms of saving the Armington property for the artists living there, including the daughter of Bill Struever. With the help of developers and preservationists, they completed a financial plan and learned how to navigate the challenging legal matrix of the local, state, and national tax incentives that would tie the for-profit development to the nonprofit next door (Marsh 2011). Today, Monohasset provides high-end housing as well as HUD-subsidized artist live/work units for many people affiliated with The Steel Yard.

Rockefeller says that the idea for "the Yard", took shape after September 11th, 2001. He was inspired by witnessing New York residents come together purposefully, and by his own volunteer experience at The Crucible, a treasured industrial arts institution in Oakland California:

> I kept on coming back to the fact that people need to feel like they can do something and be useful … and can be rooted in the process of making things. And so it started to kind of evolve, my looking at the Steel Yard … I started to look at economic development aspects as well, and job training things, and you know basically trying to teach people skills or offer people the opportunity to learn skills that … could also be applied toward bringing an income in (Ketten 2011).

Rockefeller and Bauta never saw the Steel Yard as a project that would add something new to Olneyville. Rather, they saw it as a facility that would support work that had been going on for centuries in the neighborhood. As Rockefeller says, "I was particularly interested in what the next successful model for industry would look like, and if it could be sustainable in a scaled-down way that focused on local markets" (Cameron 2011: 72). The Steel Yard's Executive Director,

Drake Patten, explains the challenge of getting funders, bureaucrats and arts advocates to think about how Rhode Island has always had a creative economy: "There have been, and still are, a ton of creative businesses that are not 'arts' or 'tech' businesses; we have a rich industrial heritage and we don't really understand how to incorporate it" (Patten 2010). The scaling down of industry at The Steel Yard has taken its most tangible form in the organization's extremely successful Public Projects, through which it hires local artists and artisans to design and fabricate an array of street furniture, like trashcans, tree guards and bike racks. The organization has landed 90 contracts since launching its Public Projects, including five for the Providence Downtown Improvement District, a Downcity beautification program instituted by the Providence Foundation (Cameron 2011: 73).

The Steel Yard prides itself on its slow growth approach and its successful partnerships. In 2010, the organization administered a workforce development program funded through the American Reinvestment and Recovery Act. Between July and August 2010, eight young people learned to weld, torch cut, work with jigs, and use power tools and hand tools at The Steel Yard. While earning a minimum wage, they practiced their fabrication skills making frames for trashcans that would later be sold through the organization's Public Projects program, and then designed and fabricated an intricate fence for the Yard. When they left the program, trainees had a portfolio and work samples to bring to future employers, as well as professional references from supervisors and artists/metalworkers from the Steel Yard community.

Partnering with the City of Providence Department of Art, Culture + Tourism (AC+T) was crucial to the successful implementation of The Steel Yard's workforce development initiative. Lynne McCormack, the Director of AC+T, says that the city acted as a pass through agent and streamlined the tedious reporting for the federal grant so that The Steel Yard was able to launch the program in just six weeks: "It was crazy! Some cities that had workforce programs, they turned it around. But no other cities in the country, that didn't have a program already, turned a whole program around as quickly as we did." McCormack also notes that her work as an ombudsman for the arts helped to facilitate communication between the Providence Planning Department, The Steel Yard and the EPA, who awarded the Yard a $400,000 grant allowing it to complete its brownfield remediation work in 2010 (McCormack 2011).

The Steel Yard's work with AC+T exemplifies The City of Providence's new way of doing business with arts organizations. Authentic, relationship-based partnerships are key. Long gone are the days of closed-door negotiations and arbitrarily disbursed discretionary funding.

Figure 2.3 Steel Yard Workforce Development participant
Source: Courtesy of the Providence Department of Art, Culture and Tourism.

The Department of Art, Culture + Tourism: Advocacy and Visibility for the Cultural Economy

When Mayor David Cicilline came to City Hall in 2002, he immediately began to revamp the City's approach to supporting its cultural sector. First, he charged arts policy consultant Ann Galligan with making recommendations for establishing a new bureaucratic structure for arts and culture in the City. Galligan not only met with a core group of representatives from resource-rich legacy institutions, like Trinity Rep and PPAC, she also met with members of the business and funding community, and culturally specific arts organizations. Most significantly she convened unaffiliated arts leaders from across the city's cultural landscape.

Cicilline's transparency and deliberation signaled a departure from the days of the AEDTF and top-down, closed-door cultural policy. By November of 2003, using Galligan's recommendations as his guide, Cicilline was ready to articulate a new vision for a centralized office under his direct purview:

> Through the visibility of the Mayor's Office, the Department of Art, Culture + Tourism will raise the status of art and culture in city government, signaling their importance as integral components of city life. Arts and culture will be elevated

> to a position of equal partnership, in creating and implementing public policy, and will serve as a vehicle to create collaborative working relationships across city government. The Department will celebrate the social and economic power of art, as well as the tremendous potential of creative workers, as engines of growth and development in Providence (Cicilline 2003b: Vision).

Cliff Wood, a lobbyist who had been working for the Downcity developer Cornish Associates, became the first Director of the new Department. Wood also brought in Lynne McCormack to work as AC+T's first Deputy Director. McCormack, who had worked at Cultural Affairs in the Parks Department under Bob Rizzo, had a grasp of institutional history as well as deep connections in the arts community. Her addition to the team helped AC+T to resolve some of the issues created by Galligan's overly ambitious cultural plan.

The plan was not sufficiently funded; this meant that there were few resources to hire staff to support existing (let alone expanded) programming. Taking on the mandate of the Mayor, Wood and McCormack immediately began creating partnerships with Downcity performing arts presenters and producers to ensure that the public did not perceive AC+T to be siphoning funds from the art produced by the City. Partnering with the performing arts presenter FirstWorks helped AC+T sustain high-quality multi-arts programming similar to that presented during The Convergence Arts Festival, a flagship program produced by the Department of Cultural Affairs and its nonprofit arm, CapitolArts, under Cianci (McCormack 2011).

Wood sought out Donald King, the Artistic Director of The Providence Black Repertory Company (Black Rep) to continue producing music similar to that featured at the City's Rhythm and Blues festival, also a Cianci-era program produced by CapitolArts. The Black Rep was a theatre and live music venue spun off from a public program at AS220 that had recently purchased a new building in the heart of the Downcity district. Wood was confident that with King's curatorial skills, Black Rep's production capacity, and Art, Culture + Tourism's financial and technical support, something like the world-renowned Montreal Jazz Fest could one day be achieved through AC+T's partnership on Providence Sound Session (Wood 2010).

In addition to these two key partnerships, AC+T maintained its Friday night concert series in Waterplace Park, and began to focus explicitly on bringing more performances to the city's neighborhoods. Acknowledging the myopia of the Cianci administration when it came to developing cultural programming outside Downcity, Wood told *The Providence Journal* that, "for a long time, the focus has mostly been on downtown. That's not a bad thing, especially when you're trying to market the city to outsiders. But there has to be balance" (Van Siclen 2004). It is still important to sell the city to outsiders, both potential home buyers who might purchase property and contribute property taxes to the city's coffers, and tourists who, through taxes on the city's hotels, support the Tourism Council, which in part supports salaries at Art, Culture + Tourism (Wood 2010).

Wood and McCormack established the Neighborhood Performing Arts Initiative (NPAI) using money that had been growing for nearly 100 years in the city's Edward Ely Trust Fund for performing arts in public parks. Rather than curate the performances itself, AC+T signaled its willingness to share cultural authority by sending RFP's to arts organizations and neighborhood groups, encouraging them to decide what type of arts programming would best suit their local, neighborhood audiences. As McCormack says, her relationship with the Director of the Parks Department, where the Office of Cultural Affairs was housed, was critical to the success of the program. Again, partnering was not only the best way to push things forward in Providence, it was the only way (McCormack 2011).

Today the Department of Art, Culture + Tourism, run by McCormack since Wood was elected to the Providence City Council in 2006, has a full-time staff of four, and a rotating cast of interns supporting its daily operations, but is still in need of increased support. AC+T supports FirstWorks, Providence Sound Session, the Summer Waterplace Park music series and the NPAI performances, film and television production, and any other commercial cultural events taking place in the city. The department also devoted the last several years to an intense, participatory planning process through which it has conceived a new 10-year cultural plan for Providence. The plan recommends permanently institutionalizing Art Culture and Tourism by incorporating it within the City's Planning Department (City of Providence 2009).

In addition to creating a framework to ensure the permanence of AC+T, the Cultural Plan, developed with the participation of hundreds of stakeholders in the city's cultural economy, incorporates several updated, strategic goals for AC+T and its partners. These goals include raising awareness for the cultural sector as a whole, developing the creative practices of Providence youth, fostering sustainable organizations, increasing cultural participation and bolstering neighborhood investment in arts and culture (City of Providence … 2009).[6]

McCormack acknowledges that neither the cultural plan, nor AC+T will solve all the problems of the cultural sector in Providence. The department is still figuring out how to balance the needs of neighborhoods with the demands of the city's core arts and entertainment district. What is certain about McCormack's approach is that like her predecessor, her management style is shaped by her identity as an artist, and her belief in the power of art and culture to transform communities. She acknowledges that although strategic thinking and long-term planning have always been central to AC+T's operations, many of the tactics that the department used in its infancy were incredibly risky. She says the office will continue to take

6 McCormack anticipates the incorporation of AC+T will proceed smoothly with City Council approval and that initiatives like the new federally-funded study shaping investment in key transportation corridors will be more and more in sync with existing and future growth of the city's cultural sector, which by 2007 accounted for nearly $111 in revenue and 3,000 jobs in the city (Americans For The Arts 2007).

risks, as all great artists do, as it moves into its second decade: "If you ask my staff they'll tell you I'm still opportunistic, and we still take a lot of chances and push the envelope like crazy, but it is shifting" towards a more balanced, long-term approach to cultural management and development (McCormack 2011).

Moving Forward, Looking Back

Cultural development is circuitous and at times random. What factors today seem predestined and productive within systems that support cultural change may once have been seen as the vestiges of corruption, vice and lethargy. By understanding that the valences assigned to these factors in Providence's history have not been static over time, but have shifted with the larger tides of economic and social change, scholars of post-industrial development can better understand and analyze the small post-industrial cities and their complex, context-specific conditions.

By tracing a path from Providence's early industrial history, and the infrastructure that made its efflorescence possible, to the development of historic preservation as an engine for economic development, and finally to arts and culture as the driving forces energizing the historic spaces preserved and the people who inhabit them, I hope to have shown that Providence's post-industrial cultural economy is the product of both serendipity and foresight, risk-taking and strategic planning. Without moments of benign neglect, such as those precipitating the coalescence of the College Hill preservation coalition as well as the AS220 collective and the Fort Thunder coalition, the official cultural apparatus housed at City Hall would never have become quite so responsive to issues centered around availability and preservation of historic space for, and accessibility to, the arts. Yet without the planning, agitation, and coalition-building spearheaded by Providence's bureaucrats and cultural elites, the germinating cultural buds born from benign neglect would never have been able to bear institutional fruit like WaterFire, AS220, and The Steel Yard.

Not all of Providence's successes can be attributed to the ingenuity of its resident culture makers and bureaucrats – many stem first from the city's small size and its strategic placement between larger marketplaces along the Eastern seaboard. Providence's size, in particular, has facilitated the development of a dense web of collegial affiliation in the cultural sector cemented by immaterial exchanges and platonic social ties. Additionally, the density of institutions of higher learning in Providence has made it possible for Providence to capitalize on and retain highly skilled culture workers who then become a part of this network.

Providence's story shows that no city can survive and prosper by foregrounding a simulated cultural front, but that such a front is sometimes necessary to create the public support needed to foster institutional stability. As we can see in this post-industrial cultural economy, the cart, preservation and infrastructural redevelopment initiatives, had to come before the horse of a robust cultural economy.

Bibliography

Americans for the Arts. (2007), *Arts & Economic Prosperity III: The Economic Impact of Nonprofit Arts and Culture Organizations and their Audiences: National Report* (Washington, DC: Americans for the Arts).

Brett, R. (1976), *Temples of Illusion: The Golden Age of Theaters in an American City* (Bristol, RI: Brett Theatrical).

Brown University. (2006), *Slavery and Justice* (Providence, RI: Brown University).

Cameron, K. (2011), "Rising from the ashes", *Metropolitan Magazine*, January 2011, New York, 70–74.

Carnevale, L. (2011), Co-Director Providence Partnership for Creative Industrial Space (Personal communication 25th February 2011).

Cicilline, D. (2003a), Inaugural Address, 26th, Jan., 2003, Providence, RI.

Cicilline, D. (2003b), Vision for Art & Culture in Providence, 24th November 2003, Providence, RI.

Cook, G. (2010) "AS220 at 25: A lively oral history of Providence's artistic Mecca", *The Providence Phoenix.* [Online] Available at http://providence.thephoenix.com/life/106719-as220-at25/?page=1#TOPCONTENT [accessed 4 March 2013].

City of Providence Department of Art, Culture + Tourism. (2009), *Creative Providence: A Cultural Plan for the Creative Sector* (Providence, RI).

Crenca, B. (2011) Founder and Artistic Director AS220 (Personal communication 1st September 2010).

Donnis, I. (2000), "The Safari Lounge wins a surprise reprieve," *The Providence Phoenix*. Downtown Confidential, 10th–13th February 2000.

Duany, A. and Plater-Zyberk, E. (1992), *Downcity Providence: Master Plan for a Special Time* (Providence RI: The Providence Foundation).

Department of Art, Culture + Tourism. (2009), *Creative Providence: A Cultural Plan for the Creative Sector* (Providence, RI).

Fink, L. (1981), *Providence Industrial Sites Report, Rhode Island Historical Preservation Commission* (Providence, RI).

Galligan, A. (2003), Recommendations to the Mayor for the Office of Art, Culture and Tourism.

Goldstein-Plesser, G. (2010), "Keeping it real: searching for significance, authenticity, and integrity in Three Providence Place conflicts" (BA Thesis in Urban Studies) (Providence, RI: Brown University).

Greenfield, B. (2004), "Marketing the past: historic preservation in Providence, Rhode Island". In M. Page and R. Mason (eds), *Giving Preservation a History: Histories of Historic Preservation in the United States* (New York: Routledge).

Hocking, P. (2010), Former Director of the Brown University Swearer Center for Public Service (Personal communication, 8th August 2010).

Jerzyk, M. (2009), "Gentrification's third way: an analysis of housing policy & gentrification in Providence", *Harvard Law Review* 3 (2009), 413–29.

Ketten, J., Salkind, M. and Sakash, S. (2011), *Who Made Us Creative? People, Place and Power in Providence*, New Urban Arts, Providence RI (7th April 2011).
Kuehl, M. (2000), "The Safari Lounge and determinism," *Providence Journal*, 1st April, Editorial, 7B.
Leazes, F.J. and Motte, M.T. (2004), *Providence, the Renaissance City* (Boston: Northeastern University Press).
Marsh, C. (2011), Director Communityworks Rhode Island (Personal communication, 12th July 2011).
McCormack, L. (2011), Director Providence Department of Art, Culture + Tourism (Personal communication, 13th January 2011).
Patten, D. (2010), Director The Steel Yard (Personal communication, 7th September 2010).
Providence Journal Staff. (2011), "Whither Safari Lounge?" *Providence Journal*, 15th March, Editorial, 6B.
Providence Plan. (2007), Olneyville Neighborhood Profile. [Online] Available at: http://local.provplan.org/profiles/oln_main.html [accessed 4 March 2013].
Providence Plan. (2011), "Providence Plan Releases 2010 Census Information and Maps," 5th April 2011. [Online] Available at http://provplan.org/news/updates/the providence-plan-releases-census-2010-information-and-maps/ [accessed 4 March 2013].
Providence Preservation Society. (2002), Industrial Sites and Commercial Buildings Survey 2001–2002. [Online] Available at http://local.provplan.org/pps//index.asp [accessed 4 March 2013].
Rickman, R. (1999), *African Americans on College Hill, 1950–1979* (Urban League of Rhode Island, The National Trust for Historic Preservation's Antoinette Downing Preservation Fund for RI, and The Rhode Island Committee for the Humanities, Providence, RI).
Rhode Island Historical Society. (2011), Historical Walking Tours. [Online] Available at http://www.rihs.org/events_walking_tours.html [accessed 4 March 2013].
Rosaldo, R. (1989), "Imperialist nostalgia", *Representations* 26, 107.
Ruth, S. (2011) Founder and Artistic Director Community MusicWorks. (Personal communication, 7th September 2010).
Schupbach, J.S. (2003), "Artists downtown: capitalizing on arts districts in New England" (MA Thesis in Urban Studies and Planning) (Cambridge, MA: MIT).
Smith, G. (2000) "Residents, artists clash over shopping center," *Providence Journal*, 22nd November, News, 1C.
Stern, M. and Seifert, S. (2008), *From Creative Economy to Creative Society. The Social Impact of the Arts and the Rockefeller Foundation* (Philadelphia, PA: University of Pennsylvania).

Tannenbaum, J. and Allison, M. (2006), Wunderground: Providence, 1995 to the present: Providence poster art, 1995–2005; Shangri-la-la-land: [exhibition September 15 2006–January 7 2007]. (Providence, RI: Museum of Art, Rhode Island School of Design).

Tsui, B. (2005), "In Providence, faded area finds fresh appeal", *NY Times*. [Online] Available at http://www.bonnietsui.com/articles/new-york-times/in providence-faded-area-finds-fresh-appeal/ [accessed 4 March 2013].

Van Siclen, B. (2004), "Creative energy: he marshals the arts to fuel a more vibrant Providence", *The Providence Journal*, 18th July, Arts, B-01.

Werth, Alexander (2010), "Historic preservation and the politics of downtown renewal in Providence, Rhode Island" (BA Thesis in Urban Studies) (Providence, RI: Brown University).

Wood, C. (2010) Former Director, Department of Art, Culture + Tourism (Personal communication, 16th September 2010).

Woodward, W.M. (1981), Downtown Providence, Rhode Island Historical Preservation Commission, Providence RI.

Chapter 3

"Lofty Artists" vs. "El Oro del Barrio": Crafting Community and a Sustainable Economic Future in the *Paper City*

Myrna Margulies Breitbart

Introduction

> When I was a child growing up in Holyoke I used to stand on the canal bridge and watch green dye come out of one side of a factory, while red dye came out from the other. Where they met was a belly-up sucker. Anything that can be born from that reality will happen from the mills.

This comment was made by an artist attending the first meeting of a new creative coalition formed in Holyoke, Massachusetts in 2001. Metaphorically it captures the vastly different approaches to the cultivation of a new creative economy that continue to proceed along parallel yet dissimilar paths. One emphasizes the need to build upon Holyoke's demographic diversity and the aspirations and talents of existing residents, while the other emphasizes the importation of an entrepreneurial "creative class".

José Colon represents the former perspective. He is a long-time resident of the city, an artist, and the director of El Arco Iris ("The Rainbow"), an after-school art program that serves many children from lower-income households. Colon expresses excitement as he walks the streets of Holyoke envisioning places to put local art. He does not fantasize a single "arts district" but rather an entire city saturated with art. He believes this process could begin with the embellishment of an ugly concrete wall along one of the three canals in the city that runs through an impoverished neighborhood. "Faces of Holyoke", the name he gives to the mural he envisions, would be created using images of objects of importance that residents of many cultural backgrounds would provide. Ideally, the objects would come from peoples' homes as an expression of their identity and heritage. Colon imagines an older French Canadian woman showing him a precious vase that he would take a picture of and then draw onto the wall. He talks of what it would be like for this woman to pass by this painted wall and see a symbol of her heritage displayed for others to learn from.

> You're driving by and you're French Canadian or Jewish or Puerto Rican. I don't care. You're in Holyoke and there's something that actually identifies who you are … It's up there [on the wall] now. I'm no longer just a person who drives by that you are never going to know (Interview 2001).

Complementing Colon's vision is that of Mary Jo Maichek, a storyteller who performed at Heritage Park, the Boys and Girls Club, and many venues in and around Holyoke. More than a decade ago she expressed the dream of using the creative arts to build bridges between the city's older and newer immigrant communities through the construction along the canals of something like a big dance centre offering Latin dance lessons and classes in dance across the board.

> I would make it beautiful … really go for it … right downtown so it is walkable, and accessible, with a lot of affordable classes … [I would] do outreach in the streets … have cars going around with signs and loudspeakers saying what it is, just like the Blues Brothers …. Around the canals we could really do a lot …

Maichek shared many ideas for enlivening the city through the arts but expressed concern that "they don't make it all gentrified". "I'd like to see it come from within," she said, "because this community has so much to offer, it is just amazing." Claire Williams, a flute player who grew up in Holyoke, also pointed to its "rich heritage" and expressed resentment about preconceived ideas that many people from outside hold about the city. Appreciating the many forms of culture present, she praised venues like the Canal Gallery because "they bring in artists from all segments of the city's population to display their work".

The same year I spoke with these residents, a different vision for a cultural economy was shared in a meeting held in a newly renovated café on the ground floor of one of the largest 19th-century mill buildings in Holyoke. An entertainment-marketing consultant hired to promote space to artists facilitated the meeting. He began by revealing his long-term intention to,

> mutually produce a wide scope of events from Gallery shows to Charity Events … work within the system for appropriation of the millions of grant dollars available from countless agencies … [and] provide an eclectic network of resources that will bring together talent, venues, media, production and internet applications.

The goal of these efforts, he went on, was to "lure wealth back to Holyoke" by making it a "more interesting, vibrant, and safe place". This was something, parenthetically, he felt local government had failed to do because of what he described as a "preoccupation with Holyoke's poor and an over-abundance of housing subsidies". The owners of the mill also spoke of their strong belief in Holyoke's ability to "make a comeback" due to its location on the I-91 corridor, its proximity to other attractive arts-connected communities[1] and its unique and beautiful architecture and setting along the canals. "We want [it] to develop as an urban area", followed one attendee [a curious remark given Holyoke's well-known industrial history]. "Holyoke", he continued, "is the only urban area in western Massachusetts with a grid and industrial buildings. It is *waiting to be populated* [my emphasis]." Striking about this comment is the point of focus on the inanimate structures and environmental attributes of a *potential* city, made for *potential* residents and perhaps a few tourists. Invisible is the existing population – not only the well-established Puerto Rican population, but also the older descendants of European immigrants who had worked in the mills and built Holyoke's reputation as a premier industrial city over more than a century.

As the meeting progressed, the owner of a nearby art gallery who had also inherited the much smaller mill in which it was situated, asked the artists in attendance to "see the city as I see it". He had been networking within Holyoke for many years to promote the arts, and asked the assembled artists to look down on Holyoke as if from a plane, view the amazing canals, and imagine a proposed Canal/Arts Walk alongside. "There's a lot of space to fill", he said, "and no end to how many people can start businesses here". In what he surmised to be support for these sentiments, the entertainment-marketer added his own vision for the mill and the water and canals that surround it. With "our efforts", he said, this area would become known as a "safe haven", a positive respite from "crime".

During the introductions that followed, two artists who currently live in Holyoke commented that while their parents had actually grown up in the city, chosen to leave, and probably contributed to its decline, they had decided to move back in. Another visual artist felt obliged to add that she actually """lives in downtown", that you "get used to it", and that it is "not so bad". Two other local artists, also women, spoke of the negative reaction, even resistance, they get when they mention they are from Holyoke. They spoke of how difficult it is to get their friends from the upper valley to visit them. "I always wondered", said one of them, "Why not Holyoke?" When the meeting reconvened after a refreshment break, one of the local artists asked a pointed question of the organizers: "What plans do you have to bring a more diverse group of local people in to this group?" The question provoked a quick and defensive response from the facilitator who said that they were "trying to hook into the local music scene" but that the "real problem" was that in trying to bring an economy to downtown you have to "pay people and you need people who can pay". The people of Holyoke, he said, "can't pay".

The contrast between this vision of the role that a cultural economy might play in the regeneration of a former industrial city and the earlier notions of building upon the cultural assets of a diverse population, frames the story that follows. Holyoke is not unique as a city in search of a new economic engine to replace

---

1 Nearby Northampton, Massachusetts was once named the number one small arts community in the country in a book by John Villani entitled *The One Hundred Best Small Art Towns in America* (1994).

its once vibrant manufacturing base. Yet few studies highlight the important class, racial and ethnic dimensions of the debates and practices surrounding the creative planning process. The parallel tracks that have evolved since these early meetings in Holyoke now incorporate visions of technological and entrepreneurial development. Combined, these approaches to regeneration enable a particularly useful exploration of the contrasts between more mainstream creative economy models and the potential that might emerge from more sustainable and inclusive creative economic and community development practices. Such an examination provokes questions that get at the heart of how the problems and assets of smaller post-industrial cities are understood, and revitalization approached, by the city, outsiders, newly arrived artists, young techie entrepreneurs and various cohorts of Holyokers whose paths rarely cross. It identifies the forces of resistance these divergent narratives spawn, and in the spirit of a "people's geography", engages in a complicated search for the alternate functions that art, cultural creativity, and innovation play in the lives of marginalized urban dwellers, or *might* play in a very differently conceived urban development strategy.

"There are Only Three Cities in the World, Holyoke, Paris, and New York"

When Belle Skinner (daughter of silk magnate, William Skinner) uttered these words in the early 20th century, Holyoke, Massachusetts, which came to be known as the "Paper City", had more millionaires per capita than in any other U.S. city (Dunn 2005). By 2010, it was the sixth poorest city in the country.

The Boston Associates, a group of prominent Boston industrialists who already held interests in several cotton mills in Lowell, Massachusetts in the East, and Chicopee, Massachusetts in the West, founded Holyoke in the mid 1840s. They sought to build one of the first planned industrial cities in the U.S. by harnessing the power of the Connecticut River's Hadley Falls through the construction of a series of canals. The idea for the city originated with George C. Ewing, a traveling salesman who sold scales, and knew of the mills in Lowell, Massachusetts and along the Blackstone canal in Rhode Island. After he convinced several Boston businessmen to join him in the venture, and their agent carefully persuaded farmers of Ireland Parish to sell their farms, the Associates developed a plan for a wholly new city of mills that would replace the agricultural land with a dam, three levels of canals that total 4.5 miles in length, and at least 50 water-powered factories (www.thelivingmuseum.org). When the water flow was measured in the driest year of 1847, it was found that it could produce 30,000 horsepower sufficient to power 550 mills (Thibodeau 2006: 4).

Holyoke soon became a city of immigrants. Many male workers were Irish, though young women and men were recruited from farms all over New England, and as far away as Quebec. Initially the Irish came in search of employment and they constructed the dam and the mills, living in the lower wards near the factories and the canals. Mill owners lived up the hill. Successive immigrant groups, including French Canadians, Germans, Polish, Italians, Portuguese and English moved into the city where they encountered tremendous discrimination and poverty. Those who arrived last were met with distrust, while residents who were more established moved out of the lower wards and up the hill into better-resourced neighborhoods. Diversity according to ethnic background and social class, along with craft exclusivity, hindered labor organizing, though labor unions remained strong in the city for some time. Ethnic diversity also divided neighborhoods and defined residents' everyday cultural worlds (trends that continue to the present day).

Holyoke became known more for its production of high museum quality archival paper than for its textiles as a result of shortages of cotton during the Civil War. Unlike the textile industry, run by absentee owners, the paper mills were developed using local capital (Green 1939: 79). Joseph Parson's mill produced more than two tons of paper per day and was the largest writing and envelope industry in the U.S. in the 19th century, earning huge dividends for its investors (83). By the turn of the century, Holyoke had a more diverse economy. Some mills still produced fine silk, linens, cotton, and a range of products that included trolley cars, books, and bicycles. There was also a thriving machine and tool industry that was organized on a smaller scale and sparked inventions to support the larger industries (81).

Population doubled by decade throughout the late 19th century reaching its highest level of around 65,000 in the 1920s (Destination Holyoke: 1). By 1928, Holyoke was reputed to be one of the wealthiest cities in the U.S., hosting numerous full-functioning live theaters, music and dance halls, and a spectacular Opera House with touring companies from New York City. Then, as now, a variety of ethnic festivals were held throughout the year and rival sports teams competed.

Post-Industrial Holyoke

The decades from the 1960s through the 1980s were not good ones for Holyoke. Thousands of residents who could no longer find jobs left as more manufacturing firms closed in search of cheaper labor in the south or off shore. Abandoned factories and residential properties in need of repair, or no longer generating a profit considered sufficient by their owners, became the target of arsonists. In the decade of the 1980s, Holyoke acquired the unfortunate title of "Arson City" and dangerous fires with loss of life became an almost weekly occurrence. Residents led marches in protest of building owners who were paying people to burn their buildings down, some with dire consequences.

Like many similar small industrial cities in New England, Holyoke did not benefit much from the economic upturn in the 1990s, and is now a net exporter of jobs in the region. It is estimated that 45% of its residents work outside the city, and higher skill jobs within Holyoke tend to be filled increasingly by non-residents. The city's economic base moved squarely into the service sector, bifurcated by pay and skill levels, following decades of deindustrialization. One of the largest current employers, the Ingleside (Holyoke) Mall, built in 1979, contributed to the decline of downtown business. Other significant employers include Holyoke Community College, which also moved its main campus from a downtown location (though it recently opened a new adult education center in a renovated downtown fire station), Holyoke Hospital, and the newly renovated Holyoke Health Center that is downtown and serves the health needs of a significant proportion of the low-income population of the city and the region.

Puerto Ricans became U.S. citizens in 1917 and migrated to the Pioneer Valley in successive waves starting in the 1940s, the late 1960s, and again in the 1980s and 1990s. Many original migrants were forced out of Puerto Rico by the disastrous economic effects of Operation Bootstrap.[2] They arrived in the Northeast as migrant farm workers just as tobacco fields in nearby Connecticut were diminishing. In addition, unlike the situation of early in-migrations of Irish, French-Canadians or Polish who arrived in Holyoke when labor was in great demand, Puerto Ricans arrived just as the manufacturing jobs were leaving and the infrastructure was literally going up in smoke. Much of this migration was local, as the construction of Interstate 91 bifurcated primarily Puerto Rican neighborhoods in nearby Springfield Massachusetts in the early 1960s, forcing many residents to leave. Both the percentage and absolute number of Puerto Ricans in Holyoke continues to grow and many white residents of Holyoke have since left or are growing old in place. The city has a population today of about 39,000 and close to 50% are Puerto Rican. This number represents one of the highest concentrations of Puerto Ricans in a city outside of Puerto Rico (http://quickfacts.census.gov/qfd/states/25/2530840.html).

In 2010, Holyoke was ranked the sixth poorest city in the U.S. and 43% of the city's households had an income of $25,000 or less, compared to 25% for Massachusetts and 29% nationally. Per capita income for Latinos in the city is about one quarter of what it is for white non-Latino residents, and the percentage of those living below the official poverty level is around 35% and growing (http://factfinder2.census.gov/faces/tableservices/jsf/pages/productview.xhtml?src=bkmk). Obstacles to economic achievement for Latinos result from discrimination, insufficient political representation, and a lack of proficiency in English, due in large measure to an insufficient number of ESOL (English as a Second Language) programs. Additional barriers to employment and social mobility include a lack of access to transportation, childcare needs and inadequate funding of schools. Holyoke's economic challenges are augmented by its location in a larger region known far and wide for its liberal life style and quality of life.

2 Operation Bootstrap was a U.S.-sponsored program that invested millions of dollars in encouraging factories to relocate to Puerto Rico for cheap labor and to escape federal taxation. This shifted the economy there from agriculture to manufacturing and tourism. The new jobs never compensated for the loss of rural agriculture and industry. This led by the 1960s to massive unemployment and the migration of many Puerto Ricans to the U.S. in search of jobs.

The city of Northampton and town of Amherst, which house several elite private colleges and the University of Massachusetts, lie just north of Holyoke, separated by the Holyoke Range, or what some refer to as the "tofu curtain".

Neighborhood residents in the four lower, mainly Puerto Rican, wards of the city abutting downtown are anxious to move forward having survived the tumultuous 1980s. This was a time when law suits were brought against the city for attempting to restrict Puerto Rican access to housing through the rezoning of residential areas to industrial – tactics that led to abandonment, disinvestment and even arson. The neighborhood agenda now consists of increasing business along Main and High Streets, and expanding a range of social, language, service and youth programs. Residents would also like to beautify existing, and develop new, infrastructure, recreational facilities and parks. They demand safer streets and access to quality affordable housing, as well as the cultural assets to reinvigorate neighborhood life. Community development corporations such as Nueva Esperanza, Nuestras Raices, Enlace de Familias and the Community Education Project (among many others) focus their agendas on improving access for all Latino residents to higher education, better health care, good jobs, housing and improved environmental quality. As will become clear below, the litany of challenges that these organizations, and Holyoke residents in general, face is accompanied by a significant number of assets that include a strong sense of family and a vibrant cultural environment with talented residents. There is also a deep appreciation for improved aesthetics, which residents take as one important measure of the quality of urban life.

Setting the Stage: City-led Economic Development and the Cultural Turn

For several decades following the post-war period of accelerated deindustrialization and population loss, Holyoke's politicians and planners followed the path of similar cities in their effort to retain manufacturing. They put numerous incentive packages together to lure new manufacturers into outmoded spaces or vacant lots. More recently, an emphasis has been placed by the city on downtown revitalization, dealing with surplus vacant property and developing more activity near the mall and the Interstates. In the last city administration, efforts also concentrated on promoting a new image for the city and developing the workforce. Disposal of city owned property due to foreclosures and encouragement of development were other high priorities.[3]

3 While Holyoke's 11 million square feet of industrial and manufacturing space may look abandoned, 10 million of those square feet (about 90%) are actually occupied by a variety of small manufacturing enterprises, services, and now some spaces for creative enterprise, though much of the space is underutilized. The city also owns 54% of the acreage in Center City (Massachusetts Technology Collaborative and John Adams Innovation Institute 2011: 2–18, 2–20).

Attention to the possible role of artists in Holyoke's regeneration accelerated in the 1990s. This emerging vibe was driven largely by publicized examples of artists moving into former warehouses and factory-lofts in districts such as Soho in New York. Around the 1970s, a group of people involved in local theater and heritage tourism formed an entity called "Passport Holyoke" to draw attention to Holyoke's unique industrial history and promote all of the city's diverse cultural, educational and recreational resources (http://www.passportholyoke.org). In 1976, the U.S. Department of Parks constructed Heritage Parks in eight downtowns across Massachusetts. Holyoke's Heritage Park was built in 1984 and placed near the first level canal. It now shares this site with a Children's Museum, the Volleyball Hall of Fame and an historic carousel that was once in a nearby amusement park atop Mount Tom.

By the 1990s there was a significant "buzz" in New England about the role of the arts and culture in the regional economy and numerous statewide think tanks convened to explore "best practices". These meetings and the ideas that emerged did not escape the attention of the city planner for Holyoke who began to import practices from cities such as Providence, Rhode Island, Lowell and Worcester, Massachusetts. Among these ideas was the creation of a mixed use Arts Industry Overlay District, which, at the time, employed an inclusive definition of the "arts" and did not require the bulk of an artist's income to be earned from the arts in order to access space. Also incorporated into this vision was the important goal of linking mainly Puerto Rican residents living in abutting districts to the city center.

Once the Arts Industry Overlay district was formally designated, divergent paths towards a new cultural economy surfaced. Explaining this divergence, however, is not as simple as drawing a line between local government and community-based initiatives. As the following story reveals, there have been times when efforts to rejuvenate Holyoke's downtown and the canal area have drawn both sides of the political and social spectrum together, and other times when differences have been more visible.

The Cultural Economy Meets the "Paper City": Take One

Holyoke's master plan from 1999 identifies the historic mill and canal area as a potential asset due largely to a consultant's report that focused attention on what it called the "19th century remnants" found in the city's downtown canal district. "Architecturally significant structures" and the presence of the waterway are seen in this report as a potential backdrop for new commercial and residential revitalization in spite of the deteriorating physical infrastructure. Proposed in that master plan are a Canal Walk, public markets and retail incubators, factory outlets, a brewpub, artists' lofts and workspaces and performance venues.

Canal Walk/ArtsWalk

Of these early ideas, the Canal Walk captured the imagination of some residents, and funding was secured to begin a participatory public design process. A Boston architectural firm hired by the city to produce the design built upon the special enthusiasm of youth and included them in the early visioning process. Indeed, the firm worked closely for more than a year with a coalition of youth organizations to support their efforts to generate a Canal Walk design. The Massachusetts Cultural Council also funded many brainstorming events that brought youth and adults together. The firm's eventual design reflected ideas generated through these well-attended exercises.

As the design process moved forward, the idea emerged for a once-a-year ArtsWalk to encourage residents to imagine how the arts could improve their quality of life. This initiated one of several moments of convergence among the varied constituencies in the city. An ArtsWalk committee was formed to plan a once-a-year festival along the site of the future Canal Walk. Greater Holyoke, Inc., a nonprofit organization that promotes economic revitalization, led this effort. The goal was to highlight the creativity of local artists and raise money to support local organizations that provide services to Holyoke youth.

Figure 3.1 Canal model created by Holyoke Youth

Figure 3.2 Youth parade to ArtsWalk

ArtsWalk was sustained for three years (1999–2001) with limited funding from the Massachusetts Cultural Council's Economic Development Program.[4] The original steering committee comprised an eclectic mix of representatives from city, community and youth organizations. It also included Holyoke's city planner, a representative from Career Point (an organization that works with local residents to ready them for the workforce) and the owners of some factory space for artists.[5]

The first year of the ArtsWalk Festival devoted considerable time to showcasing the arts and culture within Holyoke. Local artists, youth organizations, and performers were identified and recruited to participate. Additional tasks included the solicitation of local food vendors and the identification of sites and times for performances and parades that would move through many neighborhoods.

4 These well-attended events attracted a large and diverse number of residents. Yet, a recent newspaper article on the burgeoning creative economy and new artists attracted to a former mill building erases these events from memory. Its focus, a recent ArtsWalk that took place along the newly constructed Canal Walk in May 2012, is touted as a "first for the city" (Ward-Wheten 2012).

5 I was asked to join this committee to help identify and recruit talented Holyoke residents to participate in the ArtsWalk festival because of my ties to El Arco Iris, an afterschool art program in Holyoke where several Hampshire College students interned.

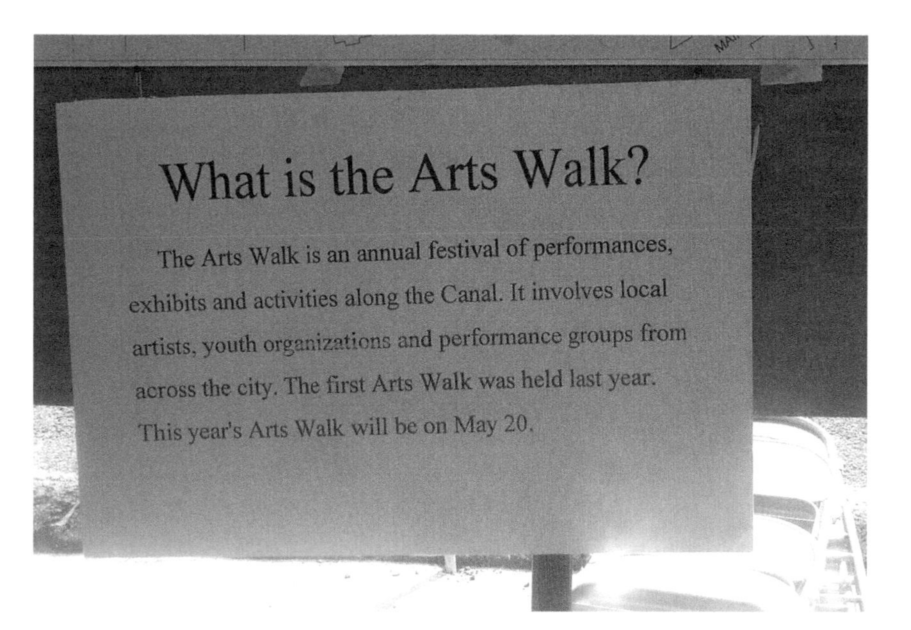

Figure 3.3 Sign displayed at ArtsWalk

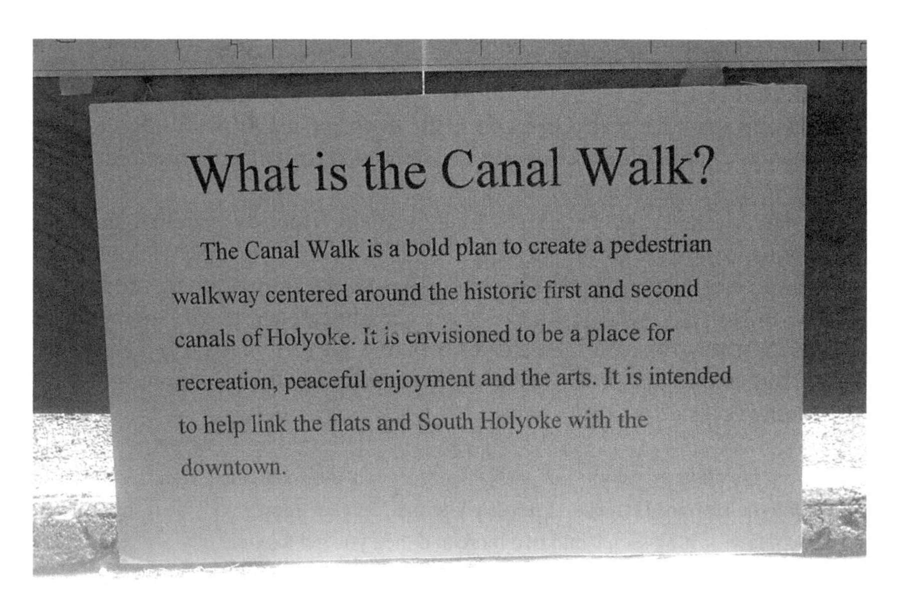

Figure 3.4 Sign displayed at ArtsWalk

Figure 3.5 Arts Corridor and Canal Walk Map

Permits were secured and money was raised to match grants and advertise. The original goals were to bring all youth and community organizations, creative arts organizations and adult residents with arts-related talents together to foster new alliances and spark the dream of a constantly evolving and aesthetically pleasing Arts Corridor that would contribute to neighborhood revitalization and provide a connection to downtown.

In spite of rain, the first festival drew many Holyoke families and the participation of several community-based organizations to view the work of local artists and performers. Among them was Tom Keeler, a performing artist who grew up in Holyoke, worked for the water department, and played drums for a group called Moonshine Earl. He spoke about how his part time life as a musician brought him into contact with so many different people in the city, "friends for life now, through music".

> There are always people out there that see you play and then come over. "Oh, you're from Holyoke? I never knew that. I never knew you." Next thing you know … it blossoms into two more and then you get together.

Having lots of drums eventually brought Keeler to the idea of creating his own authentic miniature replicas of drums used in real bands made out of old plastic film containers and other recycled materials. These were displayed at the ArtsWalk Festival.

Figure 3.6 Tom Keeler's drum sets

> Being born and raised in Holyoke I am proud of the history … Some great musicians, athletes and artists have come from Holyoke. It's my vision to be part of the history of Holyoke through my artwork. I also hope that art-related activities increase for all city youth to enjoy!

A second festival a year later focused even more attention on the idea of creating a uniquely Holyoke program of changing exhibits and interactive cultural activities along the proposed Canal Walk. In both years, members of the ArtsWalk steering committee contributed considerable time to organizing the event, encouraging resident participation and evaluating outcomes. Tee shirts and other souvenirs with the ArtsWalk logo were also produced to keep the idea of a rejuvenated Holyoke in residents' minds after the yearly event was over.

The third and final year of the festival took an approach that fundamentally changed the event. The entertainment marketing consultant mentioned earlier joined the planning committee hoping to draw the attention of outside artists to one of the largest mill buildings, and to use the event to "create a scene" that would attract more people from the larger region. In a generous gesture, he offered a huge parking lot next to the mill as the site for the event. Moving ArtsWalk from a string of locations along the canal, and within abutting neighborhoods, to a single location shifted the ethos of the event. More emphasis was placed on selling vendor space and recruiting "professional" artists from the larger region than on outreach to Holyoke residents and organizations. Reactions to the new format were instantaneous. Young people who had been actively engaged in prior years, described ArtsWalk 2001 as "bootleg" (a.k.a. "lame"). They complained of being harassed by vendors who quelled their curiosity about the artwork, and they claimed that they were "treated like we were going to steal something". The prices of items such as ice cream and French fries also went up considerably, and because there was less local outreach, there was a noticeable decline in the number of people attending the event from Holyoke.

As if to illustrate further the emerging divide, two organizations of artists subsequently formed. A *Creative Coalition* was started by the arts marketer to promote artist live/work space and generate a "buzz" regionally about the arts in Holyoke. A second group called the *ArTes Alliance* was fashioned by early ArtsWalk organizers, the director of planning, local artists and a number of Latino organizations. Its purpose was to promote new cultural initiatives and harness the existing talent and cultural resources in the city to promote collaborative community development. This included the use of art to enhance the built environment and the creation of opportunities for resident artists to connect with schools and businesses. Both efforts were designed to increase awareness regionally of Holyoke's cultural assets.

After early Canal Walk design charettes, the *Holyoke ArTes Alliance* and *Creative Coalition* dissolved and there was a hiatus in organizing until The Friends of Canalwalk and C.R.U.S.H. (Citizens for the Revitalization and Urban Success of Holyoke) formed around 2008. The latter group maintains an active event website and is comprised of people who describe themselves as wanting to "live, work, and have fun in a healthy, successful and revitalized former industrial city" (http://holyoke.ning.com/). Among other things, C.R.U.S.H. sponsors "Parties with a Purpose" where people gather to discuss important local issues, such as the likelihood of gentrification, while also enjoying some form of recreation such as a block party. Recently the group initiated a project to restore Pulaski Park, an Olmsted designed park. They also sponsored a series of "Bring Your Own Restaurant" (BYOR) nights, where an outdoor site is announced along the canals via the web and people show up in random fashion with chairs, tables and food to share.

The Holyoke Community Arts Inventory and Map

Documentation of the early ArtsWalk and Canal Walk design process involved the compilation of information about residents with local talent and arts-related industries in Holyoke that operate largely under the radar. The idea of building an alternative regenerative cultural economy – one that focuses on developing the capacity of residents – emerged from this work and led me to three collaborative projects, all of which combined to challenge Holyoke's image as a dying industrial city. The process of involving Holyoke residents in an ethnographic exploration of their *own* cultural and creative resources had multiple aims.

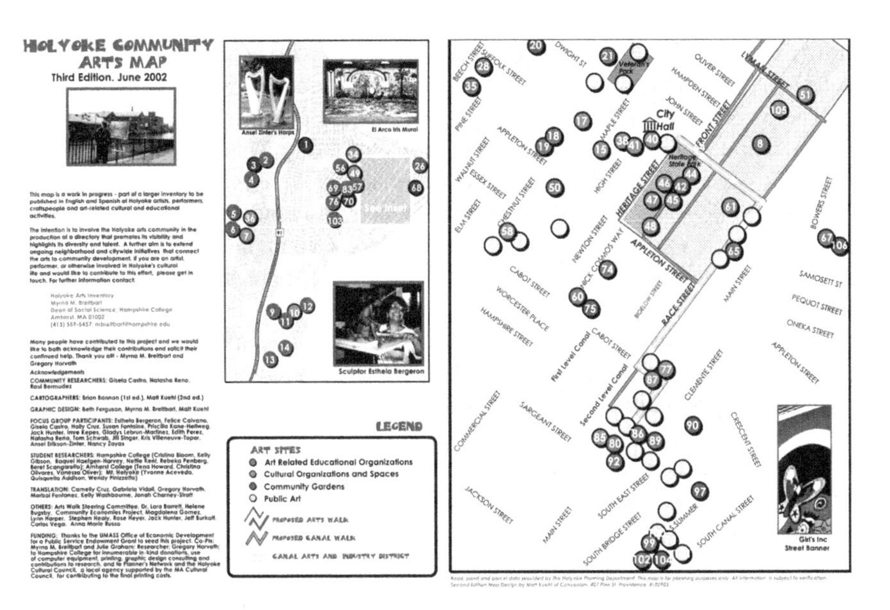

Figure 3.7 Holyoke Community Arts Map 2002 – English version

The first was to examine the meanings that art- and cultural-related activities currently play in residents' lives and might play in the future. A second goal was to use this information to develop inventive and collaborative community economic development strategies that build upon residents' talents and needs.

Since its inception, the project has resulted in two versions of a Holyoke Community Arts Map. Both are printed in English and Spanish, and highlight cultural institutions and spaces, arts-related educational opportunities, community gardens, murals and other forms of public art (Lauer 2001).[6]

A Community Arts Inventory, which I designed as a participatory action research project to generate residents' creative visions for the city, was also intended to draw attention to talent within Holyoke, and spark useful alliances to

6 This was followed in 2009 by a Holyoke Community Arts Calendar, produced with a colleague, Mari Castañeda, to highlight the history and vibrancy of the Latino community and draw attention to the very rich cultural resources and community organizations in the neighborhoods abutting downtown. The calendar includes key arts and cultural events, community information, and profiles of community members such as a graphic artist who directs an arts-based youth program, and organizations such as Kimbombo theater group, a program housed at the Community Education Project that teaches literacy and English as a second language through script-writing and performance of plays that address compelling health issues in the community.

further collaborative planning.[7] The research began with neighborhood outreach and a series of small focus groups with a diverse group of local artists.

The picture of arts-related activity that emerged provides a stark contrast to mainstream definitions of creative work. None of the residents interviewed lived in the loft-style mill buildings and the vast majority were members of extended families. While some worked in isolation, much art production in Holyoke was conducted as a *social* enterprise within the larger community. Some residents earned small incomes from their creative work yet all appreciated the recognition they received for their talents from others in the community and beyond.

The methods used to identify residents involved in arts-related activities began not with professional art association rosters, but rather, with referrals from established community-based organizations and social networking (not of the computer variety). As residents recommended people to speak with our definition of "artist" expanded to incorporate activities that would be otherwise invisible. One example is Maria Ortiz, who bakes for various family and community celebrations to supplement her income from babysitting and a school cafeteria job.

> I dedicate myself to the perfection of bizcochos – cakes made of flour, eggs, butter, and a flavor[ing] …. When I am making a bizcocho I am expressing what I feel and I would love the opportunity to express myself in an exhibition, because I do believe this is a talent.

Maria Salgado-Cartagena, another resident and artist, recalls how she "has been writing for as long as she can remember". As a young girl coping with strict parents and the absence of youth programs in the city, writing became a way to express feelings. With inspiration from her mother, who wrote letters, Salgado-Cartagena saw writing as a way to negotiate the two seemingly different worlds of being white or Puerto Rican in Holyoke. She said she considers creative writing a "life skill".

> Working as an artist has allowed me to become more active in my community, realize that I do have some rights as a member of this community. There's like a ripple effect because once you have self esteem, once you have cultural identity, it allows you to become a much more involved person.

With grandchildren who are now third or fourth generation Holyoke residents, Salgado-Cartagena is more committed than ever to revitalizing the city in ways that build the artistic and income earning capacities of all Holyokers.

After many similar conversations, it became clear that residents, whose talents contribute in meaningful ways to community and family events, celebrate their talents *as artists*. In addition to reinforcing the need for a much broader

7 Many people contributed to this project, most especially my co-researchers Gregory Horvath and artist/resident Gisela Castro, as well as Jessica Payne.

definition of the arts, these discussions brought into view complex livelihoods that can be difficult to measure. Elaborate cake decorating, catering, musical performances, DJing, and small craft production provide supplementary income for many Holyokers. In some cases, however, fear of jeopardizing government transfer payments or the lack of material resources and space inhibit these artists from developing and promoting their talents further. "I learned to cook helping my grandmother in Puerto Rico", said one woman. "She had a kiosk and to survive she sold all kinds of food. In Puerto Rico you can set up a little business where you want and support it yourself by making different kinds of food. Here, it's more complicated, you have to seek permission, and you need a lot of money."

Many of the talented residents we interviewed engage in complex relations of mutual obligation with family and friends, exchanging artistic practices for practical assistance when needed. Several artists within the Puerto Rican community apply their talents to playing music at community events or making requisite party favors, clothing and decorations for gatherings such as quinceañera celebrations (15th birthday parties for girls). The community-building aspects of art production are not restricted to the most recent immigrant communities, however. Nearly all of the artists we interviewed placed value on creative work as a way of bringing diverse communities together. "My feeling", said one playwright, "is that the arts *build* community … when people are involved in a creative process, they're vulnerable. And through that vulnerability comes a bonding and deeper listening to each other." Another musician and piano accompanist, who grew up in Holyoke, and has worked to restore the old Victory Theatre, envisioned this space not as the exclusively commercial operation it once was, but primarily "as a wonderful place for artists in Holyoke to … perform, a community performing arts space."

As our conversations progressed, a community arts sector located both inside and outside the nexus of the marketplace – in studios, performing arts centers, and public space – but also in the "everyday space" of households, basements, schools, churches and other family and community contexts, gradually revealed itself. Supporting an extension of the definitions of both "artist" and "cultural space", this included many large cultural events in public venues outside any designated "cultural district" (e.g. an in-city farm, neighborhood parks and city streets). Several residents we spoke to noted how much of the cultural production in Holyoke is hidden and subordinated to negative media about drugs and crime. In spite of this, they harbored dreams about using art to bring the Latino and older immigrant white populations of the city together in new spaces that they hoped the city would provide for music and art production. Concern was also expressed about how policies to promote a cultural economy might contribute to gentrification; especially as the rising property values in nearby Northampton drive artists to Holyoke. As one artist expressed it, "What we want here in Holyoke is Noho [Northampton] without the attitude!", rising property values and displacement.

A database of residents involved in arts-related activities, along with some visual documentation in both Spanish and English, was compiled from these discussions and subsequent networking. Preliminary findings from more in-depth interviews were then presented at a public forum in November 2002. This led to a discussion about how to proceed with arts-based development that could draw upon rich local talent and an unmet desire of many residents to access training in such fields as graphic design, digital music production, photography and other art forms.

Conversations with Holyoke residents and initial inventories of local talent reveal the importance of cultural expression, especially to the growing Latino population. This is reflected in the way several community-based organizations incorporate cultural development into their work. Nuestras Raices, a community-led "agri-cultural" development organization formed in 1992 by farmers who migrated from Puerto Rico, is significant both for its broad definition of environmental justice, and for its recognition of the important cultural basis of economic development and educational outreach. Its forward-looking agenda also recognizes the economic potential in establishing regional links with Latino populations up and down the Connecticut River. Since its formation, the organization has developed a network of community gardens across the city and has started a large 35-acre farm (La Finca) on vacant land along the Connecticut River. To build on cultural ties to the land, and train residents in farming and other related income-earning opportunities, farmers share equipment and collaborate in selling their products through weekly farmer's markets and a yearly Festival del Jibaro that draws visitors from the larger region and features music performances and crafts.[8] Food products are also sold to restaurants and at other cultural festivals. A number of food-related businesses have been incubated at the organization's headquarters on lower Main Street, which includes a commercial kitchen, and at one time housed a restaurant run by neighborhood residents, as well as the El Jardin Bakery, which has since moved but continues to sell organic bread around the Pioneer Valley. In addition to promoting small businesses, Nuestras Raices has responded to the neighborhood's desire to utilize vacant lots to create public spaces for socializing. It did so by beautifying the urbanscape outside its own offices on Main Street with a public plaza that is filled with plants reminiscent of Puerto Rico, a fountain and a large colorful mural.

In response to the dearth of economic opportunity for local residents, Nuestras Raices cultivates other forms of enterprise development as well. A former tenant rented a space for an exotic fish terrarium business, Marine Reef Habitat, which designed aquariums for offices, residents and commercial businesses. Nuestras Raices also maintains ENERGIA LLC, a socially responsible energy efficiency company that supports resident job training and career development in the green

8 It should be noted that not everyone in Holyoke is enthusiastic about including sustainable agriculture in the city's future planning. On one recent blog about the Victory Theater renovation, an anonymous resident wrote, "Now get rid of the third world socialist farm(s). The future can't be built on "stone knives and bearskins" (http://ramblingvandog.blogspot.com/2008/07/inside-victory-theatre.html).

economy by providing upgrades for residential and commercial property. Also located on site are after-school programs that address environmental justice issues by engaging youth in community mapping projects that identify toxic areas that contribute to a number of health-related environmental problems, such as juvenile asthma.[9] More recently, grants have been solicited to allow Nuestras Raices to expand its role in promoting healthy lifestyles and combating obesity and diabetes.

The Cultural Economy Meets the "*Paper City*": Take Two

Canal Walk 2

More than a decade lapsed from the initial participatory Canal Walk planning charettes and ArtsWalk events to the actual start of the Canal Walk construction. During that time, many of the original design ideas were lost[10] and many of the youth who had been involved in the first phase of the project moved on. Once revived, a different engineering firm was placed in charge.[11] Phase One, along the first level canal, across from Heritage Park, the Volleyball Hall of Fame, and the Children's Museum, now provides a brick walkway and some seating.

Phase Two, scheduled to begin in summer 2013, focuses on the 2nd level canal, with connections to Phase One and the rehab of an abandoned railroad bridge into a pedestrian walk. Phase Three, with no specified date for completion, is supposed to connect the neighborhood of South Holyoke to downtown (one of the key desires expressed by residents early on in the planning process).

Decisions about what sections of Canal Walk get built, and when, are driven in part by the city's ability to secure easements from the owner's of property alongside the water. They also depend on the city's goal of developing downtown as a cultural tourist destination. The Friends of the Canalwalk, a committee formed in 2008, is comprised largely of artists, some residents, and property and business owners along the proposed route. The early focus of The Friends was on getting the city to move on the construction, yet it maintains interest in the review of all plans.

---

9 In one green mapping project, the youth expanded the definition of "green" to include businesses that are Puerto Rican owned and spaces that they consider "welcoming to Puerto Ricans".

10 For example, though pedestrian and bicycle use in downtown are encouraged in a recent *Center City* report to improve health and social equity, no provision for bicycles and bike racks, specified in the earlier planning charettes by the youth, is currently incorporated into the existing Canal Walk.

11 The "Big Dig" in Boston drew resources away from the Canal Walk project in Holyoke. This was a major rebuilding of the central artery and tunnel project that was completed in late 2007 and monopolized nearly all of the state's transportation budget up to that point.

Figure 3.8 Photo of Canal Walk Phase One

Like the city, the committee envisions this as a branded space similar to other river walks. While there is talk of building performance spaces where outdoor cultural events can take place, it is not at all clear who will be able to program such spaces and who will have access to them.

The Victory Theatre

The soon-to-be restored Victory Theatre sits in a newly designated Arts and Innovation District that includes the War Memorial Building, where there are many musical and dance performances, the Heritage Park complex, a recently renovated public library, Wisteriahurst Museum, and several buildings with artist studios. This theatre is the last remaining of seven live theatres in Holyoke that once included a 3,000-seat opera house predating the New York Metropolitan Opera House by six years. The Victory was built in 1920 as a venue for theatrical performances (Broadway shows often came there first), and was once able to accommodate an audience of over 1,600. The theater originally sported silk curtains and wall panels made in the Holyoke mills. There are stairways of Vermont marble, Brazilian mahogany paneling and murals from the era of the Works Progress Administration. Detailing features the best of Beaux Arts and Art Deco, including an incredible ceiling of an oval room below the balcony and individual features such as curved walls, a black marble drinking fountain with

24-carat gold glass mosaic, Tiffany alabaster glass windows and a magnificent proscenium surrounding the stage.

Older residents recall a time when downtown streets were filled with many people headed to the Victory to see vaudeville or the latest motion picture. The theater was turned into a cinema and eventually closed in 1979. Since then, dedicated people have worked to save the building from demolition. The restoration project is now under the auspices of MIFA (the Massachusetts International Festival of the Arts), with headquarters across the street in another architecturally significant building. The organization has secured substantial funding through the state and federal government's historic preservation fund programs, and through new market tax credits and grants, but is running into problems with deadlines and funding on a project estimated to cost around $28 million.

The city is excited at the prospect of having a performing arts space with the capacity to host major productions and draw many people from the region and beyond to Holyoke, where they will presumably spend money on food and lodging (an estimated multiplier of $18.8 million annually for downtown). However, what distinguishes this flagship development project from others like it in larger cities is the attention that is being given by MIFA to providing benefits for Holyoke residents. To cultivate an audience for future performances school children are brought to matinee performances held in smaller venues. At these events, actors and performers informally share stories about their career trajectories with the audience. MIFA is also exploring ways to subsidize access to performances once the Victory opens, and is using the renovation process as an opportunity to develop the interest and capacity of local high school students to pursue careers in the technical fields of theatre, entertainment and media. This latter aim involves creating opportunities for young people to gain hands-on experience, with the help of local unions, through training programs in all aspects of theatre management and technical practice (e.g. historic renovation, lighting, set design, cosmetology, marketing and sound) as well as in the "softer" fields of playwriting and directing. MIFA has already initiated conversations with the local unions, Holyoke High School, and Dean Technical High School to develop these educational, training and job creation models further. Future programming is also designed to attract the diverse cultural mix of residents now present in Holyoke. Don Sanders, the Director, has difficulty containing his excitement about these varied possibilities, in part, because of their unpredictability. "There's something in it that's so unforeseen", he said, as he recounted the fan clubs that public school children in the city formed for their favorite actors after seeing a free performance of Shakespeare's *Love's Labour Lost.* Sanders is also excited by the possibility of training young people to run the lights and sound in future productions. He even dreams of creating a musical training program for children similar to the extraordinary El Sistema in Venezuela that produced Gustavo Dudamel, the conductor of the LA Symphony Orchestra. Other possibilities include renting access to the theatre to local colleges to raise money to support local outreach and more skill-based training programs.

The Massachusetts Green High Performance Computing Center

Perhaps the most anticipated anchor in the new Arts & Innovation District was the Massachusetts Green High Performance Computing Center (MGHPCC), the largest public investment made in Holyoke in 50 years, announced by Massachusetts Governor Deval Patrick on August 9, 2010. The MGHPCC opened in 2012 and is built on the site of the former headquarters of a fabric manufacturer (Mastex) that shut down operations in 2009. This site is in the downtown canal district and is designed to both spark development in the new Innovation & Arts District and generate new growth in the Pioneer Valley ("Computing center set for Holyoke" 2010).

In March 2010, Governor Patrick designated the Holyoke Innovation District, encompassing the four downtown wards, as the 20th Growth District in Massachusetts. Twenty-five million dollars were allocated for the project, with an additional $50 million coming from a consortium of colleges consisting of MIT, Northeastern University, Harvard, Boston University and the University of Massachusetts. Two giant computing companies, CISCO and EMC worked with Holyoke to establish the center as a collaborative research facility in the life sciences, clean energy and green computing. The MGHPCC, a LEED certified green building, now houses thousands of networked computers, and is in the top 500 of the world's most powerful computing facilities. Holyoke was chosen to house this complex for the same reason it was developed in 1860 as one of the first planned industrial cities – because of its hydropower – and, more recently, its location in one of the fastest fiber optic connected areas in the state. Indeed, the project was initiated to save money on electricity and infrastructure by combining computers together at one site and utilizing cheap "green" power. The facility cost around $95 million and is operated by a nonprofit organization created by the five affiliated research universities.

A kick-off event held in October 2010 at the new Multi-modal Transportation Center in downtown Holyoke, was attended by the Governor, many state and city officials, the President of Holyoke Community College, the Director of the MA Technology Collaborative's Innovation Institute, a few residents, and some young local entrepreneurs. In an interesting juxtaposition, the new Picknelly Adult and Family Education Center was dedicated just a few days prior, to provide adult basic education classes and day care. The Governor and Director of Housing and Economic Development stressed the impact the MGHPCC would have on new investment, jobs and "the people of Holyoke" more indirectly through the development of "local assets" described as "techies" and "entrepreneurs". Also mentioned was a large factory converted to various small businesses and artists, which is expected to host innovation events. The new Director of the MGHPCC spoke of the way that "science has changed" from lab experiments and theories to "simulations" and the building of computer programs to "let you ask questions requiring complicated computation". The "customer base", he said, are universities, researchers and businesses collaborating with universities that need to process huge amounts of data. The building housing these giant computers, the audience was told, would also use state of the art green design.

Only one meeting attendee, Gladys Lebron-Martinez, a respected member of the city council and youth services manager at Career Point, an organization that helps young people with "job readiness", inquired about associated initiatives to deal with the technology divide in the city. She wanted to know how the MGHPCC would increase the capacity of lower income residents to access the education and training necessary to secure meaningful employment in these burgeoning sectors of the economy. "How", she asked, "do you convert this [MGHPCC] effort to involve our school systems and our children?" The Director of MGHPCC, John Goodhue (MIT graduate and former President and CEO of eBay) turned this question over to others on the panel. They replied that some of the new center would be for "community use" – e.g. a lab where students from the technical high school and the community college could train in telecommunications and operating systems, or where those with an interest in using simulation in science instruction at the high school and middle school levels could potentially work. The stated intention of CISCO to work with the school system and the state of Massachusetts to incorporate technology into math education, and use technology to "address urban issues", were also mentioned.[12] The MGHPCC does not generate tax revenue directly for the city since it is a collaboration among non-profit educational entities, and only employs about 15 people.

Manufacturing Young Entrepreneurs

In attendance at the launch of MGHPCC was one of Holyoke's most notable young entrepreneurs, Brendan Ciecko, described by national magazines such as *Inc*, as one of the "coolest" and most "determined" under 30s in the country. He started Ten Minute Media, a digital design, marketing and branding company when he was only 12 years old. After his freshman year at Hampshire College, Ciecko received a call from Mick Jagger's manager who said that Jagger had seen his work and wanted to hire him. He left college to build Ten Minute Media into a thriving website development and graphic design firm for music groups and famous musicians as diverse as Natalie Cole, Katy Perry, New Kids on the Block, Schwayze, Snoop Dog and Julie Andrews (www.brendanciecko.com).

Ciecko's impressive array of clients is not the only thing that prompts Massachusetts officials to use him as the poster child for creative economy

12 In September 2012, Holyoke Community College (HCC), Dean Technical High School and the MGHPCC received a $200,000 grant to start the Cisco Network Academy, which will provide training and professional certification to students in networking. Cisco and the MGHPCC also initiated a "Smart and Connected Communities" program to place tele-equipment at HCC and (later) in the city's fire and police departments (http://www.mass.gov/governor/pressoffice/pressreleases/2012/20121114-ma-green-high-performance-computing-center.html).

initiatives; it is his vocal and material commitment to the Gateway city of Holyoke. On Twitter, Ciecko describes himself as "an entrepreneur, designer, and dweller of the post-industrial city". Besides locating his music website development business in Holyoke, he has purchased buildings in downtown, one of which he is developing for artists and cultural activities, and another, as an "epicenter" for small business development. He also created a website to draw attention to the new MGHPCC. In a recent interview and follow-up emails, Ciecko, who is still in his 20s, took the time to lay out his vision for the city of Holyoke. He talked about providing "an urban living experience" in a building that would provide an art space, a music room, classrooms and a student-operated café/gallery on the first floor. He also outlined a vision of downtown that includes a greater presence for Holyoke Community College and provides more entrepreneurial space to enable social interaction in retail and food spaces. One of the enterprises Ciecko envisions is an operation similar to Greyston Bakery in Yonkers, New York, with the motto, "We don't hire people to bake brownies. We bake brownies to hire people." Such a bakery would meet a number of needs, he suggests, including the provision of social services, access to affordable housing and health care, job training and career services.

An enthusiastic traveler with a special appreciation for former industrial cities in Eastern Europe, Ciecko speaks passionately about urban architecture and the Polish city of Lódz, which has managed to attract a good amount of foreign business investment and vastly reduced a 20% unemployment rate. Most exciting to him is a $250 million Euro construction project to build Manufaktura/Andels Hotel, which he describes as "the coolest factory conversion I've ever seen". After drawing similarities between Lódz and Holyoke, down to negative self-images, Ciecko speculates about whether one of Holyoke's largest factories might undergo a similar conversion, and whether vacant lots between buildings might be transformed into Green Thumb-style parks and gardens. Another dream he shared is to build an Architecture Center in Holyoke to honor George P.B. Alderman, the designer of some of the city's most beautiful historic buildings and mansions.

Ben Einstein, another young entrepreneur and recent graduate of Hampshire College who focused his undergraduate studies on commuter-aided design and fabrication, also drew the attention of state and city officials when he announced his intention to purchase a 4-story, former steam and plumbing supply building in Holyoke with Ciecko and launch a "Maker College". The for-profit space had dual goals. One was to provide a machine shop to give fee-paying members access to computer-controlled milling machines and lathes, 3-D printers and water-jet cutters to fabricate metal and other materials into tangible consumer products. Another was to start a "business accelerator" to launch product-based businesses that Einstein and other prospective investors would retain a share in ("Making Maker College" 2011).

Einstein's vision of a Maker College is compatible with the evolving ideas behind the newly designated Innovation District yet departs from this vision in some fundamental ways. While much attention is currently focused on how to help the city capitalize on the MGHPCC and attract firms interested in computing, research and technology development, Einstein's "Maker College" is focused squarely on reviving manufacturing, if on a smaller, more malleable scale than that fashioned in Holyoke's earlier paper and textile mills. The idea presumes a role in the new economy for small-scale manufacturing and, in many ways, references the growing movement in cities such as Detroit, of "hacker space" cooperatives filled with so-called "kitchen-table industrialists" who focus on small runs of products that others design. These spaces generally provide cooperatively owned machinery that the public can access, and are meant to encourage the development of small businesses among those with little capital (Giridharadas 2011). Unfortunately after several months of trying and failing to lure a sufficient number of venture capitalists to invest in these ideas in Holyoke, Einstein moved to Boston where he anticipates attracting more interest.

Another effort to direct the attention of entrepreneurs and venture capitalists to the city came in the form of "IdeaMill = Entrepreneurship + Design + Urbanism + Technology", an event held in Holyoke in October 2011 to provide a forum for "industry leaders and visionaries" to tell their stories of success. The event was meant to encourage collaborative profit-making endeavors through the design of creative industries in new media, music, "viral marketing" and other up and coming technology-based fields.[13] The gathering is emblematic of the continued, if slow, movement into canal mill buildings of a growing number of engineering, design, marketing and technology businesses for whom the existing population of the city remain largely invisible.[14]

The South Holyoke Revitalization Strategy and Central City Vision Plan

In 2008, a regional non-profit affordable housing agency, in collaboration with Nueva Esperanza, a local community development organization, conducted a comprehensive survey of the mainly Latino South Holyoke neighborhood. The aim was to encourage the city to "transform the … neighborhood into a more desirable place to live and work" and build on its assets (South Holyoke Revitalization Strategy 2008). Recommendations were made to address the

---

13 A panel entitled "Urban Defibrillator – How to revive a city: a discussion about urban design, start-ups, the arts and more" tended to propagate the image of technologists and designers as the life saving "medical staff" ready to revive Holyoke. With a continuing focus on the physical infrastructure, one twitter entry prior to the event referred to Holyoke as "Venice bones with a little high-tech." The Mayor himself now refers to Holyoke as "Digital City" in an effort to follow the path of nearby Pittsfield, MA, which is trying to rebrand itself as the "Brooklyn of the Berkshires".

14 The architect-owner of Open Square (formerly the large Lyman mill along the canals) has already successfully attracted such businesses, and there is evidence that the city is now drawing more attention of media entrepreneurs such as Steve Porter, an internationally known DJ, producer, and musician, who has restored another mill building.

"negative image" of the area and expand housing options to introduce more income diversity (especially for residents whose economic situation improves). The need for better linkages to downtown, better code enforcement, relevant job training and the construction of a community center were also noted. Education and sustainable economic and cultural enterprise development are the foci of many community-based organizations in this part of the city. However, this report was quickly overshadowed in August 2009 by another special report from a consulting firm commissioned by the city, and funded by the Massachusetts Department of Housing and Community Development.

The report, *Holyoke's Center City Vision Plan: A Plan for the Heart of Our Community*, resulted from five months of "stakeholder" interviews and public workshops, with the objective of creating a "sense of community ownership" over the city planning process. It highlights the four census tracts that make up the downtown and aspires to reconcile "conflicting views", though these are never specified in the final document (City of Holyoke 2009a). Much of the focus is on improving points of physical access, especially streetscape and architectural features that announce one's arrival into the city. A desire to "capitalize on the unique characteristics of Holyoke's urban neighborhoods" and to reuse historic resources within a combined Arts & Industry Overlay District along the canals is also mentioned. The report goes on to suggest a variety of improvements to the built environment and several "neighborhood initiatives", such as the renovation of local parks and new community gardens, transit, jobs/commerce and arts/history. Particular suggestions are made to develop key thematic and geographically distinct "nodes" around government activity, learning and transportation. One example is the "Cabot and Main Street" node, with mixed-use development and improvements to Main Street facades. The intention is to "create a sense of place" in the South Holyoke neighborhood and improve "connectivity" between this neighborhood and downtown. No discussion is included in the plan of the social/cultural barriers to achieving these goals or of the importance of capacity building among residents with respect to accessing education and workforce development, top priorities for community residents. The *types* of new jobs desired are not specified in the report though a section on the "enhancement" of neighborhoods speaks of the need to prepare places to be more "inviting" to residents in a range of income levels, and to encourage home ownership and the introduction of more moderate income housing. No mention is made of how to preserve affordable housing and prevent displacement, should that become an issue.

With respect to the arts, all three proposals in the *Center City* report suggest that new cultural activities be focused between the 1st and 2nd Level Canals (closest to High and Maple Streets in downtown). This was apparently based on a "community process" that identified an existing artistic community within buildings along these canals. It is suggested that this district be extended further into downtown to encompass the renovated public library, the Victory Theatre and another performance venue. The report does not discuss the primarily Latino neighborhoods between the 2nd and 3rd level canals where many artists live in rental apartments with their families. While the *Center City* report suggests "new events that bring **Holyoke's** diverse population together", it does so without acknowledging several Latino-sponsored festivals that are already designed to do just that.

The Innovation-based Economic Development Strategy

Shortly after the Center City Vision Plan, an Innovation District Taskforce was set up to complete a strategic plan. The resulting *Innovation-based Economic Development Strategy for Holyoke and the Pioneer Valley* outlines actionable strategies to attract private investment and regional assets to the city and region (http://www.innovateholyoke.com/index.php/download_file/view/3/173/). Several elements of this report pick up on the earlier Center City Vision plan, especially those pertaining to improvements to the built environment such as the design of more flexible work and live spaces near downtown Holyoke and the enhancement of existing sites for new business. The Canal Walk is one of several infrastructural projects expected to draw people and businesses to the city and downtown. A new Transit Oriented Design District focuses attention on the the restoration of Amtrak train service to New Haven and New York. This project involves faculty and students from the University of Massachusetts's art, art history and architecture departments in a plan to renovate the now vacant historic H.R. Richardson train depot (most recently used as a machine shop to rebuild engines). The *Innovation* strategy also proposes that Holyoke become a "test bed" for innovative energy production and urban agriculture.

With respect to the creative economy, the report suggests an Industry Overlay District and an Arts/Innovation Overlay District. Within the latter, priority is given to incorporating more middle income housing[15] and developing sites and the streetscape within one-two blocks of the newly constructed MGHPCC and the proposed train station. This Innovation District is smaller than the older combined Arts/Industry district. It hovers around institutions downtown but does not extend down to Latino neighborhoods near the 3rd level canal. Any reference to "capitalizing on the unique characteristics of Holyoke's urban neighborhoods" (present in the Center City Vision Plan) is missing.

The absence in this report of reference to the existing population in the city other than in the context of speaking of its deficiencies (e.g. high unemployment; inadequate education and skills, incomes considered too low to support a downtown retail revival) is striking. While entities such as an "Innovation Entrepreneurship Center" are proposed to advance the digital technologies and

15 A Center City Housing Initiative Program that follows sets out a plan to "expand the diversity of housing stock" within the Innovation District by providing substantial tax incentives for developers to rehab properties as multi-unit market rate housing (www.holyoke.org/cchip).

attract new residents, most of the jobs explicitly mentioned for Holyokers are urban agricultural positions that the report presumes can "be filled by the unemployed". The report also recommends growth in the urban agriculture industry cluster, with coordinated marketing and outreach (including access to healthy food) to the over 300,000 Hispanic residents in the region. The target population is said to have "underutilized skills and business potential," yet Nuestras Raices, the community-based organization that is almost wholly responsible for the development of sustainable agriculture, community gardens and food-related small businesses in Holyoke, is not explicitly mentioned within the list of possible "Collaborative Partners". Also recommended are efforts to increase young people's awareness of manufacturing as a "viable career".

Envisioning Holyoke: Future Scenarios

More than a decade ago, at one of the first gatherings of artists in Holyoke, the proprietor of the 600,000 square foot former Lyman Mills spoke of the role that he believed artists could play in the rejuvenation of Holyoke. He emphasized the compactness of the city and the beautiful architectural details of the physical plant. He said he believed that new people could be attracted to live in Holyoke "without displacing the existing population" or reducing the city's diversity. Some residents are questioning this possibility.

The new Mayor, Alex Morse, who drew a significant vote from Latino constituents and opposed a casino as the solution to economic woes in favor of building a new base of green technology and arts-based enterprise, initially supported the demolition of a low-rise public housing project located in downtown near the new Canal Walk and Open Square. Many residents critiqued this decision as an example of "poor people removal" reminiscent of the early years of urban renewal (http://hush.fluxmass.org/2012/05/10/response-to-mayor-morses-statement-on-lyman-terrace/#.T7Zk1r_bbPR). Some also saw the decision not to renovate long-neglected structures that residents want to remain in, as an example of a building conflict between the Mayor's desire to attract a new "creative class" to Holyoke and the right of an existing population to continue to occupy downtown neighborhoods.[16]

Discussion of Holyoke's "weaknesses" in the *Innovation-based Economic Development Strategy* refer to "entrenched poverty and issues related to residential development" in Center City as a "threat" to economic development and growth of the retail sector (section 3-2). The report recommends that residents there be linked to "opportunities, jobs, education/skills development and entrepreneurship," utilizing the nearby Five Colleges, improvements in STEM (Science, Technology,

16 It is estimated that more than 50% of Center City population is under the age of 24 and 83% identifies as Hispanic (Massachusetts Technology Collaborative and John Adams Innovation Institute, 2011: 3-2).

Engineering and Math) education and various regional and talent development initiatives. However, it notes the inadequacy of K-12 education in Holyoke as a real obstacle to achieving these goals. The report further suggests that the success of the Arts/Innovation Overlay District requires a revision of the city's incentives and permitting process prioritizing sites for building rehab and site clearance (4-14).

> Strategies should be prepared for the viable redevelopment of all sites and properties. These strategies can include promoting and enhancing the marketability of existing buildings already on the market; assembling adjacent parcels to form larger tracts, more amenable to successful redevelopment; the selective razing of properties *whose structural deficiencies, overall condition or physical character leave them antithetical to overall goals of this district* [my emphasis]; fostering public-private partnership with specific redevelopment goals and incentives.

The "obstacles" to regeneration noted here, along with the high cost and formidable challenges of renovation, may explain why the Mayor took the initial position he did on downtown public housing. However, he recently reversed this position saying that he "cannot in good conscience support any demolition of Lyman Terrace – total or partial – until our citizens have ample opportunity to have their voices heard regarding the community needs there." Morse urged the Holyoke Housing Authority to be "more responsive to the immediate needs of Lyman Terrace's tenants" for more green space, sufficient parking and a community center, and for any future plan to "be consistent with a long-term vision for our downtown as a diverse, densely populated, vibrant, and prosperous place." "Growing our population", he continued, "must entail keeping people in Holyoke, not forcing them out" (www.masslive.com/news/index.ssf/2012/08/holyoke_mayor_alex_morse_asks.html).

Holyoke's effort to spur economic revival by cultivating entrepreneurial innovation and a creative economy mirrors patterns in many smaller post-industrial cities. One can read this progression from ArtsWalk 1 to Idea Mill as a hopeful sign of revitalization-to-come or as impending gentrification. Though moving along parallel tracks, these very different approaches to creative economic growth sometimes converge to reinforce a sense of hopeful possibility. More often than not, the separate tracks remain unaware of each other's existence.

Uncovering "Hidden Pearls"

Participants in the early Holyoke Community Arts Inventory identified a myriad of positive impacts that they believe would come from a more serious consideration of culture and the creative arts as one aspect of economic development. Most residents concur that there is real talent in the city, something that Pedro Rodriquez refers to as "el oro del barrio", the gold of the neighborhood. Like others, however, he believes that this "gold" can take effort to find. "We are like a hidden pearl that you have to search for", he says.

Today some residents worry that redevelopment efforts might threaten an emerging sense of community, pushing locals to use creative expression only as a form of resistance to their own marginality, or as a means to cope with geographic and cultural dislocation.[17] The will to search for the "gold" from within Holyoke is hard to find in current redevelopment documents, which sometimes present the city as an unpopulated stage set comprised of romantic mill facades waiting for a new creative or middle class to move in. What would it take to begin to fuse these parallel creative tracks?

One starting point is to expand the definition of the creative arts to enable residents who are not currently defined as professional artists or creative entrepreneurs to access the resources needed to expand their skills and develop more community-building and income-generating opportunities. Massachusetts is considered a leader in its support of creative industries, and reports over 14,000 creative industry businesses in the state, employing around 80,000 people. These businesses are defined as entertainment-based industries, advertising (including digital), and a variety of design-related professions. Governor Patrick recently launched a Creative Economy Initiative and appointed a Director to work with such businesses and artists. With such a limited definition of creative industry, however, few of the talented residents of Holyoke will be counted or provided with the funding and training support necessary to access needed capital and the space to develop their capacities.

Smaller industrial cities hit especially hard by the out migration of industry were once dependent for jobs on a single enterprise. Today, economic diversification in cities like Holyoke is a sought-after norm. Just as broader economic development strategies aim to attract a diversity of well-paying jobs, efforts to develop a creative economy must involve more than a mix of heritage tourism, flagship developments and artist lofts on the one hand, and a community-based arts scene on the other. While each may individually enrich the quality of life, neither alone addresses the need for residents to participate as full cultural citizens and attain educational and economic opportunities that provide higher standards of living.[18]

17 "Having most of my family in Costa Rica can be stressful", said Maria Isabel Ramirez, a weaver and dressmaker. "These [art-related] activities are a therapy. I really enjoy myself … This work connects me to Costa Rica, where many people practice these arts. They also connect me to people in Holyoke. The work doesn't only benefit me; it benefits the whole community."

18 Indeed, there is a continuing skepticism in Holyoke about basing the city's future on its past. Until the Victory Theatre renovation began, the city seemed intent on razing as many historic buildings as possible. Now there seems to be recognition of the importance of retaining a sense of place and history, though not to the exclusion of reinvention and a serious consideration of how the city's assets might be used to fashion a different future.

Among many things, diversification of the local economy requires further exploration of Holyoke's position within the larger regional economy. This should include the building of a more extensive network of Latino businesses that connect residents along the Connecticut River's so-called "knowledge corridor" from Hartford to Holyoke and up to Vermont. Professional market research commissioned by Nuestras Raices in 2006 suggests that there is a strong unmet demand for cultural activities among the over one million Latino residents in the Connecticut River Valley corridor. The report further suggests that families would be willing to travel up to an hour and a half to visit (and spend money at) events perceived to be safe and family-oriented. The Center City Vision plan makes no mention of such potential.

Greening Holyoke

Another area targeted for economic growth in Holyoke is the green economy. Holyoke is designated as one of 11 "Gateway" cities in Massachusetts, called "Gateways" because they share in common an industrial past, are the entry points for many immigrants, and experience real economic challenges. In 2008, the state passed the Green Jobs Act to help promote job creation and workforce development. The act encourages the use of recycled materials and green energy (often the water power available in these cities). It also fosters less toxic production methods and products. Weatherization programs for residences and businesses are considered to be well suited to cities that have an older building stock. Holyoke and Springfield are targeted for funding to support these conversions and the training of local residents in related fields.

Holyoke's new urban renewal plan *Connect Construct Create* (www.holyokeredevelopment.com/tag/connect-construct-create/) includes a focus on bringing new investment in "green" development to the city, and there are currently several on-going efforts to build new small creative businesses to foster sustainable ways of living and making a living.[19] The Massachusetts Economic Development Incentive Program (EDIP) is supposed to stimulate business development and job creation through three-way partnerships between the state, companies that want to expand and a city or town. The City of Holyoke's Economic Development Department administers this program for the region. An Environmental Integrity Company, funded in 2009, invested $650,000 to retain existing employees and committed to adding 12 new positions with preference given to Holyoke residents. A marketing video titled *Holyoke: Green Business Grows Here*, developed by

19 At the Centro Agricola, attached to Nuestras Raices, formerly unemployed young men and women aged 17–21 take part in a green jobs training program and learn about a new community-based business enterprise that is being formed to promote design work in the area of green housing and commercial development. This is in addition to the other small sustainable businesses at this location and at the 30-acre farm down Main Street mentioned earlier in this chapter.

the Holyoke Gas & Electric Department, the Pioneer Valley Railroad, and the Holyoke Economic Development & Industrial Corporation (HEDIC) highlights the advantages that Holyoke offers for modern manufacturers and green business and biotechnology initiatives, especially with the new MGHPCC (City of Holyoke 2009b).[20] The owner of the large mill complex, Open Square, has moved far along in meeting this goal. The seven-building complex is now the largest net zero energy mixed-use development, generating 50kW of energy through two hydroelectric generators. This represents more energy than they consume (www.opensquare.com).

Neighborhood Regeneration

In the *Holyoke Community Arts Inventory* a number of residents said that they lacked adequate public spaces and sufficient access to performance venues, materials, equipment and the means to sell their crafts. Miguel Rodriguez said that when he was small he used to sing all the time, bring music books back to his house and keep singing until his family "finally told me to stop". His manager discovered his talents by accident when he heard him singing in his basement. As an accomplished musician, Rodriquez plays all over Massachusetts but travels to New York for recording sessions. More local performance spaces and a recording studio that could be accessed affordably would make it possible to do more work in Holyoke.

Similarly, Gisela Castro, an artisan who caters and produces crafts for many Holyoke family celebrations, finds it difficult to locate the art supplies she needs in the city. She and the women with whom she shares her culinary skills value the opportunity to both teach and learn from others about traditional dishes from different parts of the world and, like so many other residents, her involvement in activities through the arts, "make Holyoke feel more like home". When we spoke a few years ago she longed for access to a commercial kitchen and more affordable space where children and adults can converge to share talents, learn and produce together. Rather than consign the arts to galleries, Castro wanted to see large art spaces on busy street corners in people's neighborhoods.

According to Maria Salgado-Cartagena, a poet and creative writer, the demand for community space derives in large measure from the housing situations of many residents.

> A lot of the [Holyoke] community lives in buildings where you rarely have enough space for you and your family .... If we had a community building where there was a darkroom then people who take photos could use that space ....

20 To my knowledge, no mention was made in this video of Nuestras Raices' long-term efforts to bring more job-seeking residents into this green economy and to improve the important but less tangible impacts of green infrastructural development, such as parks, plazas and cultural events that contribute in key ways to improving the quality of daily life.

> The biggest luxury is space. And there is no space for a lot of people, including myself. This is a big issue.

Many musicians cutting CDs, caterers, artists and craftspeople currently work out of basements and cramped rental apartments shared with their families. "We need space," said José Ramirez, "and people need to understand it's one of our most important things."

Standard policies that promote the development of small live/work studios in a downtown Innovation District will not address the needs of these residents for whom the separations between art and daily family life do not exist, and are not necessarily seen as culturally desirable. What residents say they need are flexible and accessible *public* communal art and design spaces that comply with health laws and building codes (e.g. commercial kitchens, sound-proofed practice and performance spaces etc.), and that can be flexibly scheduled for use by different groups of people.[21]

There was a time in the early to mid 20th century when different ethnic groups in Holyoke shared cultural spaces. One example is the former Valley Arena, built in 1926 in the middle of Holyoke's working class neighborhoods. The space was run as a boxing arena until 1944 when it was converted to house wrestling, performances, dancing, and Big Bands, including Glenn Miller and the Dorseys. Each event apparently appealed to a different audience and was scheduled for a different night (http://boxrec.com/media/index.php/Valley_Arena_Holyoke_MA). A return to this idea of developing more flexible uses for venues through creative scheduling could contribute to the economic vitality of Holyoke and help to address imbalances in residents' ability to access cultural, economic and educational space. Expanding ideas such as a "Maker College" to include spaces and times when local high school youth, young adults in search of an income-earning, and recent college graduates could come to acquire new skills, incubate products, and market their ideas could also begin in small ways to generate new career possibilities.

Building Capacity through Education/Supporting the Aspirations of Youth

In the late 19th century, many pointed to education and the public schools as the answer to "acute problems of community living" in Holyoke. Interestingly, they emphasized science and math and its ties to manufacturing (Green 1939: 126). Holyoke currently ranks 330th of 332 Massachusetts school districts, an obstacle that the new young Mayor and recently elected State Representative Aaron Vega, are profoundly aware of, and vow to address (Jonas 2012).

---

21 Many people interviewed also see a real potential for neighborhood-based theatrical productions that enhance literacy and political involvement in the city through the subject matter of the plays. Kimbombo is already doing such productions, though it must often search for spaces in which to rehearse and perform.

A real challenge is thinking about how the educational system can raise not only the scores of Holyoke youth on standardized tests but also keep young people in school by awakening their desire to learn. When Holyoke youth see underpaid service jobs as the only career options, there is not a lot of incentive to move through the system. Exposing them to careers in growing fields such as digital music, media, graphic, industrial, web, fashion and other forms of design will not automatically entice youth to complete school or result in a job, but it can engender more interest in learning. This is happening in Lawrence, Massachusetts, where a Community Development Corporation, Lawrence CommunityWorks, rehabbed a vacant parochial school into a state-of-the-art after-school learning center where young people build skills and college-bound portfolios in a number of creative fields (see Chapter 11). There are many people from Holyoke and the surrounding colleges, who could mentor young adults if subsidies were available for educational resources and spaces to house creative learning incubators. If programmed to invite learners to interact and share skills, such spaces could also generate a more collaborative atmosphere than is characteristic of more typical incubator spaces. Holyoke also has the capacity to revive manufacturing on a smaller scale, and schools can push this agenda forward by providing more applied background training in those STEM and art- and media-related fields that attract young people's interest. This could be facilitated through collaborative partnerships with area colleges, entrepreneurs and nascent businesses, as noted in the *Innovation-based Economic Development strategy* report.

The Politics of Change

Neoliberal planning places value on areas of a city that are deemed marketable and will attract outside investment. Given the levels of economic disinvestment that many post-industrial cities like Holyoke have suffered, encouraging place marketing can quickly over shadow a concern with generating more heterogeneous forms of *placemaking* that address structural inequalities and build the creative and technological capacity of existing residents. Amidst this economic reality and political and ethnic divides, simplistic approaches to producing a creative economy can similarly view arts and technology solely as "imports", and the physical infrastructure, mill buildings and canals as little more than ambient backdrops for tourism.

Diálogos, a recent book on planning in cities with large Latino populations, describes how tensions arise between "professional values of rationalism and universalism, and … places characterized by heterogeneity, flexibility and contingency (Rios, Vazquez, Miranda 2012: 15)." Given the economic and social realities, they believe that, "Improvisational, novel, and culturally derived uses of space that foster community building" require "equally creative and flexible forms of design, planning and regulation … (15)". Such challenging places, they say, provide a real opportunity to develop new democratic models for planning that can result in "a collective sense of belonging, authorship, and power" (204).

Mustering support for such a planning agenda requires the sustained political participation of residents.

There is some evidence that the political wind may be shifting with the surprise election in November 2011 of Alex Morse, a bi-lingual, 23-year-old resident who graduated from Holyoke High School in 2007, and was the first in his family to attend college, completing a dual major in Urban and Africana studies from Brown University. In his acceptance speech, delivered in both English and Spanish, Morse said he would become Holyoke's "main marketer" and that he would get residents and outsiders to focus on the city's "opportunities and potentials rather than its problems". Morse is on a crusade against resignation and wants to bring pride (and a middle class) back to the city. He is working already to capitalize on the new MGHPCC and a budding spirit of entrepreneurship among young residents and artists.[22] In April 2012, he announced a new young Director of Planning and Economic Development, Marcos Marrero, who grew up in San Juan, Puerto Rico and earned a dual Masters in Public Affairs and Urban and Regional Planning from the Woodrow Wilson School at Princeton. Echoing Morse's optimism, Marrero shared in the belief that,

> Holyoke is a unique place where diverse urban, suburban and rural neighborhoods can coexist; where its residents can have multiple transportation alternatives and job choices; where different ethnicities can coincide and share. People from our region and beyond want to experience our cultural offerings and Holyoke is poised to become the creative economy capital for the region, taking the lead on arts, innovation, and technology (Plaisance 2012).

In June 2012, the Holyoke City Council voted to support Mayor Morse's request to mobilize the creative sector by creating a new position in the city for a Creative Economy Coordinator.

Morse's election raises the important question of whose vision will prevail in the city, and how inventive, experimental, and inclusive this post-industrial

---

22 Announcing a fundraiser in fall 2011 in a campaign press release to MASSLIVE.COM, Morse invited residents to an evening where "we will, together, create a one-of-a-kind 'community cinema experience' at one of Holyoke's overlooked gems, the Wauregan Mill Building, in the canal district." Residents were directed to bring their own chairs, tables, and concessions, and participate in a "flash cinema" where they would "converge on the Wauregan", and transform the inside of this now vacant ground floor of a Mill building into their own Hollywood/Holyoke-style while watching an Academy award nominated film: Street Fight, which followed the mayoral campaign of Cory Booker in Newark, New Jersey. The campaign announcement went on to say that,

> Cities are transformed only when citizens (and the leaders they support) participate and reclaim them. This is a night where we, as citizens, create the future we want to see in Holyoke. Its up to us to participate in our city's renaissance and rebirth. (http://blog.masslive.com/campaign_dispatches/2011/10/alex_morse_for_holyoke_mayor_campaign_offers_simultaneous_fund-raiser-movie_night.html).

city is willing to be in crafting its future. Creative industry and cultural economy practices built solely upon an elite and imported "creative class", however young and energetic, along with traditional economic development metrics that ignore the needs of an existing one-third of the population living under the poverty level, must be augmented with policies that build the educational capacity and opportunity for all residents. One population cohort that cannot be overlooked is the category of young adults who, like Mayor Morse, grew up in the city and are committed to its future.[23]

In the last year, Holyoke has seen an immense upsurge in new events[24] designed to draw attention to cultural production that cuts across ethnic lines. Like the earlier ArtsWalk festivals, these gatherings bring visibility to, and celebrate, the work of a wide range of talented residents. The attention being given to the planning of these activities is a hopeful sign that suggests the untapped potential of inclusive creative economy planning and a way for Holyoke to begin to craft a distinctive path to future development.

At *Holyoke Points of View*, a recent exhibit of painters, someone remarked at the way that visual artists were able to "see the city differently from everyone else", capturing its uniqueness and beauty in small slices of time and space. Any sustainable vision plan must build on this uniqueness and see Holyoke's notable diversity as an opportunity on which to capitalize. Ten years down the road Holyoke may well have the capacity to share useful lessons about a post-industrial

---

23 This commitment is well expressed by Holyoke resident Jesse Maceo Vega-Frey, who, while aware of the challenges posed by living and growing up in the city, also sees opportunity:

> "Having one Latino parent and one white parent would be a tricky endeavor anywhere, but in Holyoke, where the lines between those worlds are so divided, the balancing of those two realities has always been a complicated line to walk. Add to that a non-religious and politically radical upbringing, an Ecuadorian – not Puerto Rican – dad and a non-Irish mom from Ohio, I didn't fit squarely into any of the boxes laid out before me. This identity I've grown up with has resulted in feeling a deep connection to Holyoke and all its communities as well as a deep isolation from them at the same time. In truth, I feel like I can hold and embody all the dynamics at play in this city in my own life, in my own body, and in my own mind and flow – or stumble – between them with some degree of integrity and grace. I think this dance between connection and isolation, between belonging and alienation, has everything to do with my artwork. This city and my relationship to it continues to be a fundamental driving force of my creative endeavors (quoted in Thibodeau 2006: 128–9)."

24 Examples include La Noche de San Juan (a celebration of Puerto Rican Arts in Holyoke) held in Heritage Park in June that includes parades and prizes, and Holyoke Point of View held in April, a whole series of arts and cultural events to benefit the public library. The latter included exhibits as diverse as the Puerto Rican Cultural Project and "A Walk Through Holyoke – 1982", an exhibit of the never seen photos of Jerome Liebling (www.holyokepov.org/). This photo exhibit was followed by hands-on photography workshops with an after school program at Girls Inc.

regeneration that supports a multi-dimensional and vibrant creative economic base. To accomplish this, it may still be true, as one artist suggested, that "Anything that can be born from [Holyoke's] reality will happen from the mills" and draw upon the city's rich heritage.

References

City of Holyoke. (2009a), "Holyoke's center city vision plan: a plan for the heart of our community", Report prepared by Vanasse Hangen Brustlin, Inc.

City of Holyoke. (2009b), *Annual Report for Fiscal Year 2009* (Holyoke: Office of Planning & Development, Economic Development Department).

Computing center set for Holyoke. (2010), *Daily Hampshire Gazette*, August 10.

Dunn, J. (2005), *Holyoke: The Belle Skinner Legacy* (South Hadley, MA: Flats Press).

Giridharadas, A. (2011), "Meet the makers: the kitchen-table industrialists", *The New York Times* May 15, MM50.

Green, C. (1939), *Holyoke Massachusetts: A Case History of the Industrial Revolution in America* (New Haven: Yale University Press).

Jonas, M. (2012), "Holyoke hope", *Commonwealth Magazine* 11 January (Winter).

Lauer, M. (2001), "Holyoke maps its gems in cultural directory", *The Union News* 27 December, H3.

"Making Maker College: Northampton man plans Holyoke fabrication shop business accelerator". (2011), *The Daily Hampshire Gazette* 21 March.

Massachusetts Technology Collaborative and John Adams Innovation Institute. (2011), "Innovation-Based Economic Development Strategy for Holyoke and the Pioneer Valley" (Report prepared by HDR Decision Economics, Boston, Ma).

Plaisance, M. (2012), "Marcos Marrero named Holyoke director of planning and economic development", *The Republican* 25 April.

Rios, M., Vazquez, L. and Miranda, L. (2012), *Diálogos: Placemaking in Latino Communities* (New York: Routledge).

South Holyoke Revitalization Strategy. (2008), Prepared by HAP Inc. and Nueva Esperanza for the South Holyoke Revitalization Coordinating committee, March. Available at http://www.holoyke.org/index.php?option=com_content&view=article&id=52<emid=59 [accessed 16 August 2012].

Thibodeau. (2006), *Destination Holyoke: Immigration and Migration to Holyoke* (Holyoke, MA: Wisteriahurst Museum).

Villani, J. (1994), *The One Hundred Best Small Art Towns in America* (Santa Fe: John Muir Publications).

Ward-Wheten, S. (2012), "Holyoke artists, businesses spur downtown renaissance in canal district", *Daily Hampshire Gazette* 22 May.

Websites

http://factfinder2.census.gov/faces/tableservices/jsf/pages/productview.xhtml?src=bkmk [accessed 15 March 2012].
http://holyoke.ning.com/ [accessed 10 March 2012].
http://hush.fluxmass.org/2012/05/10/response-to-mayor-morses-statement-on-lyman-terrace/#.T7Zk1r_bbPR [accessed 15 August 2012].
http://ramblingvandog.blogspot.com/2008/07/inside-victory-theatre.html [accessed 15 August 2012].
http://www.brendanciecko.com [accessed 12 June 2012].
http://www.masslive.com/news/index.ssf/2012/08/holyoke_mayor_alex_morse_asks.html [accessed 30 August 2012].
http://www.opensquare.com [accessed 16 August 2012].
http://www.passportholyoke.org [accessed 15 August 2012].
http://www.thelivingmuseum.org [accessed 12 March 2012].

Chapter 4
From Post to Poster to Post-Industrial: Cultural Networks and Eclectic Creative Practice in Peterborough and Thunder Bay, Ontario

Alison Bain and Dylann McLean

Introduction

Within the paradigm of the creative economy, the drivers of economic success have shifted for cities of all sizes (Lorenzen et al. 2008). The global rise to prominence of the "creative city" has placed culture, cultural facilities, and cultural labour centre stage as essential components of community development initiatives and competitive urban economies (Bain 2010). This chapter contributes to a growing body of literature on the cultural economy of smaller cities (Nelson 2005; Bell and Jayne 2006; Waitt and Gordon 2009; Lewis and Donald 2010; Van Heur and Lorentzen 2012) by identifying eclectic creative practice as a key, yet under-appreciated, resource that can help to build a more socially inclusive future in former industrial cities.

Peterborough and Thunder Bay are two mid-sized Canadian cities in southeastern and northwestern Ontario that capitalized in the 19th century on railway and waterway linkages to grow into industrial, service, and administrative centers serving considerable hinterlands. In the contemporary context of significant deindustrialization, both cities, despite their geographic separation, have become interlinked through membership in the Creative City Network of Canada and their shared reliance on a Richard Florida-inspired creative capital model of urban-economic development.[1] The Floridian influence can be seen in the top-down diffusion of recent municipal cultural planning strategies, which we contrast in

1 Such a model privileges knowledge, creativity, and commodified differences as mechanisms for urban renewal. It is reliant upon the attraction of a broad "creative class" of professionals who are employed in business, law, engineering, science, healthcare, architecture, design, and the arts who are said to share similar approaches to complex problem solving and a common work ethos that values individuality, difference and merit (Florida 2002, 2005). Florida argues that these iconic citizens of the knowledge-based economy are attracted to cities that offer the "3Ts" of technology, talent, and tolerance.

this chapter with a longer history of grassroots arts mobilization rooted in artist-run centers and a landmark Supreme Court of Canada ruling that made postering a constitutional right for all Canadians. These examples of spaces and tools of cultural community building through informal social networks are used to argue that unregulated, eclectic expressions of local creativity have a valuable role to play in challenging the homogenizing influence of mainstream cultural planning policy in smaller, post-industrial cities.

Networking the Cultural Economy

A fundamental feature of the post-industrial cultural economy is its reliance on networks to sustain it. In his foundational research on the rise of the network society in the information age, Manuel Castells (1996, 2000: 5) explains that while networks are a traditional form of social organization, they are now increasingly complex and empowered by new communication technologies to better cope with "flexible decentralization" and shared decision-making. Inspired by the work of Castells and other sociologists, scholars of the cultural economy have illustrated the essential role performed by elaborate networks of business and personal contacts in cultural production in large urban centres (e.g., Norcliffe 2003; Stolarick and Florida 2006; Vinodrai 2006; van Heur 2009). Such networks can (re)confirm business and social relations through formal and informal collaborations, establish new contacts, facilitate co-operation and social connectivity, obtain information, produce knowledge, and escape the perceived limitations of a smaller locality. While much cultural economy research and many government reports tend to focus on formal cultural networks that can produce value chains with quantifiable economic outputs (e.g., indicators of firm productivity and innovation, cultural sector employment and career trajectories, per capita investment in culture and cultural infrastructure, cultural event audience participation and sponsorship, and performance measures), informal networks of creativity should not be neglected as they do more than just help to anchor cultural workers in local arts scenes (Bader and Scharenberg 2010).

This chapter adjusts the focus away from larger cities to smaller, former industrial ones, to consider some of the tensions that exist between formal and informal social networks of cultural information exchange, the participation they encourage, and the kind of creativity they support. The argument is made that the smaller, post-industrial cities of Peterborough and Thunder Bay are unique in the opportunity their informal social networks of cultural production afford to draw more people into artistic practice. For as Lewis and Donald (2010: 49) assert in their new rubric of creative city potential "collective action is easier" in smaller cities because "the bonds of familiarity and trust facilitate consensus and collaboration".

Urban Industrial Legacies in Ontario: Peterborough and Thunder Bay

The City of Peterborough is located in south-central Ontario, 130 kilometers northeast of Canada's largest financial and cultural centre, Toronto, and 280 kilometers west of the nation's capital, Ottawa. For thousands of years First Nation peoples lived in and moved through this area of rolling till-plain, lakes, rivers, drumlins, and mixed-coniferous hardwood forest (Bain and Marsh 2012). In the 1800s, British settlers were attracted to the area by the availability of land and water for farming, logging, transportation, and power, and radically transformed the landscape over the ensuing two centuries. In the 19th century, investment was made in railway and canal infrastructure. From Peterborough, rail lines extended in all directions to facilitate trade with developing cities around Lake Ontario, but the Trent and Otonabee Rivers never became a major transportation corridor because it took over a century to complete construction of the locks and canals (Tatley 1978). Attracted by the rail and waterway linkages, industries such as Quaker Oats (1902), Canadian General Electric (1892), Outboard Marine Corporation (1928), and The Peterborough Canoe Company (1893) were established and flourished, along with a tourism industry based in the resorts and cottages of the Kawartha Lakes (Jones and Dyer 1987). The road system was gradually improved and expanded into a twinned highway, rendering the Trent Severn Canal obsolete for commercial (although not recreational) purposes, and surpassing the railway system in transportation importance.

In the contemporary period, all that remains of the founding industries in Peterborough is Pepsi-Quaker Tropicana Gatorade, heading up a food processing industry that includes Minute Maid, Baskin Robbins, and Unilever Canada, and a downsized General Electric. In addition to food processing, metal fabrication and automobile parts are other key manufacturing employment sectors. From 1986 to 2006 the proportion of the labour force over the age of 15 employed in manufacturing declined from 21% to 4% (Statistics Canada). In absolute terms the number of manufacturing jobs in Peterborough decreased from 10,835 to 3,235.[2]

Over the last few decades, local decision-makers have been forced to diversify Peterborough's economy, shifting it from manufacturing to service industries. Regional education, health, and government-related functions have become important local economic drivers, with Trent University, Sir Sandford Fleming College, the Peterborough Regional Health Centre, and the Ontario Ministry of Natural Resources as significant employers. In 1999, with the establishment of the Greater Peterborough Area Economic Development Corporation (GPAEDC), an arm's length economic development and tourism agency for the city and county, biotechnology research was targeted as a new strategic area of economic growth. While biotechnology research is now widely

2 Over the same time period, the percentage of primary industry employment remained fairly constant at approximately 4% while trades, transport, and equipment operating employment decreased slightly from 19% to 17%.

celebrated as "the focal point, the catalyst for a bright new future" in Peterborough (http://www.dnapeterborough.ca/DNA%20Focus%20Trent%20insert%20final%20proof.pdf), the cultural sector also has its political supporters who are seduced by visions of creativity-based economic growth. In 2006, cultural workers represented 2.2% of the local labour force. With 1 in 75 residents professionally employed in the arts and culture sector, civic leaders began to recognize the contribution of cultural workers to local economic development and quality of life with a moderate increase in per capita public spending on culture.

Twenty hours' drive north and west of Peterborough, along the shores of Georgian Bay and Lake Superior, the City of Thunder Bay is located at the natural junction between Lake Superior and the Kaministiquia River. While it is the largest city with the strongest and most diversified economy of the hinterland communities in northwestern Ontario, it shrunk by 4% in population size between 1986 and 2006. The longstanding heartland/hinterland relationship that exists between northern and southern Ontario dates back to the arrival of French and the fur trade in the 17th century. Fort William, now in the east end of Thunder Bay, was formerly a Hudson Bay trading post. The period of urban economic growth that began with the Western Wheat Boom continued until the First World War and largely shaped the industrial landscape of Thunder Bay with the building of railroads, and grain elevators (Di Matteo 1993). The growth in the grain transportation sector was followed by a wartime economy focused on munitions and ship-building that took advantage of the city's Great Lake's location. Thunder Bay became Ontario's largest port with a 2 million tonne grain storage capacity. Present day Thunder Bay was formed by a Provincial Bill in 1969, and is the amalgamation of the City of Port Arthur, the City of Fort William, and the Municipalities of Neebing, Paipoonge, and Shuniah.

Today, with an international airport, Thunder Bay functions as a regional transportation hub. It is located at the intersection of major railway and highway routes that cross Canada as well as run north-south from the Gulf of Mexico to the Upper Midwest of the United States. Confederation College and Lakehead University help to make Thunder Bay an educational center for northern Ontario. Despite the city's medical, administrative, and business services hub functions, the skyline is still dominated by the pulp mills and billowing smoke of the forestry industry. Bowater Forest Products is the city's largest private sector employer while the AbitibiBowater mill continues to produce 330,000 metric tonnes of market pulp and more than 230,000 metric tonnes of newsprint each year (http://www.abitibibowater.com). With fragility in the pulp and paper industry due, in part, to a decline in demand for newsprint and the concomitant expansion of truck shipment rather than train and boat shipment, Thunder Bay's position within the forestry and shipping industries has weakened. To address this economic precariousness, the city has sought to attract more knowledge-based industries and has developed a cluster of firms specializing in molecular medicine research and DNA analysis (Thunder Bay Community Economic Development Commission 2009). The cultural sector is another area of the economy the city is seeking to strengthen.

As of 2006, Thunder Bay had 1,295 people employed in Arts, Culture, Recreation and Sport (2% of its labor force).

Road, rail, and water linkages were the hard infrastructural investments central to the growth of Peterborough and Thunder Bay as industrial cities. In both of these smaller cities transportation connections have gradually been eroded. Trucking and bus services have encroached on freight and passenger rail services and boat shipment. In the post-industrial era, the connectivity that has helped to sustain these places is one that is under-written by computer-mediated communication technologies. It is the soft infrastructural investments in information and knowledge networks that have come to matter.

Color by Number: Post-Industrial City-Making through Municipal Cultural Planning

Like other smaller cities in North America, Britain, and Europe, Peterborough and Thunder Bay have struggled to survive within the parameters of the new post-industrial economy by reinventing themselves as places that are attractive to new businesses and residents alike (Nelson 2005). That reinvention often involves an increase in the size and significance of the cultural sector to the local economy. Investment in culture is widely touted as the way to revitalize cities and to diversify the local economy through the formal channels of municipal cultural planning.[3]

The oft-cited cultural planning consultant Franco Bianchini (1993) defines cultural planning as an integrated approach to urban development in which material and immaterial cultural resources are strategically leveraged by municipal policy-makers to create social and economic development opportunities for a community. More recently, Bailey et al. (2004) describe cultural planning as a way to engage "with the lives of people who live in the city", and to improve the quality of life for local residents rather than just a mechanism for regenerating the city through culture. Such a focus on everyday lived realities seems to be an interpretation of cultural planning that has not been widely adopted in practice by policymakers in Ontario's smaller cities.

In Ontario, over a decade after cultural planning made in-roads in European policy circles, the Ministry of Culture responded to the gap in policy tools and knowledge base by establishing a working group to advance cultural planning in the province (Brooks-Joiner and Stasiuk 2005). The working group organized 10 provincially-funded Municipal Cultural Planning forums between 2005 and 2007. The forums were designed to attract municipal staff, elected officials and local cultural, business, and community leaders with the intent to foster a broader

3 An increase in demand for cultural amenities and services in smaller Canadian cities is based in part, Duxbury (2008) suggests, on retirees moving from large urban centres and expecting similar levels of cultural infrastructure.

vision of culture and cultural resources as the foundation of a new economic development strategy. In this policy road show, case studies of "arrangements that work", in which successful partnerships were established and/or cultural resources documented, were used to present cultural planning as an inclusive, long-term, silo-breaking, and network building activity that all municipalities should engage in (Brooks-Joiner and Stasiuk 2005). Examples of this best-practice approach to cultural policy development can be seen in the Municipal Cultural Planning Incorporated's (2010) recent publication *Cultural Resource Mapping: A Guide for Municipalities*. In this online publication, both Peterborough and Thunder Bay are showcased as valuable case study examples. Peterborough is commended for integrating cultural resources into local government planning and decision-making. Peterborough's Arts, Culture and Heritage Division used data from an arm's length utility company to map opportunities in the historic downtown where sprinkler upgrades in vacant second-floor heritage buildings could facilitate their transformation into live-work space for artists. Thunder Bay is praised for developing a succinct cultural mapping project work plan that describes key objectives, rationale, project scope, benefits, and outcomes as a tool for better informing the decision-makers, partners, and the public. Despite these rather specific minor examples from smaller cities, much of the provincial cultural planning policy recommendations appear to be prescriptive, scaled-down replicas of large city strategies with minimal modification to better meet the needs of a smaller, post-industrial context. Emphasis, for example, is repeatedly placed on the need to support the use of cultural planning for economic development, cultural tourism, and branding and the need to strengthen cultural clusters and cultural resource management.

Back in the spring of 2005, the City of Peterborough hosted one of the initial regional forums on Municipal Cultural Planning. The Director of Community Services (who as the manager of the Arts, Culture, and Heritage Division single-handedly holds the city's cultural policy reins of power) was involved in the initial provincial working group. He packaged Peterborough to provincial decision-makers as a leader in municipal cultural planning. Peterborough's self-identified leadership claim hung on the assertion that it was one of the first cities in Ontario to undertake a cultural mapping exercise (yet it still does not have a cultural plan). The 1995 map was a preliminary quantification exercise that identified 134 arts, culture and heritage organizations, 185 cultural businesses, and hundreds of cultural workers within a 15-kilometer radius of the city. Despite the rudimentary nature of the mapping exercise, the Director of Community Services maintains that it has helped to begin to shift local attitudes toward culture as reflected in an increase in the municipal cultural budget by 60% between 1995 and 2001. While political support for culture may have expanded slightly, the power to plan and manage this sector remains exclusive and hierarchical with minimal participation from Aboriginal, immigrant, youth, or industrial worker groups. Hosting the forum, then, was part of an on-going self-affirmation strategy to augment cultural-sector buy-in from local decision-makers and to formally facilitate regional networking.

Through workshops, presentations, and conversations Peterborough city managers and cultural administrators exchanged ideas with their counterparts from neighboring towns and municipalities over a wide catchment area. This formal networking opportunity, available to those who registered, can be seen as a valuable preliminary mechanism for establishing new contacts and for opening up future channels of communication and collaboration across administrative layers in the management of the cultural economy. For example, the current young Supervisor of Cultural Services in Thunder Bay obtained her bureaucratic and policy training by helping to organize this forum in Peterborough. She is now helping to coordinate a cultural mapping[4] exercise in Thunder Bay as a precursor to the development of a municipal cultural plan. It is these kinds of formal and informal social, management, and policy networks that underpin the cultural economy, linking places as seemingly geographically separated as Peterborough and Thunder Bay. Such linkages, rather than bringing new residents to the decision-making table, involve the same kinds of people, in the same kinds of management roles, employing the same kinds of policy methods.

At a wider geographic scale, another significant by-product of the organization of the Municipal Cultural Planning Forum was the formation of the Municipal Cultural Planning Partnership (MCPP). MCPP is a coalition of provincial government agencies, municipalities, cultural service organizations, and post-secondary institutions that "have come together out of a shared belief in the power of culture to transform local economies and communities" (www.ontariomcp.ca). This mandate is intended to position MCPP as a leading promoter of cultural planning in Ontario.

In the Canadian context, municipal cultural planning is usually overseen by a cultural policy advisor and steering committee who, in turn, may out-source significant portions of the process to consultants. At the center of the cultural planning process are two key activities: cultural mapping and community participation in the development of a cultural plan. Building on the assessment of a municipality's arts and culture resources, the cultural plan through community consultation is intended to identify what is unique about a place and what could augment its cultural distinctiveness. In practice, the unique and diverse characteristics of a place seldom get adequately showcased, particularly if it involves highlighting a city's industrial legacy or its working-class history (which are often too readily dismissed as out-dated or irrelevant in a post-industrial, knowledge-based economy). A comprehensive cultural plan is intended as a public statement of a community's recognition of the arts as an essential service that contributes to the economic growth and liveability of a place. But who crafts a city as economically viable and liveable, and how this is accomplished, should be of greater concern. For the most part, it is the needs and lifestyles of middle-class

4 Cultural mapping involves the systematic identification and inventorying of local arts, culture, and heritage resources and the subsequent mapping using Geographic Information Systems (GIS) software (Evans and Foord 2008).

professionals, arts and cultural workers, and university graduates that cultural planning exercises privilege rather than the quality of life requirements of former industrial workers and their families.

At present, there is a complex array of local cultural planning mechanisms in place in Ontario municipalities. Back in 2005, the Ontario Ministry of Culture identified insufficient access to models, best practices and expertise, as well as low levels of awareness as to the value of cultural plans amongst senior staff, elected officials, and the community, as key barriers to widespread adoption of cultural planning in the province (Brooks-Joiner and Stasiuk 2005). To address some of these perceived barriers the Ministry of Culture ran the Municipal Cultural Planning Forum and helped to establish the Municipal Cultural Planning Partnership. These state-led formal cultural planning promotional and networking initiatives have turned the cultural plan into a trendy, yet generic, policy visioning document designed to support a creative capital formula of urban-economic development. In practice, cultural plans may do little to enhance local employment opportunities for non-artists or to improve the quality of life for working-class residents.

The Creative City Network of Canada

On the national stage, The Creative City Network of Canada (CCNC) has played a central advocacy role in disseminating information about cultural planning, particularly to smaller cities. Initiated in 1997 out of Vancouver's Office of Cultural Affairs and established as a not-for-profit organization in 2002, CCNC's goal is to connect Canadian professionals working in municipal cultural service delivery (www.creativecity.ca). The CCNC's membership base is made up of 107 regions, municipalities and organizations. As Table 4.1 illustrates, 44 of CCNC's members are mid-sized Canadian cities with populations between 50,000 and 500,000, 36 member communities have populations between 10,000 and 50,000, and 14 have populations under 10,000. Only four member municipalities have populations over 500,000 (e.g., Mississauga, Toronto, Vancouver, and Winnipeg). These statistics suggest that the CCNC is most widely used as a cultural planning resource by small- and mid-sized cities in Canada. Yet the cultural planning resources that many of these smaller municipalities rely upon are drawn from the experiences, methods, and best-practices of larger urban centers (in Canada and around the world) that have the greater financial and human resources to direct towards cultural sector development.

Table 4.1 Mid-sized city membership in the Creative City Network of Canada

Mid-sized cities	Province	Population (2006)	Population change (2001–2006)	Labour force 15+ (2006)	Art, Culture and Sport (ACS) employment (2006)	ACS employment as percentage of labour force (2006)	Artist-run centres (2010)
Abbotsford	BC	123,864	7.2%	65,645	1,050	1.5%	0
Barrie	ON	128,430	24.0%	72,030	1,645	2.0%	0
Burlington	ON	164,415	9.0%	92,590	2,770	0.3%	0
Burnaby	BC	202,799	4.6%	109,545	3,600	3.2%	0
Chilliwack	BC	69,217	11.0%	32,255	765	2.1%	0
Coquitlam	BC	114,565	1.5%	62,900	1,855	0.0%	0
Fredericton	NB	50,535	6.2%	28,840	1,065	3.6%	1
Guelph	ON	114,943	8.3%	66,380	1,785	2.6%	1
Halifax R.M.	NS	372,679	3.8%	212,880	7,590	3.5%	3
Kamloops	BC	80,376	4.0%	44,605	No data	No data	0
Kelowna	BC	106,707	11.0%	57,305	1,745	3.0%	1
Kitchener	ON	204,668	7.5%	116,820	2,605	0.0%	2
London	ON	352,395	4.7%	191,555	5,015	2.6%	1
Maple Ridge	BC	68,949	9.2%	37,665	930	2.4%	0
Markham	ON	261,573	25.0%	144,735	4,325	0.2%	0
Medicine Hat	AB	56,997	11.0%	32,355	560	1.7%	0
Moncton	NB	64,128	5.0%	35,755	965	2.6%	2
New Westminster	BC	58,549	7.1%	34,255	1,220	0.3%	0
Niagara Falls	ON	82,184	4.3%	44,405	955	2.1%	0
North Okanagan R.D.	BC	77,301	7.1%	39,400	815	2.0%	0
Oakville	ON	165,613	14.0%	92,285	3,475	3.0%	0

Ottawa	ON	812,129	4.9%	456,480	19,970	4.0%	6
Peterborough	ON	74,898	4.8%	38,350	1,120	3.0%	1
Port Coquitlam	BC	52,687	2.8%	30,380	760	0.2%	0
Prince George	BC	70,981	-2.0%	40,870	650	1.0%	0
Red Deer	AB	82,772	22.0%	51,005	1,010	1.9%	0
Regina	SK	179,246	0.6%	102,625	3,195	3.0%	2
R.M. of Waterloo	ON	478,121	9.0%	272,530	6,115	2.0%	0
R.M. of Wood Buffalo	AB	51,496	24.0%	33,790	380	1.0%	0
Richmond	BC	174,561	6.2%	92,475	2,870	3.0%	0
Saanich	BC	108,265	4.4%	59,785	2,200	3.0%	0
Saint John	NB	68,043	-2.3%	64,175	No data	No data	1
Saskatoon	SK	202,340	2.8%	114,025	3,030	2.0%	3
St. Albert	AB	57,719	8.7%	46,270	No data	No data	1
St. Catharines	ON	131,989	2.2%	68,925	1,830	2.0%	1
St. John's	NL	100,646	1.5%	53,175	1,895	35.0%	2
Strathcona County	AB	82,511	15.0%	49,240	1,120	2.0%	0
Surrey	BC	394,976	14.0%	210,935	4,200	1.0%	0
Thunder Bay	ON	109,140	0.1%	56,550	1,295	2.0%	1
Trois-Rivieres	QC	126,323	3.2%	63,135	1,570	2.0%	0
Victoria	BC	78,057	5.3%	45,270	2,440	5.0%	0
Waterloo	ON	97,475	13.0%	56,265	1,820	3.0%	0
Welland	ON	50,331	4.0%	25,990	430	1.0%	0
Windsor	ON	216,473	3.5%	108,240	2,235	2.0%	2

Source: Statistics Canada.

The CCNC supports its primary networking goal through electronic media, a website, a listserve, and a newsletter. Contact between individuals and municipalities is also supported by annual conferences and workshops focused on knowledge sharing and "best practices". The CCNC has published and widely disseminated two toolkits on *Cultural Planning* and *Cultural Mapping* (2007) that are supported by how-to workshops. The use of toolkits, workshops, and an on-line discussion board has served to streamline and to standardize the cultural planning process, continuing the ideals of "good governance" and "best practices" that have become central features of neoliberal economic platforms (Griffths 1998).

The cultural sector, much like other sectors of municipal economies, "depends heavily on an ability to manage itself" and "the development of sustainable national networks that facilitate knowledge exchange supports this" (Duxbury 2008: 40). The CCNC functions as an inter-city and inter-organization co-operation network through which cultural policy knowledge is circulated amongst a select group of cultural administrators and planners, many of whom have the substantial annual membership fees paid by taxpayer dollars and are in a position to influence urban public policy.[5] The restrictive nature of the membership structure, places the CCNC in a regulatory position to control the flow of information and access to its resources. Thus the CCNC and its membership may in fact help to sustain the garret and silo model of policy development and implementation by restricting access to resources to decision-making elite with particular occupational training and employment purviews. Were a more horizontally integrated model of municipal cultural planning implemented, a broader array of local residents, non-experts, and non-artists would have the potential to participate in decision-making. Such diverse input could help to challenge the use of jargon and break down boundaries among professions; it would also enable categories of knowledge to be questioned and the policy production process to be reconfigured.

For decision makers in mid-sized cities in Canada, the CCNC is lauded as a valuable source of current cultural policy information. With its on-line community of support, it is the transfer of best practices and problem-solving knowledge that provides the CCNC network with its added value for members.[6] The applied knowledge sharing process that the CCNC facilitates between municipalities and

5 The annual membership fees for CCNC are on a sliding scale dependent upon municipal population size: $3,675 CDN for a population greater than 500,000; $2,100 for a population between 300,000 and 500,000; $920 for a population between 100,000 and 300,000; $395 for a population between 20,000 and 100,000; and $265 for a population less than 20,000.

6 Listed here are a small sample of the many guides and toolkits available on the website for members: Increasing Cultural Participation: An Audience Development Planning Handbook For Presenters, Producers, and Their Collaborators (2001); Capitalizing on Cultural and Heritage Tourism (2002); Evaluating and Achieving Through Performance Measures (2003); Facility Planning and Advice (2003); The Art of Renewal: A Guide to Thinking Culturally About Strengthening Communities (2005); Celebrations Toolkit: A Practical Guide to Planning Celebrations and Special Events (2006).

organizations is designed to minimize risk and to increase economic efficiency in cultural sector governance. We argue that by foregrounding an economic imperative, minimizing risk-taking through policy emulation, and potentially fostering a disconnect from cultural and working-class labor, the toolkit approach to creativity is potentially stifling of genuine innovation in the cultural sector. Artist-run centers offer an alternative networking model that suggests possibilities for a more socially inclusive and less hierarchical approach to cultural planning that might have greater relevance to smaller, post-industrial cities and their non-artist and working-class residents (Markusen and Johnson 2006).

Arts Mobilization through Artist-Run Centers

Artist-run centers (ARCs) are "a visible network of galleries and museum-like spaces run by artists" (Bronson 1983: 33). These non-profit arts organizations were first established in Canada in the 1960s. They were a constructive critical response to the power-brokering of the traditional arts system and the perceived lack of opportunities this institutional model afforded for presenting contemporary artwork and for networking with other artists. They were established at a time when the arts and creative labor were not thought to be a foundation for economic growth. A half-century later, there are over 150 artist-run centers operating in communities across the country that are variously supported by a small membership fee and grants from municipalities, provincial arts councils, the Canada Council for the Arts, and private-sector sponsorships. Each artist-run center works with its board of directors and membership to organize a unique program of free events (e.g., exhibitions, lectures, workshops, symposiums, performances, screenings, broadcasts, and artist residencies) that showcase the often non-commercial work of emerging national and international artists through a peer-juried submissions process.[7] In addition to providing a venue and the necessary technical and promotional support for artists to present their work, artist-run centers also pay contributing artists for their creative labor, and may publish critical exhibition texts to foster debate. They usually seek to build ties within and to the local community through their programming.

The history of arts mobilization within the City of Peterborough dates from the establishment of the artist-run center, Artspace. Founded in 1974 by the expatriate contemporary American painter David Bierk, Artspace provided an alternative to the museum model of the Art Gallery of Peterborough. Some of the first artist-run centers were established in many of Canada's largest cities: Vancouver, Toronto, Montreal, Calgary, Edmonton, Winnipeg, and Halifax. Peterborough's artist-run center was unique at the time in that it reflected a rural experience, with a membership base drawn, in part, from surrounding smaller townships.

7 Some artist-run centers have also developed facilities to support creative production in the fields of video, new media, photography, and printmaking.

As the number of members grew, Artspace moved out of temporary exhibition space in reclaimed storefronts, into a retrofitted historic building, Market Hall, in the heart of the downtown.[8] Several decades and six addresses later, Artspace has an active, independent-minded membership base with approximately 30% of its members living in the surrounding rural hinterland. Located in yet another converted storefront, Artspace remains a place of cross-disciplinary intellectual exchange both materially and virtually, deliberately seeking to expose local artists to ideas and discourses that are circulating nationally (http://www.artspace-arc.org). The main gallery hosts six exhibitions annually, the incubator "mudroom" affords limited workspace for its members, the multi-media resource center provides reference materials in its library, and the website offers a visual archive of events and links to other artist-run centers across Canada as well as important local cultural resources. In the words of a previous Director (July 8, 2003), Artspace tends "to program stuff that is less safe, and sometimes a little more on the risky side" and "to pick people that are sort of a little more trouble-making". By gravitating towards critically challenging voices, and by not shying away from controversy or sharpness in its exhibition programming or community events planning, Artspace has remained a vital place of creative dissent.

In 2003, in the lead up to the municipal election, members of Artspace collaborated with other community arts organizations to form the Peterborough Arts Coalition (PAC) to organize the city's first all-candidate's mayoral debate on arts and culture issues. On the night of the debate, one of the key arts activists in the city explained that: "The Peterborough Arts Coalition was founded to raise public awareness about issues in the Arts, Culture and Heritage sector, to foster discussion, information and ideas about how to build a more creative city" (October 23, 2003). Members of the Coalition were cultural workers (e.g., painters, dancers, musicians, actors, curators, arts administrators, filmmakers, writers, editors, and directors) who were "looking for a Mayor who will put the arts and artists high on the agenda in the coming term" (October 23, 2003). The Arts Vote Peterborough debate was initiated "in the belief that it is part of a dialogue, a conversation, a work in progress, where we are all united in our efforts to make Peterborough an outstanding creative city" (October 23, 2003). A panel of Coalition members hastily assembled on the stage of the restored Market Hall in downtown Peterborough, invited debate on 10 questions that are listed below. These questions illustrate that for many members of the Peterborough Arts Coalition, a creative city is understood to be a socially inclusive municipality that supports the cultural labor of amateur and professional creatives of all ages, and ethnicities through policy and funding

8 Not long after the move, Artspace had an exhibit in which an artist quilted a depiction of a woman menstruating. The work of art was hung in a large window above the main commercial artery in the city. Local city residents phoned the mayor to complain about the "sexual depravity" of the subject matter. A flurry of editorial debate ensued in the local newspaper. In response to complaints, the City withdrew its financial support and Artspace lost its space, a piece of its history that few local artists have forgotten or forgiven.

priorities. It is a municipality that is politically interested in growing arts audiences through the development of cultural literacy among non-artist residents.

1. As Mayor, would you undertake the development, approval and implementation of a cultural plan for Peterborough? And how would you work with local non-profit, professional arts organizations to develop a cultural plan?
2. How, and by how much, can you imagine increasing direct investment in non-profit professional arts organizations?
3. What degree of support would you provide to an artist-directed, multi-purpose cultural building project?
4. How would you support the work of professional artists living and working in Peterborough?
5. How would you encourage youth to become involved in the arts? And how would support and nurture up-and-coming artists?
6. What steps would you take in the arts, culture, and heritage sector to promote cultural diversity and social inclusion?
7. How would you work with the Peterborough Arts Coalition, DBIA, the Greater Peterborough Area Economic Development Council, and the Chamber of Commerce to market and promote the creative side of our city?
8. What do you see as the role for the arts, culture, and heritage sector in the downtown?
9. Would you support a public art program? And what could you do to make a public art program a reality?
10. What advice can you give us on building an effective relationship with City Council? And how can thinking and action on the creative sector be integrated in all departments and levels at city hall?

In formulating these questions, it is clear that the Peterborough Arts Coalition partially appropriated creative city rhetoric and deployed it within a political forum to direct attention to the perceived challenges and opportunities in the local arts and culture sector. Interestingly, many of the mayoral candidates arrived for the debate armed with a hardback version of Richard Florida's (2002) *The Rise of the Creative Class* ready to parrot the creative city script that celebrates a mythical combination of artistry, inclusivity, competitiveness, and business-friendliness. The questions posed are practical and specific, but also have a social justice bent to them, which suggests that Richard Florida's work would be an inadequate resource for composing meaningful responses.

For the contributors to this edited collection, questions five, six, and seven are of particular interest. Although somewhat idealistic, the questions indicate the potential they see in informal networks and eclectic creative practices, to foster an alternative economic, political, and community development base for the city that prioritizes inclusion and collaboration. This alternative vision is concerned with the social impacts and potentialities of the arts, and seeks to strengthen local cultural resources by building bridges between artists and non-artists, as well as between the public, the private, and the non-profit sectors.

The quick assemblage of the Coalition was both its strength and its weakness. It demonstrates the potential of the local arts community to temporarily unite, and illustrates the energy, critical thought, and innovative problem-solving that can come from informal networking. However, without on-going investment and support from local civic leaders, such grassroots initiatives struggle to sustain themselves and resonate with a wider audience outside of the immediate arts community. Following the municipal election, momentum dissipated, and the Coalition dissolved. Over nine years later, many of those optimistic questions remain unaddressed. Individuals elected to office responded instead within the conventions of municipal politics with the formation of an Arts Culture and Heritage Advisory Committee (ACHAC). They also participated in formal municipal cultural planning networks and established a public art program. When the presence of cultural professionals in a small city is primarily framed as a catalyst for economic growth, the social and creative community development possibilities are likely to be under-realized. To merge these two agendas would require, at minimum, stronger policy ties between municipal cultural planning and social planning agendas and greater recognition amongst local decision-makers of the (often-unquantifiable) social impact value of the arts.

In Thunder Bay, as in Peterborough, the cultural planning rhetoric of the Creative City Network has also trickled down to impact grassroots arts activism. Like Artspace, Definitely Superior was established in 1988 as an alternative to the more mainstream public Thunder Bay Art Gallery. Its curatorial program has a regional, national, and international focus with an emphasis on showcasing emerging artists in its 2,000 square feet of exhibition space. In addition to its three gallery spaces, Definitely Superior maintains a reference room (with an archive of resource material related to submissions, job postings, grants, and proposals). It also circulates a quarterly electronic newsletter to members, detailing upcoming events and grant deadlines, calls for submissions, and job openings. The information circulated through its membership network allows local cultural workers to remain informed and connected to a cultural pulse that beats at multiple spatial scales. Where the mandate of Definitely Superior differs from Artspace is in its emphasis on youth arts mentorship and its closer alignment with municipal cultural planning policies. Take, for example, the 2011 exhibition *Urban Infill – Art in the Core*, a multidisciplinary art and performance event, sponsored by a local Business Improvement Association, which is in its fifth year. The exhibition placed 350 artists in 15 different downtown locations for four days in March. The intent of the Definitely Superior organizers was to use "creative cities partnerships" to temporarily fill empty spaces in the north downtown core with art as a means to create a "unique niche of an urban arts and entertainment district" (http://my.tbaytel.net/defsup). While such an event can find new uses for former industrial spaces, and bring art to a wider audience on a temporary basis, it exploits a conventional arts-led-redevelopment formula that provides little in the way of alternative to more conventional forms of the creative economy.

Artist-run spaces like Artspace and Definitely Superior strive to be autonomous spaces of creative engagement and possibility for the cultural community at minimal expense to cultural workers or local residents. They bring together people with diverse skills and backgrounds to exchange ideas that might not find a forum for expression in more commodified and structured conventional art world settings. It is important to caution, however, that the radical potential of these spaces to support non-competitive forms of sharing and sociality and openness to difference is limited by their bureaucratic organizational, programming, and funding structure (Bromberg 2010). Their outreach and socially transformative agenda is also limited by their disconnect from working class and/or older city residents. Despite these structural constraints, locally rooted artist-run spaces remain valuable nodes of connection and creativity in smaller cities.

Postering the Post-Industrial Landscape

The local cultural outreach performed by artist-run centers and other arts organizations in cities of all sizes is often supported by postering. Despite advances in computer-mediated technologies, the poster remains an important means of communication. The poster is often heralded as "*the* democratic medium" (Aulich 2007: 8) through which the majority of ordinary people can disseminate information and build consensus (Gianniotis 1999). At its most basic, the poster is a standard sheet of paper transformed through a printing process that uses typography, color, and/or image to convey information to the public about an event, a service, a product, or an idea (De Jong et al. 2008). Like working-class labor and industrial infrastructure, the poster is also an often taken-for-granted feature of everyday urban landscapes in post-industrial cities. Ephemeral creations that fade and peel over time, posters are part of our "collective unconscious" that can function as "a metaphor for memory itself" (Timmers 1998: 241). The continued appeal of posters, particularly to the non-profit sector and cultural organizations, is their affordability, as well as ease of reproduction and physical deployment.

In Peterborough, between 1995 and 1997, there was an ongoing dispute with downtown merchants about putting up posters advertising local arts events on utility poles. Local arts practitioners recount stories of being chased down alleyways by parking attendants and business owners when they tried to put up posters advertising cultural events. In the mid-1990s, a city by-law stated that postering on public property was a crime. One musician, Kenneth Ramsden (a.k.a. Reverend Ken), a fixture in the Canadian music scene, was charged on two occasions with contravening the City of Peterborough's postering by-law by affixing posters to publicly owned hydro poles to advertise upcoming performances of his band.

Politicized and supported by the local arts community, Ramsden fought the charges that were laid against him and took his legal battle from this small city all the way to the Supreme Court of Canada.

Ramsden's legal appeal, Ramsden v. Peterborough (City) [1993] 2 S.C.R. 1084, concerned the constitutional validity of a municipal by-law prohibiting all postering on public property. At issue, was whether the absolute ban on such postering was an infringement of guaranteed freedom of expression under section 2(b) of the Canadian Charter of Rights and Freedoms. A Justice of the Peace determined that the by-law did not violate the Charter and convicted Ramsden. The appeal to the Provincial Court, which was dismissed, was based on the determination that postering on utility poles can be a potential safety hazard to workers climbing them, traffic hazard if facing and distracting oncoming motorists and cyclists, and visual blight on the urban landscape that may become litter. Despite these safety and aesthetic concerns, it was affirmed that postering is "a historically and politically significant form of expression" that "conveys or attempts to convey a meaning" and, in so doing, "fosters political and social decision-making" (http://www.hrcr.org/safrica/expression/ramsden_peterborough.html). The Supreme Court of Canada ruled that an "absolute prohibition of postering on public property … prevented the communication of political, cultural and artistic messages" and therefore was an infringement of freedom of expression (http://www.hrcr.org/safrica/expression/ramsden_peterborough.html). According to this ruling on the Ramsden case, postering is a constitutional right and Canadian cities must make available adequate space for it to occur (Duncan 2004).

Peterborough has found a productive alternative to its former ban on postering on public property. The city now has municipally funded and maintained events boards on secondary streets downtown and events kiosks on the corners of the main thoroughfare. It has set a national precedent for other communities. This street furniture dedicated to postering is now widely used by a diverse range of city residents and non-artists. Local cultural workers use the boards and kiosks to publicize a range of different events in the arts community (e.g., studio tours, gallery openings, live music gigs, book and CD launches, auditions, and plays). These street-side posters, with their written solicitations for participation and involvement, convey a sense of the diversity, breadth, and inclusiveness of the arts community.

In both Peterborough and Thunder Bay, postering is an important means by which the cultural community talks to itself and to a diverse public who may not enter more formal cultural institutions. It is a largely unregulated system of information exchange. The unexpected and provisional moments of exchange that postering affords can seed unforeseen informal alignments and eclectic creative possibilities.

Figure 4.1 Postering events board in Peterborough, Ontario

From posters, like those that are depicted in Figure 4.1, the seeds of networks that sustain the cultural community and allow it to regenerate are sown. A local resident might: respond to an open casting call; volunteer at a local cultural event like the Peterborough Folk Festival; attend a performance by local musician Washboard Hank; participate in a community discussion about the economic crisis at the public library; place their child in a Performing Arts Summer Camp at Market Hall; become an entertainer at the Busker's Festival at Peterborough Square mall; or feast and exchange ideas at a week-long contemporary Aboriginal art event.

While postering has been little studied by cultural economists of the Floridian ilk, who are interested in "innovation-generating interactions", we would emphasize that postering is an important "distribution channel" for artists and non-artists alike that helps to foster an inclusive creative milieu (Stolarick and Florida 2006: 1815). It is a form of grassroots activism, advertising, and creativity that is vital to the cultural economic development of smaller, former industrial cities.

Nurturing Eclectic Creative Practice in Smaller Cities

Eclectic comes from the Greek word *eklegin*, meaning to select. In its common usage, eclectic means to borrow freely or to combine elements from various sources (styles, genres, methods, approaches, techniques or systems) – a technique that is well illustrated in the practice of postering. The term "eclectic" affords a useful linguistic contrast to the competitive specialization of cultural labor in larger artistic centers. In bigger, more established art markets it can be difficult for emerging cultural workers to break out of the anonymous mass into the various arts scenes; to do so usually requires building a reputation of creative worth through networking, the creation of a specialized product, regular exhibitions or performances, and critical reviews. In smaller cities, where the cultural framework is less rigidly established, the pressure to compete with other artists and to conform to art world expectations is diminished.

Peterborough and Thunder Bay provide many cultural workers with a sense of escape from the stresses associated with living and working in larger urban centers that are strongly identified as artistic. Some cultural workers describe a sense of resentment towards, and alienation from, the hip coolness of fast-paced, big city art scenes where an excess of stimulation and cultural activity can also undermine individual creativity. The decision to settle in a smaller city can be a deliberate choice to produce physical and psychological distance from the perceived arrogance, conventions, and critical judgments of commercialized, big city arts scenes. Such distance acts as a buffer; it affords a safer and less competitive space within which to experiment with different modes of creative expression. It can playfully blur disciplinary and professional/amateur boundaries in ways that may be harder to justify or to economically sustain in a larger city.

A lower barrier to cultural sector participation, affordable material and supportive psychological space to experiment with new skills, and access informal networks and mentors, suggest that smaller cities like Peterborough and Thunder Bay have the potential to draw more people into arts and cultural occupations who may not have any formal training. This potential is enhanced, we suggest, in former industrial cities where a shared appreciation for an apprenticeship model of learning can build bridges between the trade and the arts professions. Whether an aspiring plumber, electrician, musician, or dancer, informal or formal apprenticeship opportunities help to impart and refine skills through practice and feedback (Cave 2000).

In many smaller cities, the informal social scaffolding of familiarity and trust is in place to allow beginners to join experts, and to "gradually learn to become experts in their own right" (Griffiths and Woolf 2009: 559). In Peterborough, for example, there are grassroots events with a large community following that explicitly nurture interdisciplinary creativity and collaboration across the amateur-professional divide. In the *8-to-8 dance project*, musicians create musical backdrops for original dance choreography. In the *24-hour theatre project*, local visual artists create works of art that inspire playwrights and actors.

In Thunder Bay, the Symphony Orchestra combines professional, student, and amateur musicians. It also collaborates with a variety of local popular and folk musicians and the occasional artist, author, and comedian to attract a wider audience for special concert events. By performing in community halls and varying the repertoire, the orchestra is able to avoid elitist perceptions and to build an audience from across the socio-economic spectrum. Taken together, this encourages creative collaborations in Peterborough and Thunder Bay among artists of different skill levels that, in turn, help to make art more socially accessible and bring it to a more diverse audience.

The artistic communities in Peterborough and Thunder Bay can be described as inclusive, informal and built upon many collaborative and multi-disciplinary ventures. They are communities composed of many overlapping communities of practice that may share resources, space, audiences, and events. Although smaller cities may offer fewer professional development opportunities for cultural workers than larger cities, they can be unique, we maintain, in the creation of a supportive and respectful environment for informal learning in which people are encouraged to experiment with and acquire new skills without the need for formal training. The spirit of eclectic interdisciplinarity can be seen in the following comments that describe how different local cultural workers in Thunder Bay self-identify. The owner of an art gallery (September 23, 2009) explains that "I'm known as a singer/songwriter and then I'm also known as a sculptor and painter … I'm in so many different mediums. I used to be in theatre too and I've written plays and directed them." A recipient of the Governor General's Award for Children's Literature and Illustration (September 28, 2009) states: "Primarily I'm known for writing and illustrating children's picture books … But I also paint and am working on submitting an application for a sculpture competition". Equally eclectic in her interests, the director of the Community Arts and Heritage Education Project (September 18, 2009) says that she is "a cross disciplinary artist. I'm an arts administrator. I taught at university and college. I develop programming. I do arts education … So, you have this really interesting cross-weirdness of people who have lots and lots of skills."

The phrase "eclectic creativity", as used in this chapter, celebrates the value to be found in messiness, experimentation, unboundedness, and playfulness – qualities that characterize grassroots cultural labor practices in the mid-sized cities of Peterborough and Thunder Bay. These qualities associated with risk-taking and freedom can often be supported in smaller cities because the costs of living are lower and the boundaries to arts community entry are more porous and less strictly regulated by individuals and organizations. To the individual who is struggling to survive day-to-day, such qualities may initially seem inefficient, indulgent, or intangible – the antithesis of cultural economic governance strategies; but they are key to accessing the force of creativity and using it to foster positive social changes that can ameliorate the everyday lives of local residents, strengthen community ties, and open lines of communication between residents and decision-makers. As the art historian Carol Becker (2009: 85) has written: "we must recognize that at the core of creativity is a blend of the new, the revised, the rethought, and the reimagined all attempting to manifest the *what* through endless permutations of, and debates around, the *how*" (Becker 2009: 85). For Becker, play, with its abandonment of utility and meaning, is central to the creative process. Becker (2009: 56) explains that in art making, play "can necessitate the willingness to break rules, trample conventions, and simply allow the whimsy of an idea to float because it seems exciting and brings pleasure". Such rebellious and experimental playfulness and risk-taking, usually associated with art students negotiating the avant-garde edges of their disciplines, should have a productive place at the decision-making table in the management of smaller city cultural economies.

Conclusions

This chapter has been fundamentally concerned with formal and informal cultural networks in post-industrial smaller cities – their relative access, usage, and the kind of creativity they support. Formal cultural planning networks, as exemplified by the Municipal Cultural Planning Forum and the Creative City Network of Canada, with their overt economic imperative and cultural planning policy focus, often mistakenly treat creativity as if it were a precious, excludable and rivalrous human resource (Bromberg 2010). Whereas more informal cultural production networks, as exemplified by those formed through artist-run centers and postering, celebrate and support an eclectic form of creativity that is abundant, inclusive, playful, and open-ended. Both Peterborough and Thunder Bay have relied primarily on the knowledge and relationships generated through the formal networks of the Municipal Cultural Planning Forum and the Creative City Network of Canada to develop their cultural economies. While both cities have artist-run centers and have come to appreciate the value of postering, there is greater scope for municipal decision-makers in both places to nurture and to learn from the less competitive form of creativity that finds expression in informal cultural production networks.

Grassroots cultural networks in Peterborough and Thunder Bay are more visibly democratic and horizontally integrated than formal cultural planning networks. They strive to be open, affordable, and accessible to a diverse group of artists and non-artists alike. This spirit of active cultural citizenship, we argue, is a necessary counterpoint to the technical specialist cultural planning knowledge that urban decision-makers most frequently rely upon to manage the cultural economies of smaller cities. Cultural planning "experts" who are connected through the Creative City Network often share the same language, labels, and tools for diagnosing problems and identifying solutions. While this professional network might expedite the problem-identification and solving process, the stock strategies implemented tend to limit community participation and flatten, rather than augment, the unique textures of the endogenous local cultural landscape (Evans and Boyte 1992).

To better mitigate the copy and paste "me-tooism" of the creative city development model, we argue in this chapter that civic leaders of smaller cities should strive to reconnect with the grassroots networks through which local cultural activities and eclectic creativity is mobilized (Waitt and Gibson 2009). This reconnection requires more than just a verbal recognition of the value of the informal arts sector. It demands continued investment in the spaces, projects, and labour of grassroots arts and culture, particularly that which is experimental, oppositional, and interdisciplinary. It demands ongoing dialogue and collaboration among decision-makers, non-artist residents, and cultural workers that meaningfully extends beyond an election year. It also involves trust – trust that money spent on the spaces, organizations, and people that compose the cultural sector is a worthwhile investment of taxpayer dollars with the potential to make positive (perhaps even unanticipated) social impacts. If the cultural economy is something that needs to be planned and managed, then a broader range of people – Aboriginals, immigrants, youth, and industrial workers – with their different perspectives, experiences, values, and ideas, need to be more actively welcomed at the decision-making table.

References

Aulich, J. (2007), *War Posters: Weapons of Mass Communication* (London: Thames & Hudson).

Bader, I. and Scharenberg, A. (2010), "The sound of Berlin: subculture and the global music industry", *International Journal of Urban and Regional Research* 34:1, 76–91.

Bailey, C., Miles, S. and Stark, P. (2004), "Culture-led urban regeneration and the revitalization of rooted identities in Newcastle, Gateshead and the North East of England", *International Journal of Cultural Policy* 10:1, 47–65.

Bain, A. (2010), "Re-imaging, re-elevating, re-placing the urban", in Trudi Bunting, Pierre Filion, and Ryan Walker (eds), *Canadian Cities in Transition: New Directions in the 21st Century* (Oxford: Oxford University Press).

Bain, A. and Marsh, J. (2012), "Peterborough: a georegion in transition?", in Gordon Nelson (ed.), *Beyond the Global City: Understanding and Planning for the Diversity of Ontario* (Montreal and Kingston: McGill-Queen's University Press).

Bain, A. and McLean, D. (2012), "Eclectic creativity: interdisciplinary creative alliances as informal cultural strategy", in Anne Lorentzen and Bas van Heur (eds), *Cultural Political Economy of Small Cities* (London and New York: Routledge).

Becker, C. (2009), *Thinking in Place: Art, Action and Cultural Production* (Boulder: Paradigm Publishers).

Bell, D. and Jayne, M. (eds) (2006), *Small Cities: Urban Experience Beyond the Metropolis* (London and New York: Routledge).

Bianchini, F. (1993), *Urban Cultural Policy in Britain and Europe: Towards Cultural Planning* (Gold Coast: Institute for Cultural Policy Studies, Griffith University).

Bromberg, A. (2010), "Creativity unbound: cultivating the generative power of non-economic neighbourhood spaces", in Tim Edensor, Deborah Leslie, Steve Millington, and Norma Rantisi (eds), *Spaces of Vernacular Creativity: Rethinking the Cultural Economy* (London and New York: Routledge).

Bronson, A.A. (1983), "The humiliation of the bureaucrat: artists-run centres as museums by artists", in A.A. Bronson and P. Gale (eds), *Museums by Artists* (Toronto: Art Metropole).

Bronson, A.A. and Gale, P. (eds) (1983), *Museums by Artists* (Toronto: Art Metropole).

Brooks-Joiner, C. and Stasiuk, V. (2005), *Ontario Municipal Cultural Planning Inventory Project: Summary of Findings* (Toronto: Ontario Ministry of Culture).

Brunger, A. (1987), *By Lake and Lock: A Guide to Historical Sites and Tours of the City and County of Peterborough* (Peterborough: Heritage Publications).

Bunting, T., Filion, P. and Walker, R. (eds) (2010), *Canadian Cities in Transition: New Directions in the 21st Century* (Oxford: Oxford University Press).

Castells, M. (1996), *The Information Age: Economy, Society, and Culture. The Rise of the Network Society* (Oxford: Blackwell).

Castells, M. (2000), "Materials for an explanatory theory of the network society", *The British Journal of Sociology* 51:1, 5–24.

Caves, R. (2000), *Creative Industries Contracts Between Art and Commerce* (Cambridge: Harvard University Press).

Chatterton, P. (2000), "Will the real creative city please stand up?", *City* 4:3, 390–97.

De Jong, C.W., Burger, S. and Both, J. (eds) (2008), *New Poster Art* (London: Thames and Hudson).

Di Matteo, L. (1993), "Booming sector models, economic base analysis, and export-led economic development: regional evidence from the Lakehead", *Social Science History* 17:4, 593–617.

Duncan, D. (2004), "Post to post: postering issues across Canada". *Spacing Magazine*. Available at http://spacing.ca/postering/poster-canada.htm [accessed 1 August 2011].

Dunk, T.W. (2003), *It's A Working Man's Town*. 2nd Edition (Montreal and Kingston: McGill-Queen's University Press).

Duxbury, N. (ed.) (2008), *Under Construction: The State of Cultural Infrastructure in Canada* (Vancouver: Centre of Expertise on Culture and Communities, Simon Fraser University).

Edensor, T., Leslie, D., Millington, S. and Rantisi, N. (eds) (2010), *Spaces of Vernacular Creativity: Rethinking the Cultural Economy* (London and New York: Routledge).

Evans, G. (2009), "Creative cities, creative spaces and urban policy". *Urban Studies* 46, 1003–40.

Evans, G. and Foord, J. (2008), "Cultural mapping and sustainable communities: planning for the arts revisited", *Cultural Trends* 17, 65–96.

Evans, S. and Boyte, H. (1992), *Free Space: The Sources of Democratic Change in America.* 2nd Edition (Chicago: University of Chicago Press).

Florida, R. (2002), *The Rise of the Creative Class: And How it is Transforming Work, Leisure, Community, and Everyday Life* (New York: Basic Books).

Florida, R. (2005), *Cities and the Creative Class* (New York: Routledge).

Garrett-Petts, W. (ed.) (2005), *The Small-Cities Book: On the Cultural Future of Small Cities* (Vancouver: New Star Books).

Gianniotis, A. (1999), "Anti-poster laws come unstuck". *Green Left* (published online 1 December 1999). Available at http://www.greenleft.org.au/1999/387/17936 [accessed 17 December 2012].

Gray, J. (2005), "Aesthetic or ugly? Poster fight builds", *Globe and Mail.* February 12, A16.

Griffiths, M. and Woolf, F. (2009), "The Nottingham apprenticeship model: schools in partnership with artists and creative practitioners", *British Educational Research Journal* 35:4, 557–74.

Heeg, S., Klagge, B. and Ossenbrügge, J. (2003), "Metropolitan cooperation in Europe: theoretical issues and perspectives for urban networking", *European Planning Studies* 11:2, 139–53.

Heyman, T.T. (1998), *Posters American Style* (Washington, DC: Harry N. Abrams, Inc., Publishers).

Jones, E. and Dyer, B. (1987), *Peterborough: The Electric City* (Peterborough: Windsor Publications).

Lafer, G. (2003), "Land and labour in the post-industrial university town: remaking social geography", *Political Geography* 22:1, 89–117.

Lewis, N.M. and Donald, B. (2010), "A new rubric for 'creative city' potential in Canada's smaller cities", *Urban Studies* 47:1, 29–54.

Lloyd, R. (2006), *Neo-Bohemia: Art and Commerce in the Postindustrial City* (London and New York: Routledge).

Lorenzen, M., Scott, A.J. and Vang, J. (2008), "Editorial: geography and the cultural economy", *Journal of Economic Geography* 8:5, 589–92.

Markusen, A. and Johnson, A. (2006), *Artists' Centres: Evolution and Impact on Careers, Neighborhoods, and Economies* (Minneapolis: Project on Regional and Industrial Economics, Humphrey Institute of Public Affairs, University of Minnesota).

Miles, M. and Paddison, R. (2005), "The rise and rise of culture-led regeneration", *Urban Studies* 42:5/6, 833–9.

Municipal Cultural Planning Incorporated (2010), *Cultural Resource Mapping: A Guide For Municipalities* (Toronto: MCPI).

Nelson, G. (ed.) (2012), *Beyond the Global City: Understanding and Planning for the Diversity of Ontario* (Montreal and Kingston: McGill-Queen's University Press).
Nelson, R. (2005), "A cultural hinterland? Searching for the creative class in the small Canadian city", in Will Garrett-Petts (ed.), *The Small-Cities Book: On the Cultural Future of Small Cities* (Vancouver: New Star Books).
Norcliffe, G. (1996), "Mapping deindustrialization: Brian Kipping's landscapes of Toronto", *Canadian Geographer* 40:3, 266–72.
Norcliffe, G. and Rendace, O. (2003), "New geographies of comic book production in North America: the new artisan, distancing, and the periodic social economy", *Economic Geography* 79:3, 241–63.
Oldenburg, R. (1991), *The Great Good Place* (New York: Paragon House).
Payne, R. (2010), "The importance of being connected. City networks and urban government: Lyon and Eurocities (1990–2005)", *International Journal of Urban and Regional Research* 34:2, 260–80.
Stolarick, K. and Florida, R. (2006), "Creativity, connections, and innovation: a study of linkages in the Montreal region", *Environment and Planning A: Society and Space* 38:10, 1799–817.
Strangleman, T. (2001), "Networks, place, and identities in post-industrial mining communities", *International Journal of Urban and Regional Research* 25:2, 253–67.
Tatley, R. (1978), *Steamboating on the Trent-Severn* (Belleville: Mika Publishing Company).
Timmers, M. (ed.) (1998), *The Power of the Poster* (London: V&A Publications).
Van Heur, B. (2009), "The clustering of creative networks: between myth and reality", *Urban Studies* 46:8, 1531–52.
Van Heur, B. and Lorentzen, A. (eds) (2012), *Cultural Political Economy of Small Cities* (London and New York: Routledge).
Vinodrai, T. (2006), "Reproducing Toronto's design ecology: career paths, intermediaries, and local labour markets", *Economic Geography* 82:3, 237–63.
Waitt, G. and Gibson, C. (2009), "Creative small cities: rethinking the creative economy in place", *Urban Studies* 46: 1223–46.
Whyte, M. (2007), "Utility poles and free speech: citizens vie with corporations", *Toronto Star*, May 13. Available at http://www.thestar.com/news/article/213341 [accessed 1 August 2011].

Chapter 5

Learning from the Post-Industrial Transition in Northern England: Alternative Developments in the Visual Arts (1979–2008)

Gabriel N. Gee

The idea and process of cultural regeneration rapidly gained momentum in Great Britain at the end of the 20th century. Regional bodies, local authorities, and a growing number of artistic organizations recognized the shift in the role to be played by artistic creativity within society:

> The arts provide jobs; they improve the quality of life and the attractiveness of an area; they are more and more in demand as leisure-time, either chosen or imposed, expands; they are attractive to young people; and by a broader definition, which includes television, video, recorded music, graphic design and publishing, they are commercially viable, and worldwide, a multi-billion pound industry (North West Arts Board 1992–93: 14–15).

Following the election of Margaret Thatcher in 1979, conservative governments emphasized the tangible economic and social benefits to be gained from the art sector, while implementing policies to promote a free market economy and to cut public expenditure (Edgell and Duke 1991: 4–8). They championed "the economic importance of the arts" and their "effectiveness", using and promoting business models in the management of the public funded art field (Bragg 1983–84: 3; Yorkshire Arts 1988–89: 21–2; North West Arts 1992–93: 4–10).[1] In the North of England, economic regeneration had become a paramount issue following the decline of its traditional industrial sector after the end of the Second World War. Northern cities were drawn to redefine themselves in accordance with the national agenda and adopted in the 1990s a cultural regeneration model sponsored by the Arts Council of Great Britain. This model

1 On the national scale, the value of the arts as an economic asset was underlined by the publication of The Economic Importance of the Arts in Britain by John Myerscough in 1988, a report that had been commissioned by the Arts Minister, Richard Luce, to the Policy Studies Institute.

aimed to develop local cultural infrastructure through public-private partnerships in order to attract inward investment and to facilitate the shift to a service-based economy. It also advocated the role to be played by the arts in tackling social issues furthered by economic decline.

In fact, regional actors in the arts who relied on public funds were faced with a "chronic national underfunding" in the 1980s (Richardson 1984–85: 5). This was highlighted by the abolition in 1986 of the metropolitan county councils that had played an important role in arts' funding, along with local authorities and regional art associations (Richardson 1986–87: 6). Furthermore, the latter, who since the 1960s had come to manage and fund arts projects in the regions, were progressively led to inflect their policies. A privileged partner of the Arts Council of Great Britain, their financial dependency on the national body implied that they were directly affected by the ideological shift carried out by the government. Regional art associations were thus encouraged to develop a business model for artistic growth and an understanding of the arts as economic and social tools.

A number of projects emerged in northern England with the aim to articulate alternative artistic and organizational routes of development to this framework. At the turn of the 1980s, an oppositional stance responded in particular to the government's neo-liberal economic policies and its conservative vision of British identity. Collectives such as Pavilion in Leeds, Amber in Newcastle-upon-Tyne and the Liverpool Artists Workshop in Liverpool, developed a prism of strategies that reflected socio-political concerns and reactivity within the cultural field in the region. These strategies were drawn at a time when the outcome of the struggle between central government and national and regional counter-powers had not yet been decided, and they must be analysed in relation to this context. During the 1990s, artistic organizational alternatives in the North took a more nuanced profile, as actors in the northern art worlds tried to negotiate the implementation of governmental policies. A series of key values and practices associated with grass roots development nevertheless subsisted through organizations such as East Street Arts in Leeds, and the Annual Programme in Manchester.

At the turn of the new millennium, northern cities were still embedded in the transition to a post-industrial future, neither having fully discarded the industrial past, nor having quite successfully embraced the age of the leisure industry. Oppositional and alternative strategies found a new lease of life, as exemplified by the Ultimate Holding Company in Manchester and Metal in Liverpool. This chapter aims to provide a historical analysis of a set of alternative strategies to artistic production and diffusion which appeared in the North of England in reaction to the progressively dominant proposition articulated by conservative and New Labour policies in Britain. It aims to underline the means and the goals of these alternatives, while taking into account the aesthetic issues directly involved in the process.

Alternative Strategies in the 1980s: Creative Resistance from the Bottom Up

Throughout the 1980s, British conservative governments emphasized the role of the free market and the individual while advocating an ideal of British identity rooted in the Imperial Past. In the industrial north, numerous cultural agents reacted to this agenda by fostering forms of collective organization and participation that echoed the region's economic and social history. Furthermore, the negation of ethnic and gender minorities that had been reasserted on a national level came increasingly under attack by groups promoting cultural inclusion and diversity. In this context, the Liverpool Artists Workshop and Amber in Newcastle-upon-Tyne were significantly furthering a tradition of social mindedness, while Pavilion in Leeds embodied a northern form of creative resistance based on gender and ethnicity.

Pavilion was set up as a feminist photography center and collective by Shirley Moreno, Diana Clarke, and Caroline Taylor (Reid 2008).[2] It opened its doors in Leeds in May 1983 in a building taken over and renovated on Woodhouse Moor. A Feminist Art Programme had originally been created in 1981 in order to establish a new type of art center in the city:

> The aim of the gallery is to show the work both past and present of women photographers. We would like to expand this to other media where it is relevant to the issues of the exhibition. The issues will be those that arise from women's specific position in society, e.g. the representation of women as workers, as objects in pornography, as portrayed in domestic life etc. We intend to contextualise these shows by relating them to local conditions in Leeds. Furthermore, we wish to extend the gallery's function through the education programme and also through working with local communities (Feminist Art Programme 1981).

As a photography center dedicated to feminist production and thinking, Pavilion stood apart from other local and regional artistic organizations. The three women had met at Leeds University. Griselda Pollock had been recently appointed as a lecturer in the University and the questioning of the role and meanings of images furthered in Pavilion's programme and activities must be seen in the light of her groundbreaking work on feminist social art history. Similarly, the theoretical work on photographic images of John Tagg has also been underlined as influential in the elaboration of the project.[3] The programme of Pavilion had started prior

2 The following reconstruction is based on the consultation of pavilion's archive at the Feminist Archive North, located in Brotherton Library, at the University of Leeds. It was enlightened by a conversation with Anna Reid, who became the director of Pavilion in 2006.

3 John Tagg had also joined Leeds University at the end of the 1970s. Both Pollock and Tagg were members of Pavilion's advisory committee along with Stevie Bezencenet and Joe Spence.

to the opening of the space in 1983 (Pavilion 2008: 18).[4] It featured education courses in Leeds aiming "to explore the area of women's art both historically and practically" (Leeds Women Art Programme 1982). One of these courses, "Women invisible-Woman visible" developed into an exhibition entitled *Anonymous: Notes towards a Show on Self Image*, which toured regionally. The 1982 newsletter also announced other important initiatives to come, such as workshops on "Women's literature" and "Women in and out of work", as well as projects to be pursued with "girls and ethnic groups in the area".

These ambitions suggest that collective discussion and group training were central to Pavilion's development model. If photography was to be a rallying activity, it was not to be the unique focus of the group. For instance in July 1985, Pavilion held a screening of *Leila & the Wolves* by Heiny Srour,[5] as well as a L.C.U.S. Women Youth Worker's training day, a girls in need meeting, a meeting of Leeds Women anti-deportation, as well as a meeting of the Women imprisoned theatre group (Pavilion 1985). A structure of social, psychological and educational support to women based on collective exchanges between women from all ages and communities was essential to Pavilion's early development.

With regards to the promotion of photography by women, the medium was seen as more than an end to itself:

> We wanted to centre something around photography because it is one of the most ubiquitous and accessible of all the media. One of our prime concerns is the visual representation of women. Historically, women tend to be seen as represented rather than representers. And photography usually defined as something that men do and that women are the objects of is a classic paradigm of that.[6]

Pavilion's approach was more issue-based than artist-based. The space itself was equipped with a dark room to further the learning and practice of black and white photography and a convertible space, which could host exhibitions as well as workshops and discussion meetings. It could also be used as a crèche (Pavilion 1983). These facilities and most events were originally only accessible to women. Significantly, the practical and political orientation of Pavilion's activities eventually triggered the irritation of Yorkshire Arts, the regional branch of the Arts Council. Along with the city council, which had lent the building, and the University whose extra mural programme had contributed to the development of the organization, the regional arts association was the main purveyor of funds to Pavilion. In 1984, it suspended its grants. The quality of the work exhibited appears to have been blamed, as well as the "separatist" nature of the programme

4 Pavilion secured the lease of the building from Leeds City Council in 1983.

5 *Leila & the Wolves* is a 1984 docu-fiction that highlights the role of women in Middle East politics.

6 Shirley Moreno quoted in Powell (1983).

(Thompson 1993; Reid 2008).[7] In July 1986, however, Yorkshire Arts adopted a revised policy statement for photography. The turn-around resulted from "a considerable struggle" within the regional institution "with elitist notions of what constitutes 'good' practice and a recognition of the importance of an equal opportunities programme, of education and community participation" (Brookes 1986–87: 18). Pavilion was re-established as an annual client.

Throughout the 1980s, the programme reflected a dual mission: interrogating the representation of women's image and empowering women as practitioners. The first exhibition of Pavilion was thus entitled *Collective Works*, and aimed to underline the ethos of the organization. It reflected on the collective work undertaken to establish the gallery, highlighting the process of setting up a feminist network within the early 1980s Leeds environment – not a premiere in the country but certainly in the North of England. It also addressed the negotiation of funding issues and confronted traditional photographic practices dominated by male values with panels proposing alternative feminist types of practices (Pavilion 1983). These alternative practices focused on women's experiences and empowerment. They challenged the male gaze as reproduced in the mass media industry and in the history of art, and involved local community in the reception as well as production of the works.

Collective Works was followed in June by an exhibition of one of the major projects of the Hackney Flashers Collective: *Who's Holding the Baby?* The Hackney Flashers, based in East-London, had been created in 1974 and would have been a direct inspiration for Pavilion. Like the Photography Workshop, The Hackney Flashers had been set up to explore and promote community photography (Camerawork 1979). It promoted a critical use of photography as well as self-reliance in photographic users. It also encouraged enquiries into collective histories, exploring in particular women's experience of the working and domestic environment (Spence 1986: 66–74). In July, Pavilion reflected this interest in inclusive photography production through an open submission exhibition on the theme of "Feminism and Photography". Alongside the exhibition programme exploring contemporary aspects of *Womeness*, it also organized discussions about the shows, practical initiation to photography, and group discussions on social and gender issues such as the aforementioned "Women's health group", and "Working with girls" (Pavilion 1985).[8]

Although the founders left the organization in the mid 1980s, Pavilion's activism carried on until the early 1990s. It added lesbian representation to its field of investigation, with exhibitions such as *Lesbian Photographers: Don't Say Cheese, Say Lesbian* (Pavilion 1986), and *Stolen Glances: Lesbians take*

---

7 The Regional Arts Association reports merely stated: "The limitations in the 84–85 budget resulted in no grant being made to the Pavilion project" (Yorkshire Arts 1983–83: 2).

8 *Womeness* was a title of an exhibition held between the 9th of September and the 26th of October 1991.

Photographs (1991–92).[9] It also exhibited artists associated with the Black Art Movement in Britain. *Testimony* in 1986 presented the work of Brenda Agard, Ingrid Pollard and Maud Sulter (Pavilion 1986).[10] In 1992, an exhibition entitled *Keeping it Together* grouped the works of ten artists under the banner of Black women photography (Pavilion 1992). The Black Art Movement had emerged in the visual arts in Britain in the early 1980s, while tensions between afro-Caribbean communities and the government increased. The exploration of the representation and expression of singular ethnic voices and histories within a resurgent monolithic British paradigm echoed that of Pavilion's gender outlook.[11] These strategies coalesced in the aim of developing independent practical skills as well as a collective critical reflexivity on the representation of oppressed gender and communities. The heirs of 1960s and 1970s emancipation movements, Pavilion furthered a fight for recognition, which was still in the balance in the 1980s, and which must be seen in the light of post-colonial and feminist history.

If identities were felt to be a battleground, activism was also directly related to economic and geographic issues. The government's cuts in public expenditure further threatened traditional industrial and manufacture production. Northern local authorities such as those in Liverpool, Manchester and Sheffield, had embraced Municipal Socialism in the early 1980s (Green 1991: 273–91).[12] Political opposition to a hostile government was furthered on a local level by radical socialist activism. The municipal left would advocate local empowerment and local and regional responsibility in the development of economic policies, contrasting with governmental insistence on free-market reforms (Quilley 2000: 601–15)] Collective modes of organization in a northern context were thus infused with political resistance and a historical urgency. Participation and a stress on the grounding of artistic work within local communities were part of an attempt to articulate a regional opposition to a national agenda underlining individual mobility. The creation of the Liverpool Artists Workshop (L.A.W.) in 1980 by Sue and David Campbell, Colin Finn, Mark Cardwell and Pete Clarke, exemplified such resistance, advocating an approach to artistic practice based on a collective ethos:

---

9 The *Stolen Glances*, exhibition was held between December 1991 and January 1992.

10 Testimony was held between June and August 1986.

11 Furthermore, women's practices came to form a singular political identity *within* the Black Art Movement itself.

12 Municipal Socialism thrived in a number of labour controlled councils in Britain after the election of Margaret Thatcher in 1979. These "red" councils were deeply antagonist to Thatcherite policies, and aimed to base counter-powers on the local level.

Figure 5.1 Pete Clarke, Municipal Modernism, the cloth cap and the red glove, 1986, oil and mixed media on canvas

> Liverpool Artists Workshop was set up in September 1980 by a group of people wishing to work in Liverpool who saw the communal studios as a practical solution to both the financial aspects of workshop provision and also the creation of a healthy atmosphere in which to produce work. It has always been a concern amongst the members that a communal and collaborative working situation would develop as opposed to an isolated and individualistic practice (L.A.W.).

Artists' studios had been emerging in the UK since the 1970s. In the Northern provinces this included Yorkshire Artspace Society, created in 1977, Artspace Merseyside Limited which opened in 1976 in Liverpool, and Manchester Artists Studio Association set up in 1983 (Artspace Merseyside 1986; Artspace 2011; MASA 2011).[13] These organizations were formed to provide artists with cheap working space. It can be argued that deindustrialization favored the creation of artists' studios, as large disused industrial warehouses and mills in the northern regions suddenly proliferated (Cook 2003). Studios also provide a working environment. Practitioners meet like-minded individuals and find a collective spirit, which can encourage the development of an otherwise essentially solitary

---

13 The tradition of artists' studios is well established in the 20th century. However, Britain witnessed a specific resurgence and rise of collective studio from the 1970s onward.

activity. L.A.W. had a changing basis of around 15 artists who were accommodated and participated in its programme. But more than merely providing a working space, L.A.W. promoted an engagement of artists within society. As Pete Clarke recalled:

> … the difference with L.A.W. was that it was an attempt to have a more of an ideas-based studios, as a lot of them are just a mean to an end, where maybe ten to twenty artists just want to rent cheap studio space, whereas we wanted artists to share similar concerns (Clarke 2007).

L.A.W. organized "Open Weeks", aiming to bring the public in contact with the work produced within the building. It organized seminars and discussions around the work exhibited, aiming to foster a dialogue around art practices within local communities. It also ran a series of lectures, inviting well-known scholars and artists to speak in front of local audiences. Guest speakers included Terry Atkinson on "the ongoing war of position" and Griselda Pollock on "Feminism and Art in the 1970s".[14] Tony Rickaby gave a talk entitled "The class basis of art activity", while Marie Yates explored "the Woman Artist – exhibiting and teaching in a sexist society" (L.A.W.). These strategies were an attempt to contextualize the work within the specific political context of Liverpool and the national shift in economic and social policies of the early 1980s (Adams 1998: 135–6).[15] To develop art practices as social "strategies to engage in social and political debate to reach different audiences" was to try and counterbalance the negation of collective histories and values symbolically encapsulated by Margaret Thatcher's resonant "there is no such thing as society".

In Newcastle-upon-Tyne in the North East, the collective Amber similarly maintained a passionate attention to the industrial legacies, communities and lives of the northern territories. The group was set up in the late 1960s in Newcastle as a film and photography collective. Its members were attracted by the North East working class regional identity, which they began to explore through the social documentary genre.[16] Having moved from London, the group also benefited from the moderate cost of living in a former industrial region (Rigby 2008). Amber set up the Side Gallery on the derelict Newcastle quayside in 1977. The photography gallery was initially created to show the work produced by the collective. However, it rapidly began to articulate a programme commissioning and featuring the work of other practitioners who engaged with the documentation of working class communities in the North. Amber co-founder Sirkka Liisa Konttinen photographed

14 L.A.W. had a connection to Leeds University in David Campbell who studied in Leeds after having moved to Liverpool.

15 Liverpool went through a particularly acute phase of political radicalism in the early 1980s as the Trotskyite Militant tendency gained pre-eminence in the council.

16 The members of the group in the 1970s were Murray Martin, Sarra-Liisa Konttinen, Graham Denham, Graham Smith, Peter Roberts, and Lorna Powell.

the community living in Byker in Newcastle before the area was scheduled for demolition and redevelopment, John Davies looked at the Durham colliery in 1983 as the collision between the miners and the government appeared imminent, while Ian McDonald documented the traditional Quoits game played in the Esk Valley in northern Yorkshire. Amber also explored the experience of North-East communities through numerous documentary films and dramas. An engagement with live theatre in the early days led Amber to produce films where acting roles were directly informed by local communities and individuals.[17] The documentary *High Row* (1974) initiated a tradition of participatory filming that has continued into the early 21st century. The film examines the lives of seven miners employed in a small mine in Cumbria. The script was written in collaboration with the miners who played their own part in the film. In their photography, as in their film work, Amber aims to document and celebrate the working class lives and traditions of a region where ship-building and coal mining have long been amongst the major purveyors of work and identity. Crucially, the work is produced in a dialogue with the people depicted through a direct involvement of the producers within the social field. Subjects become active agents of a mode of representation which has continuously argued for a collective vision of shared experiences and histories against the tide of growing individual merit.

Like Pavilion and L.A.W., who aimed to promote a collective ethos and a broadening of production, reception and participation around the artwork, Amber opposed the conservative and romantic view that "artistic talent … is unplanned, unpredictable, eccentrically individual" (Thatcher 1993: 632). In that respect, while inheritors of 1960s and 1970s oppositional movements, they expressed a specific resistance to a historical context that emphasized an embeddedness of art production within the art market, between the artwork and its isolated and illuminated creator, severed from its social environment.

Along the Tide: Artist-led Initiatives and Negotiation in the 1990s

By the turn of the 1990s, the neo-liberal policies progressively introduced by Margaret Thatcher's governments were shaping what came to be known as a "paradigmatic change" (Heffernan 2001: 1–9). After the repression of the miners' strike in 1984, and the third national electoral success of the conservative party in 1987, oppositional stances in British society began to lose ground. During the 1990s, the art sector adopted a managerial and entrepreneurial ethos. The creation of the lottery in 1994 suddenly fed the arts with vast amounts of capital. It also furthered a top-down framing of artistic initiatives. Public funding could be

17 For *In Fading Light* (1989) which looked at the fishing industry in North Shields, Amber bought a pub and a vessel in the town, and actors lived with the fishermen on the quays as well as on the boat spending days at sea.

obtained on the condition of adapting projects to the criteria defined on a top-down basis.

The Liverpool Artists Workshop disappeared in the late 1980s when the organization was forced to move out of their premises, while Pavilion became a more generalist art agency, promoting, in particular, the diffusion and education of new technologies in art.[18] Within this context, however, an alternative model of artist-led initiative inherited from the 1970s and 1980s was still active. In Leeds, the organization East Street Arts started to investigate the conditions, functioning and role of artist studios, while in Manchester (from 1995 to 2000), the Annual Programme explored a low-cost ambitious system of artistic diffusion based on artist's networks. While capital development projects multiplied through the injection of lottery money in the art world in England during this period (Arts Council England 2002), East Street Arts and the Annual Programme exemplified an enduring pursuit of independently led artistic and collaborative strategies in the North.

East Street Arts was set up by Karen Watson and John Wakeman in Leeds in 1993. Both artists had studied ceramics in Cardiff and Sunderland, before moving back to Yorkshire (Watson 2012). The practice of ceramics required the use of a kiln, and Watson and Wakeman began to consider creating an artist studio to house the equipment. They were initially struck by the bleakness of existing visual art facilities in the city (Watson 2008). At the time, there were a number of studios in Leeds, such as Oblong Studios, Third Floor Studio, Jackson Yards, and Leeds Artspace Society. The latter, which was one of the main artists' organizations in the city, closed the year East Street Art was created. The disappearance of a major artist-led organization prompted E.S.A. to reflect on the sustainability and legacy of studio spaces. In that respect, securing the lease and use of a building appeared as an essential target,[19] though it could not be reached immediately:

> And of course, the first thing we did was to move into a very bleak and depressive building, as it was the cheapest we could find! (Watson 2008)

From 1993 to 1999, E.S.A. was based in the former East Street Mills. It aimed to increase both its programming and its visibility. As newcomers to the city,

---

18 The shift was announced In Pavilion's 1993–94 annual Report: "We are particularly keen to involve our audiences with the 'new technologies' in order to allow them hands-on experience whenever possible. We acknowledge that innovations in electronic imaging are happening with such rapidity that many people feel out of touch and bewildered" (Pavilion 1993–94).

19 A similar analysis can be applied to Liverpool, where the L.A.W. disappeared as it lost its' building on Hope street, and found itself at odds with the council's offer of relocation. This offer was subsequently taken over by *Arena* Studio, but with a more mainstream ethos. The Bridewell studio, set up in the 1970s, bought the former police station they were located in, and has been active ever since.

Karen Watson and John Wakeman were eager to gain an understanding of the Leeds art scene and to establish connections in an otherwise fragmented local art scene. E.S.A. was induced to reflect on the modalities of interactions in the art field:

> We paid for enough space to have eight artists, and we organised it so that they had studio spaces, a communal space, kitchen facilities that we could warm, a kiln room, a darkroom … In the end we had fifty spaces down there, and a big project space … But what we also did very early on: we talked a lot to the artists, we listened to what they were saying, to what they wanted, and the biggest thing we heard was that they weren't connected to anything (Watson 2008).

The group aimed to provide working facilities for artists and developed a programme to enhance artists' visibility. This meant establishing platforms for artists' work, as well as questioning artists' self-positioning and awareness of their role within society. A "Leeds Artists Network" was created to further the dialogue between the different separate visual arts entities in the city, which subsequently led to the creation of Leeds Visual Arts Forum (L.V.A.F. 2012).[20] The organization also became proactive in entering municipal decision-making boards, such as the City Council's Cultural Development Strategy. This policy involvement was induced by the historical situation in Leeds, where the council had historically favored the performance arts over the visual arts. In order to promote the visual art scene and its own activities, E.S.A. assessed that it was necessary "to become part of the political fabric of the city" (Watson 2008). This attention to the socio-political and economic environment in which artists operate has been essential to the ethos of E.S.A. Firstly, it meant getting political forces to recognize the work of art practitioners, and to participate in political decision-making. Secondly, it implied getting artists to fully embrace their practice as a professional occupation. The political implications of such strategies are crucial, for they advocate an engaged negotiation with governing bodies and individuals within the Polis as a potent conduit to reclaiming an abstractly defined creative economy.

E.S.A. eventually acquired Patrick's Studio in 2001 through West Yorkshire Playhouse, who knew they were looking for a permanent space and sold them the building for a mere £150,000. A further £1.6 million were raised to do refurbishment. By securing funds from Leeds City Council and the Arts Council Capital investment, E.S.A. became a regular funded client of the Arts Council. In effect, E.S.A. had succeeded in inserting its artistic and development model within the acknowledged and growing cultural economy of the city.

---

20 The website of the organization states "LFAF is committed to raising the profile of the visual arts in Leeds through information exchange, advocacy, the website and networking events". The organization was created in 1999.

Figure 5.2 East Street Arts and the Theartmarket, Social Club, 2007: dinner

In Patrick Studios, E.S.A. developed a program of professional development for artists. It assesses their needs, contributes to the writing of individual development plans, monitors progress, and helps artists "endeavouring to be professional in their chosen discipline" (East Street Arts 2011).[21] The strategy aims to counter the isolation that is often associated with a romantic conception of the promethean artist, and recognizes the importance of artistic exchange and networking. This organizational mode has led to numerous aesthetic projects. For instance the idea of a "Social Club" is exemplary of an ethos that offers "a platform for artists, participants, and audiences to engage collaboratively in contemporary art", and aims to narrow "the boundaries between social and art" (East Street Arts 2007: 4).[22] Social Club 2007 screened films that explored social difficulties and upheavals in Britain in the late 20th century. It featured *Housing Problems* by Athur Elton and Edgar Anstey (1935), and *The Battle of Orgreave* by Mike Figgis and Jeremy Deller (2001), which constitutes a celebrated

21 The professional development program is not a prerequisite for the other buildings held by the organization: Beaver Studios, Barkston Studios and Union 105.

22 A further introduction to the event stated that "Social club will explore contemporary art's potential to engage in and be an investigator for: social change and connecting to wider social movements; commentary on social situations; creating environments for exchange and new projects; extending the relational aesthetics/socially engaged art debate".

re-enactment of the miner's strike confrontation with the police apparatus in 1984. It also featured a *Mexican Bingo*, and a *Dinner*, co-organized by E.S.A. with Theartmarket. *Dinner* had the guests building "a den" in which to have their meal. It thereby had the art visitors working collectively on the preview "to lay the table" in order to earn the right to consume their meal.

Social Club 2008 was entitled "Multitude" and focused on our relation to objects and collecting. It made explicit references to the rise of capitalism and consumerism in the late 20th century, and featured a "non hierarchical" open submission show for E.S.A. artists where artworks could be sold. Social Club 2008 also presented a project entitled *You're never alone when you collect*, by artist Jess Wilkin, that involved a group tour of three local inhabitants' houses in Leeds. The visit looked at the collectors' eclectic range of objects and discussed how these came to be set up. It also featured a *Socialist Jukebox*, a large inherited collection of seven-inch vinyl singles. The collection was displayed on the wall of Patrick's studios where visitors could choose one item to be played before departing with the original vinyl. The guests thus constructed the musical event as they came for the social club, while the initial musical capital was shared evenly among the participants.

The *Socialist Jukebox* and *Dinner* were exemplary in exploring "relational aesthetics", which have found a large echo in artists' organizations (Bourriaud 1998; Eggleton 2007). The use of a participatory model advocating the sharing of material and ideas between the artist and an enlarged audience comes as a pertinent counterpoint to the commodification of culture in the northern provinces during the period. Within its contemporary aesthetic remit, it can be seen as subtly referring to the socio-political history of these former industrial cities. Collective organization, work and experience are decisively added to a uniformly instrumental economic value of the arts in the age of intangible financial fluxes.

In Manchester during the same period, the Annual Programme initiated an alternative path to artistic production and visibility. The programme was set up by artists Nick Crowe and Martin Vincent following the presence of the British Art Show 4 in Manchester in 1995 (British Art Show 1995).[23] Seizing the opportunity to have the national and London limelight on Manchester, Nick Crowe and Martin Vincent organized an alternative exhibition of locally based artists. The show was called "Ha!" and took place on the top floor of a warehouse in the city center. The preview on the 11th of November coincided with that of the British Art Show taking place in the main art venues of the city.[24] The party was a success with a considerable turnout, fuelled by free pizza and champagne alongside works that

---

23 The British Art Show is a major touring display of contemporary British art which has been organised every five years since 1980.

24 The main venues were the Castelfield, gallery, the Cornerhouse, Manchester city art gallery, the Chinese Arts centre and the Whitworth gallery.

were met with positive feed back.[25] It led its organizers to pursue their curatorial activities. From June 1996 to April 1997, the Annual programme was hosted in artists' houses.

> The idea was that it would be a chain exhibition. In January I will exhibit in your house, and in February you will exhibit in his house, and so on. It goes around and eventually it goes back to me and he exhibits in mine ... It was a really cheap way of getting space, obviously, but also because it had this structure, and every month there was an opening, it meant that it could keep a momentum going, locally in terms of people producing shows, but also [nationally as] it gave us the opportunity every month to send press reviews and advertise in Art Monthly (Crowe 2008).

Thereafter, the programme was pursued in various locations in the city centre of Manchester, such as shops and churches, while a gallery was briefly run on Tib street.[26] As in the case of E.S.A., the Annual Programme had developed a voluntary approach to networking, which gave visibility to their activities. It used forms of sociability which were rooted in the city, aiming to bypass the incapacity of the local infrastructure to offer opportunities to emerging artists. The reliance on artistic networks and the use of artists' present abodes and resources fostered a programme which was mostly autonomous in financial terms.[27] While based on Manchester's artists' communities, and on the city's favorable climate for artists looking for a lively social place and cheap accommodation, the programme wasn't inclined to advocate a regionalism resentful of London's pre-eminence. It specifically didn't stress the virtue of an artistic localism associated with parochialism:

> For me [the relation to London] goes back to the situation we found ourselves in the mid-nineties, in our local area. It was almost as if you could identify what you hated about your art scene, and do the opposite. If all the people around you were saying "the galleries should show local artists", you said they shouldn't. And the other thing that you found was this very backward attitude towards London, very sour. And that was genuinely not how we felt. We thought London was great, to go there, we just didn't want to live there (Crowe 2008).

25 The exhibition featured the work of Simon Greenan and Christopher Sperandio, Index; Chara Lewis, David Mackintosh, Ian Rawlinson, Michael Robertson, Martin Vincent, Roxy Walsh and Mark Harris (Greenan 2001: 16).

26 Interestingly, the Tib Street gallery was eventually relocated and became a commercial gallery: International3.

27 The series of 12 exhibitions was supported by a grant from the Arts Council, which was entirely spent on advertising the events in the national press.

As Richard J. Williams pointed out, the Annual Programme articulated a "radical pragmatism" (Williams 2000). It attempted to create a contemporary art scene in Manchester while facilitating links with London actors, such as City Racing and Bob and Roberta Smith. Its development model was itself related to the success of artist-led strategies in Britain in the 1990s. The British Art Show 4 came to be known as the crowning of the young British artists who had emerged on the London art scene following a series of do-it-yourself exhibitions in the late 1980s. The Annual Programme endeavored to create an analogical artists' network and to anticipate reception processes in the art world in order to gain visibility and recognition. Significantly, these initiatives were done at a time when most artists were unemployed. Karen Watson reflected that "When we first set up East Street Arts, the people that came to the studio spaces in the early 1990s, they were all on the dole; every single one of them except one or two tutors who were looking to get jobs. Now you would find it hard to find somebody that's out of work. That's been one of the biggest shifts over the period of time we have run the studio space" (Watson 2008). Similarly, Ian Rawlinson, reflecting on the attraction of living in Manchester in the early 1990s, suggested that a big asset was "Cheap clubs. Because most young artists at the time were signing on the dole. And that's kind of how everybody lived really. It's very different now I guess, it's much more difficult for people" (Rawlinson 2008).

The last decades of the 20th century had more or less accommodated the long-standing bohemian life of the artist that, along with teaching positions, sustained the creative desires and liberty of the actors of the visual art field. However, it was becoming more and more evident that the instrumentalization of the arts would entail an increased accountability of its practitioners. Any strategy aiming to correct or derail the objectives of government-based policies would need to engage with the pragmatics of sustainability to better assure its aesthetic and social input.

The Double Face of Cultural Regeneration: Creative Resistance within Global Flows

The Labour Party regained control of the parliament following the 1997 election. New Labour, promoted by Tony Blair, confirmed the entrepreneurial shift installed by the Conservative Party during the 1980s, while inflecting cultural policies with a revitalized social concern. The editorial of North West Arts 1998–99 annual report reflected both the continuity and change associated with the new government:

> The year was shaped, above all, by the concerns of the New Labour Government – notably its commitment to regionalism, its support for creative industries, and its emphasis on developing the potential of people, particularly via social

> inclusion policies and support for the importance of creativity in education and lifelong learning (North West Arts 1998–99: 3).[28]

At the heart of the new guidelines was a stress on accessibility, which positioned the arts and culture as a cornerstone of British society.

> Access to the publicly funded arts has become a key objective of government and the department for Culture, Media & Sport. Comprehensive spending reviews highlighted the need for greater accessibility to the arts, for both audiences and aspiring practitioners. Indeed, the Secretary of State for Culture, the Rt Hon Chris Smith MP, published Creative Britain, a collection of speeches highlighting this and other issues in order to raise their profile and to make it clear that public bodies involved in cultural provision, such as regional art board and local authorities, should take this area of their duties very seriously (North West Arts 1998–99: 18–21).

North West Arts insisted on its supportive action to groups furthering this agenda. It listed a wide range of actors who conducted projects with individuals and groups "normally excluded from arts activity".[29] The attention to a multicultural and community-based service for the arts contrasted with the 1980s stern outlook on socially-oriented endeavors. It did not, however, feature a collective ethos of aesthetic empowerment motivated by critical political reflection. The value of the economic contribution of the art sector was also confirmed, along with its role within society as a social mediator.

In Northern cities, the regeneration strategies gained further momentum at the turn of the Millennium. Gateshead completed the conversion of the Baltic Flour Mill into a centre for contemporary art (2002). Manchester organized the Commonwealth games in 2002 as it furthered the transformation of its city centre into a dynamic financial and commercial centre (Peck and Ward 2002). In Liverpool, the Tate Liverpool had been opened in 1988 in the Albert Docks. The investment in the cultural and artistic sector was pursued in the next decade, with the setting up of the Biennale in 1999 and the opening of the Foundation for Art and Creative Technology in 2002.

It is within this cultural regeneration context in Manchester that the Ultimate Holding Company was created in 2002 (Redman 2008). Based in the Hotspur Press

---

28 The New Labour policy for the cultural sector was implemented by the newly created Department for Culture, Media and Sport.

29 For instance, North West Arts 1998–99 Annual Report included a lottery award to Longsight Youth Arts to further musical and drama projects with young African Caribbean people. It registered the creation in 1998 of SACVAN, the South Asian Crafts & Visual Arts Network, and heralded the creation of the Arts about Manchester "new audiences programme", which included the provision of an administrator and education officer at the Manchester based Black Arts Alliance.

Building in the city centre, U.H.C. is a cooperative that aims to build an alternative art and design organization model based on strong political engagement:

> UHC opposes attacks against creativity and the cultural decay created by neoliberalism. We support freedom of expression in the arts, as well as the artist's freedom to create work in the public domain without the permission of any authorities ... Through planning our work on a not-for-profit basis, we seek to directly challenge commercial primacy. Through our commercial activity, we seek social change on a fundamental level ... (U.H.C.).[30]

In its support to a new instrumental positioning of the arts as a commercial vehicle for economic revitalization, U.H.C. works on artistic projects that have a political nature, and on design services that value the input of artistic thought. Jai Redman, the creative director of U.H.C. underlined the overlapping of the organizations activities:

> We had three bottom lines basically in our business plan ... One is to be financially sustainable, the other is to have a political remit so that we would make work, which is out in the community in Manchester, and challenging the local government. And it had a strong creative side to it. Those were the three lines, the artistic, the political, and the business side (Redman 2008).

The design service U.H.C. proposes to companies with ethical standards is aimed at improving the quality of visual communication in a field where the focus on the content of the message often undermines its mode of diffusion. It brings the aesthetic to the political. It also conditions the financial viability of the cooperative. The income from the design branch enables the freedom of artistic exploration, unstained by mitigation with the criteria imposed by public institutional funding.

The artistic production of U.H.C. similarly addresses the entrepreneurial turns which occurred in Manchester in the late 1980s (Quilley 2002: 76–91). It organizes charity kitchens in the street, exploiting the by-laws that enable members of the public to distribute food for free in public space. Relational aesthetics are used to counterbalance the privatization of public space in Manchester city-center, which occurs through both its commodification and an increase in surveillance control devices. It asks questions as to the hidden frames that are progressively enveloping passers-by in "neutral" public space. With *Spring Shroud* (2007), U.H.C. addressed this intrusion of private space through advertising panels. The project was realized in collaboration with comedian Mark Thomas. It consisted in a synchronized action that covered a hundred advertising boards in the city with a white veil on which one could read: "Trees breathe, ads suck".

30 *Ultimate Holding Company Statement of Artistic Process*. The first lines of the manifesto state that this introduction is "a work in progress, so don't quote us on it!".

The work effectively suggests that an ecological living being would be a more sustainable and positive presence in the urban environment than an inert advertising panel.

With *Incursions in the Knowledge Capital* (2006), UHC tackled the network of decision forces in the city concomitant to its privatization: on a wooden table, the flow chart of power brokers was imprinted and lit by three over standing lamps in the display at the abandoned Castlefield chapel. The incursion unveiled the hidden network of public and private actors and interests that it deemed responsible for the current privatization of urban space, which it so forcibly criticized. A hybrid organization functioning both in the commercial design sector and in the realm of Manchester art world, U.H.C. combines an attention to representation and an income raising activity to support its projects with a political brief exploring oppositional tactics to developments of global capitalism.

In Liverpool and Southend-on-Sea in Essex, the organization Metal has similarly been active in promoting the value of artistic thinking in society. It shares with U.H.C. a collective ethos and a belief in collaborative work as well as the capacity of artistic creativity to generate social change. It is similarly concerned with the need for artists to negotiate commercial activities and artistic freedom. Metal, which was created by Jude Kelly in 2002, is inclined to define grounds of negotiation in order to guarantee the effective impact of the projects it promotes. It functions through artists' residencies, which provide time and space to artists in a range of disciplines, who are encouraged to reflect on their work as well as produce new pieces. Residencies are nevertheless conceived within a broader perspective on the relation between art and society:

> Alongside the artistic process involved in our residencies, we are interested in how the practice of artists can contribute to, and potentially influence political and social issues of the day. Through our programme we draw out evidence for a better understanding of the artists' role in civic life (Metal Projects).[31]

The program aims to address issues within the whole prism of British society. However, it significantly tackles these from the bottom-up. Projects are developed from Metal's spaces, locations, and specific environments and history.[32] In Liverpool, Metal examined the deindustrialization process as well as the economic and social difficulties it led to. With *Cow – the Udder way*, it participated in the international ideas competition "Shrinking cities – reinventing urbanism".[33] Liverpool witnessed a considerable depopulation in the second-half of the century.

31 Metal, *Projects*.

32 Metal has two spaces, one in Liverpool and one in Southend-on-Sea.

33 *Project 7*, "Cow – The Udder Way". The project was developed by a group of architects, choreographers and a film maker: Paul Cotter, Gareth Morris, Heidi Rustgaard, Eike Sindlinger, Ulrike Steven and Susanne Thomas.

Figure 5.3 Metal, Cow the udder way, 2005

Entire neighborhoods became semi-vacant as job losses drew inhabitants away from the city center. To underline this evolution and establish a dialogue with local residents, Metal brought five cows, five calves and three farmers to a wasteland in Toxteth, which had previously been a housing area.

As its brief pointed out, the project intended to solicit local responses to the idea of a self-supplying system, which cows were seen as symbolic of. Crucially, it introduced a collective reflection on "a different agro-urban, bottom-up method for using unused urban land". This strategy relies on identifying gaps within the socio-economic texture of a city and exploring them in a constructive manner. It can be seen as exploring and using interstitial space through tactics that invest in the creative potential of left-out areas such as urban wastelands and abandoned buildings.

Significantly, Metal's Liverpool space is located in Edge Hill station. The organization secured a 25-year lease on the part of the disused building it has brought back to life. The space itself, where trains still stop by, is a potent symbol of the industrial age, which Liverpool with Manchester were once at the forefront of. Within its residency program, which contributes to the renewed vitality of the building, Metal invited the artists AL & AL to hold a long-term residency looking at the symbolical significance of the station.[34] The result was the exhibition "horsepower" which looked at "how the horse was used in art after its industrial use". It resonated with the 1830 appearance of the steam powered Rocket at Edge Hill that announced the rise of machines in the modern era.

Aside from place and history, food is an important vector used to convey the spirit of the organization.[35] Around an AGA, a heat-storage cooker, artists are invited "to share food, knowledge, experience and ideas". Cooking for others embodies the principle of exchange through which Metal aims to facilitate connections and effect transformation in our societies. This principle is extended to outdoor communities and, in particular, to schools and the education curriculum. For instance, the project King Cotton was a music theater piece whose themes evoked the embedded global cotton manufacturing trade and slave trade of the 19th century. Aside from its staging at the Lowry in Salford and the Empire in Liverpool, a community education programme was set up with a number of regional organizations.[36] The Alder Hey Children's NHS Foundation trust was one of these organizations for which a specific outreach was devised. The story was retold to the young patients of the hospital during workshops held over two months (Charnock 2008). They provided a spectacle but also the opportunity for children to appropriate the themes of the narrative revolving around freedom, love and loss through creative writing sessions. Similarly, for the project *Winter Lights* in 2006, the artist Ron Hasledon worked with Phoenix Primary School, which is located in the vicinity of Edge Hill in the Kensington neighbourhood, a district that suffered greatly from the 1970s and 1980s depopulation. The result was a 10-meter neon Polar Bear for Kensington whose design had been made by a six-year-old pupil and selected following a series of workshops on the theme.[37] It enabled the school to participate in the Liverpool contemporary art Biennale that had been created in 1999, and thus to bring children and residents in contact with the art world, which has been of such importance in the regeneration drive of the city at the turn of the 21st century.

Metal appears to be well aware of the impact of global economic changes that have led a former wealthy city such as Liverpool into decline, and the challenges that its contemporary redefinition poses to the people who witness its consequences. Its strength seems to rely on a capacity to interrogate these changes. It does not oppose them as such, but questions them within a process of communication and dialogue with those who live through them, inhabitants, communities, individuals, schools, businesses, as well as buildings, plants, urban history. It is out of this social and historical awareness that time and spaces located *aside* from the frames delimited by the dominant socio-political environments, can be created and invested with transformative benefits.

34 *Project 4*, "Al and Al".
35 *Project 14*, "Food".
36 *Project 11*, "King Cotton".
37 *Project 8*, "Winter lights, Liverpool".

Conclusion

Within the three decades that span from the election of Margaret Thatcher to the turmoil of the early-21st-century global economic crises, northern cities became engaged in a process of economic reconstruction and a redefinition of their identities. The industrial activities that had shaped their landscape and their communities for more than a century had been in decline since the end of the Second World War and were scraped aside by the liberal policies of the 1980s. Throughout these decades, a model of economic regeneration was developed and nurtured on a local and national level, advocating a shift to a service-based economy. Believed to be ideally positioned to further social transformation, culture came to play an important role within this model; it could develop tourism and attract businesses and creative classes, and it could also contribute to indigenous social welfare. The art worlds of the northern region were directly affected by the implementation of this agenda that determined its' funding and the conditions to obtain financial support.

When looking at those grassroots organizations that heralded a collective ethos during this period in the northern cities, one is confronted with a panel of stances varying according to the socio-political context and its historical evolution. In the 1980s, when the governmental policies clashed with regional resistances, organizations such as Pavilion and the Liverpool Artists workshop articulated collective values, which were rooted in a political activism adverse to the government's entrepreneurial ideology highlighting the merits of the individual over the community. In the 1990s, alternative artistic roads appeared more diverse. East Street Arts in Leeds developed a strong awareness of political matters and their importance to the practising artist articulated through a constructive negotiation process. At the turn of the millennium, it is not surprising to be able to discern the shape of a dialectical move, as some of the conciliatory stances, which accompanied the 1990s shift in British society, were radicalized in the face of overarching globalization processes. The Ultimate Holding Company devised a range of interventions and objects to address the imposition of external and private forces in the public realm in Manchester, while Metal furthers a tradition of artistic influence on society articulated through a dialogue with its diverse components.

The socio-political historical context has clearly conditioned the responses of artistic organizations advocating grassroots collective development in northern cities. So what are the common threads that bypass the specificity of these organizations' objectives and environments? One is the stress laid repeatedly on collective development, participation, and the use of interstitial space. By advocating a collective ethos, artistic groups counterbalance the renaissance of the individual aura, which emerged in the 1980s as a conservative form of postmodernism in the international art world. Such a stance has taken varied forms, ranging from politically motivated objectives to formally devising systems of development based on systematic exchange as a model of aesthetic production.

It might be noted this second component characterizes broader and no less successful alternative machinery. The 1990s Annual Programme, for instance, appears as one of the rare organizations to have flourished without financial assistance within a northern art scene devoid of commercial outcomes and largely dependent on external subsidies.

Second, an extended participation has been the concomitant element furthered by these northern groups. As we have suggested, there has been a historical twist during the period, in that the Socialist inspired inclusion strategies, which heralded an attention to local community access and dialogue have been progressively incorporated into the main national guidelines to cultural foundation (and funding), in particular following the New Labour 1997 electoral victory. This incorporation, however, has been led in parallel to an overall drive towards a conversion of past industrial economies into a service-based economy. The arts, understood as a provider of social mediation and economic assets, appeared as an updated version of the political vision that had previously neglected their inherent specificity. New Labour cultural policies reenacted the instrumental approach favoured by conservative governments in the 1980s.

A shared characteristic of these organizations has thus been a capacity to repeatedly explore interstitial spaces within the de-industrialized northern urban fabric. Bypassing top-down preplanned and instrumentalist policies, alternative strategies could locate their creative input into disused material and mental territories where they could set up their own modes of aesthetic and social development. Within the rapidly changing northern city, confronted with external global forces and the negotiation of shifting identities, the contribution of alternative aesthetic strategies based on local networking and reflections can further an empowered and creative future if it is to be based on the renewed unveiling and scrutiny of unchartered territories, to be defined and constructed by artistic agency located on earthly grounds.

References

Arts Council of England, Stetter, A. (2002), *Pride of Place: How the Lottery Contributed 1 Billion to the Arts in England* (London: Arts Council of England).

Arts Council of Great Britain. (1984), *The Glory of the Garden: The Developments of the Arts in Britain: A Strategy for a Decade* (London: Arts Council of Great Britain).

Adams, I. (1998), *Ideology and Politics in Britain Today* (Manchester: Manchester University Press).

The British Art Show 4 (1995) (London: South Bank Centre).

Bourriaud, N. (1998), *Esthétique relationnelle* (Dijon: les Presses du Réel).

Cook B. (2003), *Beyond the Endgame, Abstract Painting in Manchester* (Manchester: Manchester Art Gallery).

East Street Arts. (2007), *Social Club Review* (Leeds: East Street Arts).

Edgell, S. and Duke, V. (1991), *A Measure of Thatcherism. A Sociology of Britain* (London: Harper Collins Academic).
Eggleton, L. (2007), "Will work for food: an essay on the mechanics of the Dinner Project", in East Street Arts, *Social Club Review* (Leeds: East Street Arts).
Green, G. (1991 [1987]), "The new municipal socialism", in *The State or the Market: Politics and Welfare in Contemporary Britain*, (eds) Loney, M., Bocock, R., Clarke, J., Cochrane, A., Graham, P. and Wilson, M. (London: The Open University Press/Sage).
Greenan, S. (ed.) (2001), *Life is Good in Manchester, the Annual Programme, 1995–2000* (Manchester: Trice Publications).
Heffernan, R. (2001), *New Labour and Thatcherism, Political Change in Britain* (Chippenham, Whiltshire: Palgrave).
Myerscough, J. (1988), *The Importance of the Arts in Britain* (London: Policy Studies Institute).
Orlando, S. (2009), "Le black art dans les ann*ées* 1980: une collaboration interethnique?", in Lasalle, D. and Germain, L. (2009) (eds), *Les relations interethniques dans l'aire anglophone* (Paris: l'Harmattan), 99–115.
Pavilion. (2008), *Pavilion, Celebrating Pavilion's 25th Year* (Leeds: Pavilion).
Peck, J. and Ward, K. (eds) (2002), *City of Revolution, Restructuring Manchester* (Manchester: Manchester University Press).
Powell, R. (1983), "Setting up a British 'first'", *Arts Yorkshire* (April).
Quilley, S. (2000), "Manchester first: from municipal socialism to the entrepreneurial city". *International Journal of Urban and Regional Research*, 24, 601–15.
Quilley, S. (2002) "Entrepreneurial turns: municipal socialism and after", in J. Peck and K. Ward (eds), *City of Revolution, Restructuring Manchester* (Manchester: Manchester University Press).
Spence, J. (1986), *Putting Myself into the Picture* (London: Camden Press).
Thatcher, M. (1993), *The Downing Street Years* (London: Harpercollins).
Thompson, C. (1993), *The Pavilion: Feminism, Photography and Representation* (BA Memoir, Art & Design Bradford and Ilkley community College).
Ultimate Holding Company. (2006), *Collective Works, 2005–2006* (Manchester: UHC).
Williams, R.J. "Anything is possible: the Annual Programme 1995–2000", in S. Greenan (ed.), *Life is Good in Manchester, the Annual Programme, 1995–2000* (Manchester: Trice Publications).
Witts, R. (1998), *Artists Unknown: An Alternative History of the Arts Council* (London: Sinclair Browne).

Primary Sources

Artspace. (2010), *Yorkshire Artspace, Premises*. Available at <http://www.artspace.org.uk/about-us/premises> [accessed 1 September 2011].

Artspace Merseyside. (1986), *Artspace Merseyside Limited* [leaflet].

Bragg, M. (1983–84), "Effectiveness of the arts, President's foreword", in *Northern Arts Annual Report 1983–84.*

Brookes, P. (1986–87), "Photography", in *Yorkshire Arts Annual Report (1986–87)*, 18.

Camerawork. (1979), "Editorial", n.13, March 1979.

Campbell, D. (2008), *In Conversation with David Campbell*, interviewed by Gabriel Gee, 5 July 2008.

Clarke, P. (2007), *In Conversation with Pete Clarke*, interviewed by Gabriel Gee, 10 December 2007.

Crowe, N. (2008), *In Conversation with Nick Crowe and Ian Rawlinson*, interviewed by Gabriel Gee, 29 June 2008.

The Feminist Art Programme. (1981), *Report, News Sheet*, Summer 1981 (Feminist Archive North, Brotherton Library, University of Leeds).

East Street Arts. (2011), *East Street Arts, Workspace, Facilities, Patrick Studios*. Available at <http://www.esaweb.org.uk> [accessed 1 October 2010].

L.A.W., *Liverpool Artists Workshop* [leaflet, date unknown].

Leeds Women Art Programme. (1982), *Newsletter: The Pavilion: A Feminist Photography Centre* (May 1982, Feminist Archive North, Brotherton Library, University of Leeds).

Leeds Visual Arts Forum. (2012), *Leeds Visual Arts Forum*. Available at <http://www.lvaf.org.uk> [accessed 1 October 2010].

MASA. (2011), Manchester Artists Studio Association, History. Available at <http://www.masa-artists.com/> [accessed 1 September 2011].

Metal, *Projects* [date unknown].

North West Arts Board. (1992–93), "Partnerships and regeneration", in *North West Arts Board Annual Report 1992–93*, 14–15.

North West Arts Board. (1992–93), "Developing the arts across the region", in *North West Arts Board Annual Report 1992–93*, 4–10.

North West Arts. (1998–99), North West Arts Annual Report 1998–99.

Pavilion (1983), "Opening space", in *L.O.P.*, 29 April 1983 (Feminist Archive North, Brotherton Library, University of Leeds).

Pavilion (1985), *Newsletter*, 26 June 1985 (Feminist Archive North, Brotherton Library, University of Leeds).

Pavilion (1992), *Keeping it Together* (Feminist Archive North, Brotherton Library, University of Leeds).

Pavilion (1993–94), *Pavilion Annual Report 1993–94.*

Rawlinson, I. (2008), *In Conversation with Nick Crowe and Ian Rawlinson*, interviewed by Gabriel Gee, 29 June 2008.

Redman, J. (2008), *In Conversation with Jai Redman and Joe Richardson*, interviewed by Gabriel Gee, 22 June 2008.

Reid, A. (2008), *In Conversation with Anna Reid*, interviewed by Gabriel Gee, 1 February 2008.

Rigby, G. (2008), *In Conversation with Graham Rigby*, interviewed by Gabriel Gee, 10 March 2008.

Richardson, A. (1984–85), "Chairman's report", in *North West Arts Annual Report (1984–85)*, 5.

Richardson, A. (1986–87), "Chair's report", in *North West Arts Annual Report (1986-87)*, 6, U.H.C.

Watson, K. (2008), *In Conversation with Karen Watson*, interviewed by Gabriel Gee, 10 June 2008.

Watson, K. (2012), East Street Arts update [email 27 February 2012].

Yorkshire Arts. (1983–83), *Yorkshire Arts Annual Report (1983–83)*, 2.

Yorkshire Arts. (1988–89), "Planning for the future, the economic importance of the arts", in *Yorkshire Arts Annual Report (1988–89)*, 21–2.

Chapter 6

Halo over Barnsley: Centering the Margins in the Transformation of a Former Mining Community

Myrna Margulies Breitbart

Mum Oh dad ... look who's come to see us ... it's our Ken.
Dad *(without looking up)* Aye, and about bloody time if you ask me.
Ken Aren't you pleased to see me, father?
Mum *(squeezing his arm reassuringly)* Of course he's pleased to see you, Ken, he ...
Dad All right, woman, all right I've got a tongue in my head – I'll do "talkin". *(looks at Ken distastefully)* Aye ... I like yer fancy suit. Is that what they're wearing up in Yorkshire now?
Ken It's just an ordinary suit, father ... it's all I've got apart from the overalls. *Dad turns away with an expression of scornful disgust.*
Mum How are you liking it down the mine, Ken?
Ken Oh it's not too bad, mum ... we're using some new tungsten carbide drills for the preliminary coal-face scouring operations.
Mum Oh that sounds nice, dear ...
Dad Tungsten carbide drills! What the bloody hell's tungsten carbide drills?
Ken It's something they use in coal-mining, father.
Dad *(mimicking)* "It's something they use in coal-mining, father". You're all bloody fancy talk since you left London.
Ken Oh not that again.
Mum He's had a hard day dear ... his new play opens at the National Theatre tomorrow.
Ken Oh that's good.
Dad Good! good? What do you know about it?

Introduction

When Monty Python performed this sketch in the 1970s few would have guessed that, 30 years hence, Barnsley, England would adopt and tenaciously pursue a post-industrial regeneration scheme with culture and creativity at its core. Barnsley lies in the heart of the South Yorkshire deep-pit coal fields and was a center of resistance during the bitter and violent miners' strike of the mid 1980s. At that time, Barnsley was home to 20,000 miners, along with the Grimethorpe Colliery Band made famous in the movie *Brassed Off*. Over 140 collieries closed in the UK after the unsuccessful miner's strike, forcing more than a quarter of a million people into unemployment. The last deep coal mine in Barnsley closed in the early 1990s, leaving 17% of the population without work.

Since the mid-1990s the whole South Yorkshire region has been part of a regeneration effort designed to construct a new economic base with creative industries as the main driver.

What do you know about getting up at five o'clock in t'morning to fly to Paris ... back at the Old Vic for drinks at twelve, sweating the day through press interviews, television interviews .... That's a full working day, lad, and don't you forget it!
Mum Oh, don't shout at the boy, father.
Dad Aye, 'ampstead wasn't good enough for you, was it?... you had to go poncing off to **Barnsley**, you and yer coal mining friends.
Ken Coal-mining is a wonderful thing father, but it's something you'll never understand. Just look at you
Mum Oh Ken! Be careful! You know what he's like after a few novels.
Dad Oh come on lad! Come on, out wi' it! What's wrong wi' me?
Ken I'll tell you what's wrong with you. Your head's addled with novels and poems, you come home every evening reeking of Château La Tour ... And look what you've done to mother! She's worn out with meeting film stars, attending premières and giving gala luncheons ...
Dad There's nowt wrong wi' gala luncheons, lad! I've had more gala luncheons than you've had hot dinners!
Mum Oh please!
Dad Aaaaaaagh! (clutches hands and sinks to knees)
Mum Oh, it's his writer's cramp!
Ken You never told me about this ...
Mum No, we didn't like to, Kenny.
Dad I'm all right! I'm all right, woman. Just get him out of here.
Mum Oh Ken! You'd better go ...
Ken All right. I'm going.
Dad After all we've done for him ...
Ken (at the door) One day you'll realize there's more to life than culture ... There's dirt, and smoke, and good honest sweat!
Dad Get out! Get out! Get OUT! You ... LABOURER!
Ken goes. Shocked silence. Dad goes to table and takes the cover off the typewriter.
Dad Hey, you know, mother, I think there's a play there .... get t'agent on t'phone ...
("Oh, it's his writer's cramp!" *Monty Python* sketch #59, 1969, http://www.youtube.com/watch?v=VxL-WyC-nIg)

Facing massive unemployment and unresolved health problems, Barnsley became part of the Renaissance Towns revitalization project. This was initiated in 2001 by Yorkshire Forward, a regional planning organization that, among other things, assigned world famous architects to help individual cities imagine new economic futures for themselves. The proposals that emerged from the re-visioning process in Barnsley stunned the country and, in some cases, provoked a new round of Monty Pythonesque ribbing.

Barnsley's decision to take a more creative route to economic revitalization raises many questions. To what extent can post-industrial cities invent a wholly different future for themselves? Can they afford to be "creative" in the true risk-taking and visioning sense of the word? What forms of inclusive planning can be employed to spark new economic opportunities for existing residents? What creative assets and resources can be drawn upon to bring these visions to fruition? How should success be measured? This chapter discusses these and related questions in the context of examining Barnsley's determined and on-going response to de-industrialization.

Rethinking Barnsley, Part One

Barnsley City sits in a very old and historic borough, situated on a hillside with black soil. It is referred to in the Domesday Book of 1086. The land passed through many monks and manorial overlords before becoming a prosperous market town in the

17th century. Early industries included linen, iron, blacksmithing, winemaking and glassmaking. Coal mining expanded during Victorian times to become the most important industry by the turn of the 20th century (www.barnsley.co.uk).

Barnsley's history as a place to shop extends back to the Middle Ages when it was a regional center for the exchange of goods and services. More recently, in the mid-20th century, it housed the largest open-air market in the North of England. With around 400 stalls, it was considered a magnet for people from all over the region, and as far away as London. In 2012, the metropolitan region had a population of around 230,000 (up from around 218,000 in 2001), with the main town centre and urban area surrounding Barnsley at around 82,000 (www.barnleydevelopmentagency.co.uk/key-facts). While 99% of this population was white and born in the UK in 2001, there is some evidence that the demographics are very slowly shifting, as this percentage dropped to 97% by 2007 (http://www.guardian.co.uk/news/datablog/2011/may/18/ethnic-population-england-wales#data). Barnsley is well located between Leeds and Sheffield, and an hour's drive from Manchester. It differs from these cities and surrounding towns, such as Bradford, not only in the relative lack of ethnic diversity but also in the fact that it has no old mill buildings that once housed cotton and other textiles. Indeed, as locals will tell you, almost all of Barnsley's buildings were once underground!

Many hard feelings and hardships remain in Barnsley as remnants of the brutal yearlong miner's strike from 1984–85, and the role of the Tory government, in the guise of "Iron Lady" Margaret Thatcher, in killing the only "industry" in town.

> I remember the police charging down the main shopping area of Goldthorpe near Barnsley, knocking people to one side, including women and children. My mother was doing her shopping ... and was shaking and distressed when she eventually reached home. I was on strike for the duration and will never forgive Thatcher for wrecking the mining communities and condoning police brutality. The people of Dearne will never forget and will certainly never forgive (Stan from Goldthorpe, BBC South Yorkshire, 2009a).

These lingering emotions are so enduring that Barnsley poet, Ian McMillan, was commissioned to write a special poem to commemorate the 25th anniversary of the Miner's Strike in March 2009. In it, he applauds the local collectivism that hard times sparked while remembering the way "posh newsreaders from the south" talked about the strike as if it were occurring in some other country. This memory of class conflict and divisiveness is relived through repetition of the refrain, "But what did we learn, eh? What did we learn?"

Once the mines closed, large sums of money were spent on environmental reclamation projects to fill in the pits and restore the countryside. Huge sums of federal and European Union money were also made available to deal with the inevitable by-products of de-industrialization – poverty, declining health and overtaxed social services. Following the mine closings, skilled trades began a precipitous decline, falling at a much faster rate than the rest of the UK

(Barnsley Metropolitan Council 2002). It is estimated that about one-third of the workforce in Barnsley now travels outside for jobs, and though professional occupations have increased slightly, the rate of increase is less than half that of the rest of the country. More significant is the huge increase in part-time service work, including call centers, which one young resident of Barnsley describes as "the new pits":

> I'm 21 and wasn't involved with the strike, but my dad was. He is now a welder but his company is starting to struggle due to the credit crunch. I just wanted to express my opinion that all we have done is swap mines for call centres. Although the two occupations are not the same, they can be linked. Both have grueling hours – although mining would affect you physically whereas call centre jobs affect you mentally. To be honest, I think I'd rather be working in a pit! (James in Barnsley, BBC South Yorkshire, 26 June 2009).

Barnsley's population of young adults declined by 25–30% between 1991 and 2001, and this was accompanied by a 14% increase of people over the age of 75. The percentage of people without educational qualifications was just over 40%, and a very high proportion of residents had permanent illnesses or disabilities (25% of the population and two times the national average). Given these demographics and health issues, it is not surprising that by 2006 Barnsley had the highest level of economic inactivity within its sub region (*EKOS* Consulting (UK) Limited *2006*). Other obstacles to regeneration are purported to include a "low risk culture", lack of private sector investment, and a town centre that was characterized at the turn of the 21st century by a stagnant property market and abundant bargain stores and pubs (Rivas 2011b: 3).

While numerous schemes to address these problems were proposed, none seemed to make a difference. Many plans attempted to copy other small cities by trying to attract technology and new creative business incubators. The schemes largely failed, as there were few sites for new businesses to move to and no social networks to support their owners. Over time, a series of failed efforts generated attitudes amongst residents that tended to justify the increasingly negative media images of post-industrial life. The Manager of Remaking Barnsley, Christopher Wyatt, characterizes such popular attitudes as follows:

> Well, this is Barnsley. We really can't have that. It can't be good. It can't be nice. Because this is Barnsley. Well just put tarmac down. We'll have concrete. It will do. It'll be all right. We shouldn't waste money on it (Interview 2004).

It is into this pessimistic and challenging environment that Yorkshire Forward, a regional planning organization, entered.

In 2000, a government White Paper on urban renewal, *The State of English Cities*, provided a "new vision for urban living" that would offer a high quality of life and opportunity for all towns, cities and suburbs (Robson et al. 2000).

Regional development agencies were given stronger planning powers than local councils and this was accompanied by a substantial pot of money from federal and European sources. The White Paper inspired Yorkshire Forward to take the lead in regenerating towns and cities in the North. In 2001, £700 million were committed to more than 400 projects for the purposes of renewing coalfield and industrial sites, and to enable surrounding urban centres to "transition to a 21st-century economy".

Barnsley was not in the first phase of what became known as the Renaissance Towns Initiative to "bring lagging communities back to life". Doncaster was, and received lots of money to plan a massive regeneration of its waterfront that included a new university and marina, along with a mix of residential and leisure development. Once planners in Barnsley got wind of the initiative, according to Chris Wyatt, they "begged to be joined", and this is when he saw local attitudes begin to shift:

> Well. We can sit on our hands and we can be an ex-mining town and continue to decline, or we can think of a new raison d'être, if you like, for the town (Interview).

In 2002, Yorkshire Forward decided to invite world-class architects to join an Urban Renaissance Panel. Each was eventually assigned to assist one city or town to come up with a regeneration scheme. Barnsley was assigned Will Alsop, an internationally renowned architect known for his wildly modern buildings on stilts. As it turned out, according to Wyatt, the assignment was a "piece of genius".

Alsop approached Barnsley by car with his collaborators, John Thompson, whose partnership created the Schlossplatz Square shared by a reunited East and East Berlin, and Koetter Kim from Boston, MA. According to locals, when Alsop glimpsed Barnsley from a distance, he leapt out of his car and said, "Why haven't you told us about this place before?" Locals are purported to have said, "Who is it wants to know and what on earth are you doing here anyway?" (Wainwright 2002). Shortly after his arrival, Alsop shared his desire to "get inside people's heads and hearts" and to locate their dreams. Four participatory workshops were held with residents in February and March 2002, at which he asked residents to draw their ideas for a future Barnsley. The process caught people's imaginations and a total of 1,500 residents from all walks of life participated in the workshops and bus tours around town. Following these sessions, Alsop and associates mulled over what they heard and produced a seeding of ideas that they gave back to the community in Town Hall the end of March. He called this his "sacrificial scheme".

Alsop's Vision: Barnsley, a Tuscan Village?

Alsop's plan was premised on the idea that, "If we can make this town beautiful, people will come." He believed that Barnsley could become an important part of a network of places in Northern England that would together constitute one Super City – a place where people could live in Hull, commute to Liverpool, and go out at night in Manchester, all in one day (BBC News 2005). There would be preserved countryside surrounding these places and a good public transport system would provide the necessary links. Within this concept, Barnsley would become the principal place where people would shop.

When Alsop first encountered Barnsley, he was on a motorway. The town appeared on a hill in the distance. With all of the former coal pits filled in around it and the landscape reclaimed, he said it reminded him of Tuscany. Barnsley, he soon proclaimed, is a "Tuscan Village". Alsop believed that the town should create a modern version of Lucca's wall, with a green belt that would eventually be filled in with a "living wall" of modern buildings that would define the boundaries between Barnsley and the surrounding countryside. Streets and quarters with different primary functions would lead from this wall to a central revitalized market. Until such a time as this vision could be realized, he proposed the creation of a pulsating beaming halo of light two-thirds of a mile in diameter and more than 3,000 feet above the town, that would be illuminated at Christmas and key times during the year to draw attention to the South Yorkshire town and keep the dream of a revitalized city in residents' consciousness. On Christmas 2003 the halo was turned on for the first time.

Figure 6.1 Simulated halo proposed by Will Alsop Architects, Barnsley Metropolitan Council

Soon after Alsop Architects revealed their initial vision to the public, many humorous newspaper headlines appeared: "Will Alsop's superficial Supercities scenario", "Tuscan Vision for Barnsley", "Prepared for Takeoff", "Halo over Barnsley", "'Starchitect' to redesign Barnsley with a laser halo", and "Barnsley's Italian Job takes Shape". Residents of Barnsley may share a common past, but they do not necessarily envision a similar future. Very different views on how to move beyond its mining past emerged following the display of Alsop's colorful conceptual model of Barnsley in the future and the public showing of the film "All Barnsley Might Dream".

Various blogs and other forms of media began to reflect both a negative and positive buzz. Some residents focused solely on the halo, seeing it as nothing more than a publicity gimmick, which, it was said, was "even less likely than the prospect of the residents of Barnsley drinking Chianti and eating pasta on pavement cafes in warm sunshine". Jack Brown, a poet, author and former Labour county councilor, told *The Times*, "This scheme is the dream child of conceptual artists and semi-educated New Labour politicians who have been promoted beyond their abilities. This will turn the town into a joke" (quoted in Harrison 2003). One astronomer took issue with the effect the halo would have on the night sky:

> As a member of the Society for Popular Astronomy, I too am horrified by the "halo of light" .... It may well be that the residents living under the "halo" will have their sleep patterns disturbed and have cause for complaint .... There must be a remedy under the Clean Neighborhood and Environment Act of April 2005, light nuisance is subject to the same criminal law as noise and smells ... (quoted in *Barnsley Chronicle* 2006).

He also opposed the plan on the ground that it ignored some of the town's many practical needs, such as the replacement of old street lights, in favor of a trendy gimmick. Perhaps the most poignant of the negative reactions came from a young man who simply reflected an all too common feeling among residents that life in the North would always be grim.

> .... The North ... bleak as hell. My ... girlfriend kept banging around on about the raw beauty of the place – it's crap, it's depressing, its got nothing there but poverty and dour northerners. Full of disused coalmines and the unemployed, how's that attractive? If the Romans had any sense at all they'd have built the figgin wall further south just to keep out the brummies. They have a lot to answer for. Maybe I could get an apology from the present Italian Premier? Even Queen Victoria hated the place. She used to pull the blinds down on her royal train so she didn't have to look out on the northern city scapes (quoted in WebmasterWorld.com 2005).

Figure 6.2 Conceptual model of Barnsley, Will Alsop Architects, Barnsley Metropolitan Council

Not all the reactions to the Alsop-derived plan were negative. Many reflect the presence of creative aspirations and a more optimistic outlook on the future. One resident, J. Haigh, tells his fellow residents to "lighten up on the halo".

> Regarding the halo idea for the town hall, it seems there are a lot of killjoys in this town. Suddenly we are a town full of expert astronomers whose pleasure will be spoiled by the supposed light pollution from this halo. If your previous reports on the halo were correct, it would only be switched on for special occasions. I, for one, think it is a good idea to get Barnsley talked about. In this town it seems that every new idea is attacked by stick-in-the-muds who want everything exactly as it used to be. Give the council and this halo a chance, the astronomers will have plenty of other nights to admire the glories of the night sky (quoted in *Barnsley Chronicle* 2006).

Another resident, Mick Wilson, was apparently inspired to develop his Seurat-like painting style when a friend told him that Barnsley was "really a little Paris because of its writers and painters, sculptors and actors". He went on to say that,

> Before that, it was always those flat caps and fish-and-chips that seemed to emerge as being typical Barnsley …. It made me very iconoclastic. After all, it was in Barnsley library that I began my love affair with the Renaissance and Pre-Raphaelites (quoted in Wainwright 2002).

Still others commented on the "creative energy" behind Alsop's "mad vision", embracing the idea that Northern English towns and cities might actually welcome new ideas and erase "centuries of industrial parochialism". They saw peoples' laughter as more hopeful than mocking, and defended the aspiration and determination of Barnsley to move forward. In the mind of Steve Houghton, executive director of the Barnsley Metropolitan Borough Council (BMBC), the real key to advancing what was to become a 30-year strategic plan, was not the halo or the wall, but rather, "getting the locals to notice" (Insider Media Limited 2003).

Rethinking Barnsley 2002: Participatory Visioning and Planning

In addition to assigning an architect to major cities in the region, Yorkshire Forward established Urban Renaissance panels (URPs) to work with local communities through Town Teams. These teams were charged with creating a long-term regeneration scheme that addressed social, environmental and economic improvements. The stated goal was to "shape a way forward built upon aspirations and enhanced skills in citizenship and civic leadership" (Barnsley Metropolitan Borough Council 2002). Shortly after Alsop revealed his initial ideas in a model and film[1] seven groups of experts were sent to Barnsley to build upon this vision that had already captured a good deal of media attention.

The *Rethinking Barnsley* Weekend event involved six Urban Design Action Teams and the URP in themed workshops and site visits with key stakeholders drawn from the realms of business, planning, transportation, politics and culture. At the same time, there was a professionally facilitated participatory planning process in which the public was asked to express their hopes and concerns through "Hands-on Planning" design groups open to anyone in the town or its surroundings. The overarching goal of the weekend event was to begin to define a post-industrial future for the city that simultaneously addressed the issues of city image, social deprivation, health and environment. Many of these needs and desires for change were elicited through role-playing exercises among residents.

---

1 The film depicted Barnsley in 2025 with its wall and a glass-walled electronics and digital centre called "The Brain." It ends with a hero, the postman, who has only three letters to deliver in a totally wired city, spending his evening outside watching the halo (Wainwright 2003).

Figure 6.3 Rethinking Barnsley weekend, Barnsley Metropolitan Council

Considerable effort was made to involve young people prior to the weekend events. To drum up interest, there was a lot of publicity in the form of invitations, banners and posters, ads on the radio and in other forms of media, as well as a general email to all schools. Once students and organizations responded, the event facilitators contacted them. At the event, young people spoke of their desire for a more vibrant nightlife and public spaces in the city. Other workshops focused on urban architecture, gateway links, connections among neighborhoods, movement across space, and concerns about the centre of the city. Many participants lamented the physical eyesores that had been constructed in Barnsley during the 1960s, among them, the Council offices. "Blow up the planning department by Friday", one person shouted. "Why wait until Friday?" said another Barnsley resident (Wyatt interview 2004). One suggestion was to rebuild the new Council and planning offices as "founding stones" in the "wall" envisioned by Alsop.

The exercises addressed other issues concerning the city's image as a place with "cheap shops and pubs" by envisioning the construction a more exciting public realm with better lighting and "dramatic neon signs", water features and imaginative architecture. Participants spoke of the beautiful countryside surrounding Barnsley town, the "friendly people" and the historic market. Residents also spoke of the need to retain and attract young people, entrepreneurs and innovators, while improving the quality of life for those living in poor conditions and in diminished states of health (Barnsley Metropolitan Borough Council 2002: 20–21). With respect to the latter, residents brainstormed possibilities for sustainable landscape development with allotments, fish farms, timber production, composting and recycling. By moving from a discussion of "illness" to one of "health", residents saw the potential impact of creating more open spaces and connecting residences to them through easy-to-access trails that encourage more physical movement. They envisioned an "outward looking network" that connects with opportunities at a regional level, such as the Dearne Valley Great Park (Barnsley Metropolitan Borough Council 2002: 32).

The topic of job creation led some to suggest that Barnsley try to build upon its tradition of mining and geology by developing high-tech rock and gem cutting industries and by providing training to do machine-aided design and civil engineering. Some residents who were intrigued with how other cities were using culture and creative industries to rejuvenate, expressed interest in transforming Barnsley into a safe "cultural and creative hub" (p. 33). In total, the weekend engaged over 1,500 people in conversation. Upon completion of the event, facilitators collected the comments and ideas, summarized the results, and presented these back to the larger community in a single evening that included slides and Powerpoints. The results were eventually published and made available to the general public, in 2002, in a book entitled "Rethinking Barnsley/Renaissance Barnsley."

Figure 6.4 Barnsley Central offices

What's Tuscan about Renaissance Barnsley?

Following the *Rethinking Barnsley* visioning weekend, a 2003 Master Plan was produced that set out an implementation strategy over a 30-year period. A more specific Barnsley Growth Plan (BGP) that places the market town vision within a larger economic development framework was developed in 2007. The Vision Statement for Renaissance Barnsley lays out the basic principles upon which the new physical design proposals are based.[2] Not surprisingly (given Alsop's original vision), the design features draw heavily upon the morphology of a small medieval city, replete with a wall, a vista, a crooked street pattern, distinct districts and quarters, and a central market. The vista is depicted from an M1 motorway embankment and features the town high on a hill surrounded by countryside.

Taking the iconic wall as a starting point, the image envisioned for Barnsley can be interpreted in a modern context as a defense against sprawl, yet it was clearly not meant to keep people and activity outside the bounds of the town. As a "living" inhabited wall, it was conceived as a permeable membrane, much as the margins of ancient texts such as *Dante's Inferno* included annotations that were eventually incorporated into the main body of the writing (Anichini 2006). Medieval walls, such as those found in Lucca, Alsop's inspiration, denote the margins of a city, and are comprised of different parts and materials that often become the basis of later growth and change. In keeping with this pattern, Alsop's proposed wall of light, buildings and parkland were expected to attract new life and activity to Barnsley. According to the plan, derelict underused buildings were to be replaced by small-scale housing developments, workshops, restaurants and buildings to house new business ventures. As centrifugal forces come into play, some of the city functions located in the centre (e.g. civic buildings and the much disliked 20th-century council offices) were expected to move further out, allowing the central market to expand, much as it did when the city was issued a charter by Henry III in 1249. Another intentional feature of the plan for a sustainable future was to build up rather than out so that the basic elements of city life are within close proximity to one another. The proposed halo was a referential and symbolic feature meant to mark the boundary as both a sacred space and a space that generates great energy and inventiveness. Alsop also meant for the pulsating laser halo to signal change and draw attention to Barnsley. It was also expected to touch local residents at an emotional level and epitomize "hope for the future" (Barnsley Metropolitan Borough Council 2002).

The question of where Barnsley should place its limited resources, and just how forward-looking and risk-taking it can afford to be, is highly pertinent. The embedding of the iconography of a medieval town within Alsops' vision presents a certain irony. Ancient cities, such as Lucca or Florence, Italy presently depend

2 Developing a quality well-defined core surrounded by mixed uses (e.g. offices, housing and buildings housing other functions, such as health facilities) is key to this vision, as are quality public spaces that are accessible to all.

upon their pasts to draw economic sustenance in the form of cultural and historic tourism. As such, they face daunting challenges. Planners must negotiate a delicate balance between providing decent job opportunities for current residents, including new immigrants, and maintaining a raison d'être that connects the city to a distant past and a pre-technology creative culture and enterprise. Medieval Italian cities cannot afford to change *too much* lest they lose their referential basis. Barnsley, on the other hand, must embrace change, and yet remains constrained, in part, by a heritage that many outsiders (and even insiders) see as too remote from a creative economy-driven vision for the future.

It is tempting to debate the literal possibility of transforming a former coal mining community into a Tuscan village (as much of the media did). It is also tempting to question the authenticity of the resulting plans in relation to an actual medieval town such as Lucca, Italy. It is more fruitful, however, to consider this plan for the future Barnsley as a product of the creative imagination and to ask whether any post-industrial city has the ability to invent a future that is a far cry from its present. In this light, one can begin to see how an idea that seems to have been imported from thin air, was transformed into a modernized vision that reflects wholly new urban aspirations.

Remaking Barnsley: Implementation

The Rejuvenation of Space

One Barnsley, a Local Strategic Partnership (LSP) of citizens, businesses, non-profit service and cultural institutions, and members of local government, took on the responsibility for implementing and monitoring the progress of the various Community Plans and Neighborhood Renewal plans. These were designed to improve the borough's environmental, social and economic wellbeing, and to reduce disparities within Barnsley, and between Barnsley and the rest of the country. The overarching goals were to create a distinctive 21st-century market town at the center of the borough, and to remediate the present economic and social conditions, while developing the area's resilience in responding to accelerated change globally. Six priority areas include:

- Remaking Barnsley: Transforming the town centre and rebuilding the local economy for long-term sustainable growth.
- Remaking Learning: Improving education to raise attainment and change the learning culture.
- Workforce Development: Equipping residents with the motivation and skills to participate in a knowledge-based economy.
- Reconnecting Barnsley: Improving internal and external communication and links among cities and towns to make best use of new technologies.

- Fit for the Future: Promoting healthy lifestyles and addressing the causes of ill health.
- Community Safety: Tackling anti-social behavior and drug use (www.barnsley.gov.uk).

Implicit within these plans is the assumption that the jobs lost in traditional areas of the economy need to be replaced by knowledge-based creative industries, a revived retail market and cultural tourism. A national report by the Creative Industries Taskforce, entitled "Creative Britain – New Talents for the New Economy" influenced these priorities (2008). Endorsed by the Prime Minister and the Department of Culture, Media and Sport, the report outlines new strategies to foster future economic growth by developing capacity in the creative industries sector. A new body, *Creative England*, was created to support this goal along with a number of regional agencies. One such agency, Yorkshire Forward, estimated that around £400 million of public and private money would be invested in Barnsley's Renaissance and that approximately £31.5 million of this would come from Yorkshire Forward's Renaissance Towns programme (http://www.yorkshire-forward.com/improving-places/urban-areas/barnsley).

A number of building projects began in 2002 after the *Rethinking Barnsley* event and the publication of the Master plan and *Remaking Barnsley* in 2003. Funding for these schemes came from a wide array of sources, including Yorkshire Forward, Objective 1, the Arts Council and private investors. Several new projects planned for Barnsley included the construction or renovation of markets, centers for transportation and digital media, government offices, and both renovated and newly built cultural and retail facilities. Progress in this transformation of the built environment was swift between 2003 and 2008/9. A Transport Interchange building, a high-end office complex (Westgate Plaza One), and the Digital Media Centre (DMC) with high quality, flexible office space for new creative industries and digital or media businesses, were among the first projects to be completed as flagship architectural developments, manifesting a concern for environmental and social sustainability.

The DMC is a brand new glass-walled building that provides incubator space for creative start-ups at prices that include hot-desking options in an open-plan format, along with support services and meeting spaces that new start-up tenants can reserve. A Project and Incubation director consults with businesses and connects them to each other and to other external advisers and programmes. A new "d-space" also displays digital innovation from regional, national and international contemporary artists, designers and makers. This space and local exhibitions and events showcase new ideas, encouraging collaboration and knowledge sharing. A Business Innovation Centre (BIC) also supports the development of technology- and knowledge-based industries.

The Barnsley Civic, built in 1878 as a public hall and former Victorian theater, was variously used over time as a library, the town's college, and a venue for the first talking films. Closed for over 11 years, it was completely renovated in

2009 to serve the needs of the growing creative industry sector. Though it was not possible to refurbish The Civic solely as a theatre, European and lottery funding enabled its regeneration as a flexible hub for design and creativity. It currently houses the council's service center, Barnsley Connects, The Gallery, and The Assembly Room, a 336-seat performance space that can be flexibly used for any type of entertainment. The Panorama can host smaller events and is adjacent to The Assembly overlooking Mandela Gardens, a green space near The Civic and town centre, which is the site for events programmed by The Civic. It features a waterfall and artist-designed seats and clock. The Civic also contains retail space and a variety of studios, creative workshops and design companies, including silversmiths, joiners and architects. The 10 Creative Workspaces that are provided are meant to attract companies and artists incubated in the DMC.

By the end of 2008, it was estimated that Barnsley housed nearly 1,000 creative and digital companies and this was the fastest growing sector in town (*Insider Media Limited* 2010a). In addition to buildings centered on the promotion of creative industries, the town initiated a number of renovation and new construction projects for mixed uses, many designed to encourage people to move back into the town centre. In 2003, Yorkshire Forward invested £30 million to purchase Barnsley's Metropolitan Shopping Centre, described as a "concrete monstrosity" with 150,000 square feet of retail space. The intention was to find some private sector partners, knock it down, and spend over £200 million to replace it with a market of double the space (Insider Media Limited 2010b).

Figure 6.5 Digital design media centre

This project has yet to be undertaken, however, a £70 million mixed-use Gateway Plaza was completed in 2010 and built along an entry road to the town on the site of a derelict factory. It now houses a hotel, retail, a Health Centre, residential space, including 188 luxury flats, and 96,000 square feet of offices that include an organization to manage 20,000 council homes. In an effort to attract buyers, a campaign called "rent-to-buy" was launched in the summer of 2010, providing a no obligation chance for people to experiment with in-town living. A £40 million satellite college campus for the University of Huddersfield was also constructed in the center of Barnsley.

Planners believe that these infrastructural improvements are key to attracting a young creative demographic, and so some incorporate a special cultural or historic dimension. One example is The Beehive, an 18th-century building that now houses Lucorum, a mixed-use enterprise with a bar, rental units and a restaurant. The Elsecar Heritage Centre, built within the former ironworks and colliery workshops, is also restored to house an antique center, craft workshops and "Building 21", a 12,000-square-foot space for live music that includes the Hive Gallery, the first contemporary art venue in Barnsley.

Cultural, Consumer, and Eco-Tourism

While it might appear that *Remaking Barnsley* is almost wholly focused on invigorating the construction industry, significant attention has been devoted to programs to revive the economy and address the many social barriers to economic reinvention. Most of these intersect in some way or other with a cultural/creative economy agenda.

Barnsley Football Club is still the biggest visitor draw to town. However, more recent investments have been made in the Pennine Loop and Bridleways for cycling and horseback riding. Since Barnsley has one of the worst obesity problems in the country, the goal of improving the health of the general population through the promotion of exercise complements the goal of promoting Barnsley borough as a tourist destination. By drawing attention to the surrounding countryside and trying to generate more income and jobs through the promotion of new campsites and bed and breakfasts, planners are hoping to improve local use of the land by residents as they also draw in more tourists. In the summer of 2009, Barnsley's Secret Theatre Company hosted the only outdoor Mystery Plays in the UK at the Monk Bretton Priory, involving several Yorkshire-born celebrities (http://www.accessmylibrary.com/coms2/summary_0286-35954582_ITM).[3]

When plans were announced for a new museum to highlight the history of the town and region, this idea generated more excitement than the halo.

---

3 Originally performed in church in the Middle Ages, Mystery Plays eventually moved outdoors and into marketplaces with temporary stages. They gradually grew into large pageants that went on for days, with content considered by the church to be quite irreverent.

> I was pleased to read the report in last week's Chronicle about Barnsley Council actively seeking funds for a heritage museum, about time too! … I am confident that [such a museum] will soon outgrow its size … we must remember that Barnsley's heritage is not only linen, coal and glass. There is wire and nail making, heavy engineering and foundries, brickworks, brewing, chemicals, paper manufacturing and a myriad of small manufacturers making anything from the common to the obscure. Each of these industries, and also the businessmen and entrepreneurs, have contributed to the growth of Barnsley. Let us hope that Barnsley gets the funds required and that when the museum is established it will be well advertised and signposted and then the people of Barnsley will be able to make good use if it (Keen 2006).

This comment suggests a local pride that so many believe to be in short supply. It is also significant for its reference to a history of diversification of the economy among small manufacturers in Barnsley, and a role for preservation that extends beyond cultural tourism. Whether these lessons will be drawn upon in future planning remains to be seen.

Many outsiders find humor in the idea of selling Barnsley as a tourist attraction. For most residents the topic engenders debate about what memories should be preserved and highlighted, and what assets Barnsley should try to capitalize on in the future. The winding wheel and pithead gear of Yorkshire Main colliery, located near the Barnsley's football ground, is one monument to Barnsley's past as a center of coal mining, that all agreed to preserve. Others proposed that places more evocative of the miseries of that existence, such as the Silkstone Church, warrant attention as a reminder of the many women and children who died along with miners in a pit flood in 1838 (www.barnsleyandfamily.com). Preserving memories of Barnsley's historical contributions and sacrifice is important to residents, especially when a wholly new economic agenda based upon creative industry and technology distances them from that past.

Aspirations seemed to widen as Barnsley incorporated creative economy strategies into future planning. This was manifest, most noticeably, in the effort put into a recent competition to become the UK's next "City of Culture". Several years ago, the European Union started a competition to award one city the title of "Capital of Culture". The winner received funds and saw a boost in private investment to accelerate regeneration. Liverpool became Europe's Capital of Culture in 2008, attracting more than £800 million in economic benefits and over £200 million pounds in free media coverage (Kennedy 2010; also see Fitzpatrick: Chapter 8). Shortly thereafter, the Tory government decided to initiate a similar competition among cities in the UK, excluding London. Barnsley was among the 29 competitors but did not make the short list announced in January 2010. Eventually Derry beat out Birmingham, Norwich and Sheffield, but not before the news media shared a list of what it called "unlikely candidates lining up for the honour" (Savage 2009). In making its case, Barnsley presented more than its groundbreaking strategic planning process and progress towards reinvigorating its physical infrastructure; it cited efforts to animate the cultural sector and cultural spaces through public art, gallery art events and performances.

Promoting a Culture of Creativity through Collaboration

While many aspiring small cities have invested effort in promoting artists, Barnsley's efforts to gain entry into the larger creative economy drew the attention of local planners to a small yet growing local design sector, and to the cultivation of a stronger culture of creativity. This agenda is addressed, in part, through the reuse of underused spaces and the building or renovation of new infrastructure discussed above. There are now several structures in Barnsley that offer young start-up firms affordable places in which to incubate their businesses. Such sites offer menus of flexible services that individual practitioners and small enterprises can draw upon at little expense. Packages are generally tailored to individual needs, with a small per week charge for a receptionist and networking base (Regeneration Scrutiny Commission Report 2009: 1181–82). Marketing information is provided through an "Education Awareness Project". The Barnsley Enterprise Agency and the Barnsley Development Agency (BDA) have also established short training programs to attract "experienced people" to help out as consultants to small businesses at no cost through 2012.

Information technology and finance are key barriers to starting and sustaining a creative business (www.creativebarnsley.co.uk/aims.htm). The *Creativity Works/Creative Networks Programme* addresses these barriers by gathering mentors together to aid businesses in the preparation of action plans that include information on new product development, funding, marketing, and exporting. A *Creative Business Grant Scheme* also provides grants to cover businesses, and a *Creative Apprenticeship Scheme* acts as a broker between employers and potential employees among the unemployed, providing a subsidized and mentored six-month training and work option. Creative business networking events have been hailed as a "pillar of Barnsley's strategy towards the Creative and Digital Industries (CDI) cluster" and help to put creativity and culture at the center of an economic agenda (Rivas 2011b: 7–8).

A report, issued in 2007, indicated strong growth in the evolution of the CDI sector in South Yorkshire, with Sheffield as the fastest growing UK city region. Barnsley was signaled out for its strong expansion of CDI, especially in the arena of design and information communications technology (South Yorkshire CDI Study 2007). Much of the growth in this sector can be attributed to the cultivation of networks that support the development of creative and digital industries clusters, including Barnsley, Doncaster and Rotherham. In 2010, the BDA mapped local CDIs and found 540 active companies. The majority were micro-businesses working out of rented office space or people's homes. Of these, nearly all traded outside city limits, about half at the national level, and around 12% internationally (Rivas 2011b: 6).

A key message that planners derive from this information is that collaboration among small enterprises is important to promoting the development of "local creative ecosystems" (Rivas 2011b: 8). To this end, Enterprising Barnsley is among one of a group of sponsors (along with Northern Net, Bmedia and the DMC) that organize ad-hoc gatherings of entrepreneurs in creative fields for the purposes of sharing ideas and creating a learning environment. Barcamp Barnsley, one evening gathering held at the DMC in March 2011, invited "any creative artist, writer, blogger, technologist, or developer, geek, entrepreneur, academic researcher, gamer or investor" to participate in discussions and demonstrations of their work. One of the partners, NorthernNet, has since created a hub of companies that can compete internationally. The organization provides a Pay As You Go access to state-of-the-art editing suites and an extremely fast digital network (Barcamp Barnsley). *Creative Barnsley* also links creative people and businesses together through networks and events that feature a speaker and are held in various cultural venues. The goals are to create a "giant database" of businesses, provide workspace, and connect services to clients.

Transforming Education and Redeveloping Neighborhoods

To my knowledge, no studies have yet been undertaken in Barnsley to determine what proportion of the local population benefits from these supports to creative enterprise development. Given the challenging social statistics, however, any agenda to build the local economy must devote serious attention to improving educational attainment. To this end, a number of small former mining communities have formed the Barnsley Dearne Valley Venture, a self-managed, community-driven company located in a former primary school. This venture, along with the Barnsley Dearne Community Partnership, help the unemployed back to work through access to lifelong learning and the expansion of new employment opportunities involving further education and training. Emphasis is also placed on increasing the involvement of adults and young people in decisions related to regeneration. This includes addressing problems related to public safety, poverty and the diminished quality of residents' social and physical living environments. Community meetings to discuss these issues are held in a building that was refurbished using many recycled materials, and that now houses a youth theater.

Several other Lifelong Learning Centres have been set up in Barnsley since 2001. One £1.8 million project in the northern part of the city contains a new library and a branch of Barnsley Connects, a service center that provides electronic and physical access to local government information, a daycare, café, computer suite and multipurpose rooms for the use of the community and nonprofits. Residents of North Barnsley also created many new pocket parks and safer play areas between two residential housing estates. They refurbished a neglected pond and woodlands to increase biodiversity and set up Monk Bretton Megabytes in a renovated hairdressing salon. This space now provides the community with free computer facilities and practical training in computer design and maintenance (http://www.northbarnsley.co.uk).

Recognition of the need for a broad and inclusive approach to educational reform and training to support these local efforts has gained considerable steam in recent years. In 2010, the European Union member states published an extensive survey on the topic of creative learning and innovative teaching. The study was prompted by a growing belief that creativity and innovation are the primary "drivers" for sustainable economic development, and that schools, teachers and curricula are not sufficiently attentive to this reality, and are not providing students with an environment that encourages risk taking, critical thinking, flexibility and the license to follow ones' passions in the quest for knowledge, experimentation and competence in CDI (Cachia et al. 2010).

Prior to this report, the UK Arts Council gave several million pounds to Barnsley, Doncaster and Rotherham to boost creativity in schools, believing that "creative, innovative people are the key to the economic and cultural regeneration of the region." A group called Creative Partnerships managed the project of building partnerships between schools and creative organizations, businesses and local innovators, with the intent of transforming the way that young people learn. The goal was to cut across all areas of the curriculum and encourage young people to pursue academic pathways that meet the labor market needs of the future. The government apparently withdrew funding for this program at the end of the 2010/2011 year (www.creative-partnerships.com/about-creative-partnerships/).

Since the 2010 EU report, further efforts have been made to turn secondary schools into newly built "Advanced Learning Centres" with state-of-the-art IT and sports, and to welcome adult learners to use these facilities (Rivas 2011b: 4–5). A two-year project sponsored by the Workers' Educational Association (WEA) in 2007, "Family Learning for Social Cohesion" engages families from less advantaged communities that include Barnsley, in skill development to aid student learning and improve awareness of, and contact among, multi-ethnic cultures in the region (WEA News 2009). Further encouragement of creative learning is supported through community arts activity, including art clubs, crafts training, and informal musical and theatrical performance. This important emphasis on educational achievement has not been abandoned even though the current recession has slowed other efforts to boost the local economy through enhancements to the built environment and the pursuit of creative, digital, media and culture-related activities.

Global Recession Challenges the Barnsley Plan

Barnsley moved its regeneration agenda forward prior to the 2008 recession. Significant changes to the built environment and attention to pressing quality of life issues challenged its past imaginary and began to raise the expectations and esteem of residents. Unemployment rates fell, and noticeable job and new business growth occasioned the presence of more commerce than was present at the turn of the 21st century (Rivas 2011b: 3). The entire economy in South Yorkshire was more mixed, and in Barnsley proper, there was solid evidence of new economic activity in a variety of realms, including the visual arts, performance, software design, printing, architecture, graphic design, photography, and entertainment.

Much of this progress has slowed in recent years. Studies indicate that Doncaster, Barnsley and Grimsby have been especially hard hit by the lingering recession. Northern regions depend more on public sector jobs, and due to government spending cuts, employment prospects are more limited, especially in this area (Inman and Wintour 2010). Between 2008 and 2009, Barnsley also suffered the closure of a few large businesses, resulting in a loss of almost 600 jobs. The net loss in manufacturing jobs for the same period was 750. These closures resulted in a higher number of people claiming incapacity or accepting employment allowances (Bruff 2010: 4). Barnsley's creative and digital industries (CDI) sector growth has not compensated for these losses.

Media coverage of the economic downturn points to its uneven impact and divergent effects in different regions of the UK. This has resulted in a clear growth of disparities between people and places. For example, the unemployment gap between Hull and Cambridge widened as a result of the recession from a difference of 3.4% to one of 6.3% points between 2008 and 2009. Another study suggests that London may emerge less scathed, which has prompted some policy makers to advocate for more support services and access to financing and good quality infrastructure, for small and medium-sized firms in growing creative sectors within those small cities and towns hardest hit by the global economic crisis (Bruff 2010: 2). Some developers of newer office spaces in Barnsley are even trying to market the town to businesses that need to move to less expensive premises from more prosperous areas of the country. According to one of the developers of the Gateway Plaza, "Barnsley ticks a lot of boxes" for companies seeking to relocate and cut costs and living expenses by moving north (Insider Media Ltd 2010a).

By 2012, the design and construction of a massive new retail market in centre city, the core economic rejuvenation scheme of *Remaking Barnsley*, was held up due to funding challenges. The BDA reports that private developers and prospective retail tenants, such as Debenhams department store, are still committed to the project but, according to Chris Wyatt, "… the credit crunch has not helped, and many of the retail-based construction projects throughout England have been hit" (Insider Media Limited, 2008). Wyatt insists that this delay will not affect the overall vision, which is supposed to progress over a 30-year period, and has already infused more than £4 million into half a square mile of the city centre (Insider Media Ltd. 2010a).

The original vision for the market has nevertheless undergone change. In May 2011, a £70 million plan replaced Alsop's "Tuscan hill village" central market plan. The product of a competition, the new scheme proposed a 52,000 square foot market developed by the "1249 Regeneration Partnership". This was meant to join up with the town centre, and included a new shopping center, restaurants, a six-screen cinema, 800 parking spaces and a new indoor market that was purported to evoke the mood of an "oriental Kasbah" and "provide residents and visitors with a completely new and exciting town centre shopping experience" (Pater Mathias quoted in Gardiner 2011). Barnsley residents seemed less than enthusiastic as they wondered what an "oriental Kasbah" would look like and worried about the city centre retaining the town's dignity and identity.[4]

Given a delay in the reconstruction of this new central market, and residents' skepticism about its scope and design, the local council recognized that a sole focus on retail might be less effective than a broader-based approach that considers the role of the town centre in the day-to-day life of the community (Bruff 2010: 9). In response, a new three-season cultural program was developed to offer highly visible cultural performances in the town centre and at Sunday street markets. Such events include Gay Pride marches and various celebrations of Barnsley's small minority population to emphasize the importance of inclusivity. These activities employ both professional and community-based entertainers, and are designed to attract visitors while continuing to develop a local audience and consumer base. Neighborhoods are also encouraged to use the city centre to hold community-driven events. In response, one artist used an empty office block to create an art installation that Barnsley employed in its bid to become the UK's 2010 City of Culture (Bruff 2010: 7).

Efforts to develop an event website, branding publications and a "Welcome Programme" to attract high profile cultural events, reflect larger attempts that many smaller post-industrial cities are making to keep a positive vision of urban regeneration in the consciousness of the local population and maintain their retail sectors during hard times. Many believe that the economy that emerges from the economic downturn will look very different, and that communities that can identify and support fruitful emerging sectors will possibly recover more quickly. Given the challenges posed by the on-going recession, one might ask where and how a creative economy fits into Barnsley's future, and what can be learned about the regeneration of smaller post-industrial cities in a recession climate.

Back to the Future: Prospects for Change

The rapidity of visible change made to the urban landscape following the generation of the *Remaking Barnsley* plan was key in propelling Barnsley's regeneration forward. The infusion of funds to build a new transport interchange, several new office blocks, a state-of-the-art digital and media centre and numerous spaces for the arts – all within a span of about five years – contributed to changing local attitudes about the possibility of change; it also began to alter external images

4 Marketplace Barnsley, as this retail complex is now called, is due to be completed by 2015 with a 7-screen cinema, 35 retail units, an extended market hall and 850 parking spaces. H&M has also apparently signed on as the largest prospective retailer.

of the town and its surroundings. While Alsop's colorful model of "Barnsley, a Tuscan Village", and the publicity this generated, created a noticeable "buzz", it was the direct and indirect effects of this spatial transformation of the city centre that generated a momentum and paved the way for the more foundational development needed to further expand both the service and creative economies. Through the transformation of the built environment, the inevitability of change became visible to residents. Outside businesses that never considered the Barnsley area a prospective home, began to look at the offerings. Young people also started to glimpse future prospects for themselves and, for some, a reason to stay in school.

Nothing illustrates the attention paid to creating highly visible markers of change better than one very large series of metal sculptures that include a poem ("Full of Glory"). The sculptures are placed in the center of town and mask a substation behind the newly refurbished Civic Hall (see Figure 6.6).

Figure 6.6 Barnsley sculpture, "Full of Glory"

One can interpret this public art as a very explicit effort to rebrand the city. But who is the audience? Outsiders or Barnsley's own population? In the face of new challenges posed by the current recession, attention to rebranding still fosters debate, as illustrated by the following blog entries (www.linkedin.com/groups/Barnsley-Design-Network-3753373?_mSplash=1):

> Could we visually change and package Barnsley differently to attract greater outside investment, and create a more relevant vision of the future, but also a nod to the past. Glass blowing and mining are important to Barnsley, but are they inspiring to future generations, do they reflect Barnsley as it moves into the 21st Century? (Patrick).
>
> Personally, I'm always somewhat wary of attempts to brand towns and cities … the only people who such attempts at branding speak to, are other marketing people. In some cases, it can even damage how a town is perceived. If you remember a few years ago, there was talk of re-modeling Barnsley on Tuscan Hill villages and surrounding the town with a ring of light. This received nothing but derision from locals and made the council look out of touch. I think the really important aspects to attend to are the standard of living for residents, public services, jobs, culture and entertainment, town planning and architecture … (Andy).
>
> So what you are saying, Andy … is that we should have no ambition, no joy, no sense of identity, no pride … I'm all for anything which will stop us hunkering down and just accepting the rubbish that is being heaped upon us right now. Why not use some innovation and our own talent to do something about it? (Barry).
>
> Barry- that's exactly what I am not saying. What I question is why Barnsley or any other town would need a branding exercise to achieve these things …. What any town needs (probably Barnsley more than most) is real support and substance in growing business and jobs, building local culture and fostering pride in the area. This is where real identity comes from – not the thin veneer that branding offers … My main point is that the grass roots need to be stabilized first before a nice badge is pinned on top of it (Andy).
>
> There's a lot of work that needs doing … But the sooner we get a vision that the future generations can aspire to and work towards the better (Andrew).
>
> I believe our strongest industry could be an ideas driven economy …. It doesn't matter whether I am in Barnsley or Barcelona I can and do connect and collaborate with people across the globe on creative projects. With Barnsley investing in technological infrastructure to support super fast broadband speeds there is maybe only one thing holding us back. Us! While we always continue to see ourselves as an ex-mining town, we always will be. It all comes back to aspiration, and exploiting new knowledge economies … Let's start dreaming! (Patrick).

This recent on-line exchange among Barnsley residents suggests that they are less concerned about rebranding their city than they are with the substance of real change and with addressing the social and economic obstacles that prevent the crafting of a sustainable and economic base that can provide a range of new jobs.

Aspiring to grow a creative economy with new jobs that have meaning, require skill and provide a solid income stream, departs from earlier times when Barnsley and other post-industrial cities reacted to deindustrialization by seeking *any* form of work on the theory that "a job is a job is a job". *Remaking Barnsley 2003–2033* presents a long-term diversified strategy to revive the town's pre-mining dominance as a market. It also seeks to move beyond retail consumption to cultivate new creative industries, and defines a "successful" economy as one in which inequalities are reduced, and "all citizens enjoy a high quality of life and standard of living" (EKOS 2006: 5). The current vision is unmistakably *aspirational* in its attempt to defy both local and national imaginaries that still see little else than a defeated coal mining community or a low wage service economy. Dramatic additions to the built environment and the desire among residents for change support this vision. But can it move forward?

While the recession poses many obstacles, there may be two hopeful signs. One is indirectly related to the growing creative economy, and involves efforts to tackle underlying problems that prevent many residents from benefiting from the on-going economic transformation. The second is directly related to the evolution of the creative economy and involves the support and extension of already existing networks of people and businesses.

Regeneration efforts prior to 2008 focused on developing a digital and media-based sector to attract and retain young people. Present-day planning, of necessity, seems to be drawing more attention inward. Like many post-industrial cities whose visionary regeneration schemes are subject to the exigencies created by a global economic downturn, just as they are getting off the ground, Barnsley has had to turn its attention to smaller projects at a local scale. The Barnsley City Council is working in collaboration with the Enterprising Barnsley Programme to secure funds to help mitigate the effects of the recession for those unemployed who are still unable to benefit from jobs in cutting-edge creative sectors. The global recession, which slowed the birth of the creative economy and grand scale building, re-emphasizes the need to rejuvenate challenged neighborhoods and public spaces. It also underscores the importance of developing more sustainable planning practices, and advancing education and skill development programs for residents. High levels of deprivation, rising house prices and serious health issues, especially obesity, are continuing concerns that prompt Barnsley officials to direct more sustained attention to improving housing, health, and the general quality of life in neighborhoods undergoing decline in population and business activity. This includes the design of new residential construction to assure greater access to open space. According to Alexandra Jones, chief executive of the Centre for Cities,

> Shifting plans from building a science park to creating a public park in these places is not about giving up on growth – it's about improving the area for local residents, who should be at the heart of the decision making process (www.centreforcities.org/centre-for-cities-sets-out-a-new-approach-to-regenerate-englands-cities.html).

Returning to a planning process that is more transparent and encourages active citizen involvement in asserting local priorities is important to sustaining the momentum of this type of inwardly focused regeneration during the current recession. If successful, a renewed commitment by local planners to develop the capacity of residents, and to cultivate local creative assets through education and training, may possibly generate new opportunities and prepare residents to enter future areas of employment. The fact that social problems are so visible also means that Barnsley and other post-industrial cities remain eligible for certain outside investment programs that encourage improvements in physical health, housing and job creation.

Barnsley is trying to weather the current recession by supporting existing initiatives that promote communication and collaboration among private and not-for-profit enterprises. In addition, the city is directing more serious attention to a social economy that includes health, education and housing. There is evidence that local businesses and cultural organizations are already banding together to create a funding and resource pool. For example, the Barnsley Business and Innovation Centre, affiliated with the BDA, and the Barnsley branch of the University of Huddersfield, recently came together to advise local businesses and expand the range of services offered to them (e.g. a small business loan guarantee program and a Manufacturing Advice Service that focuses on issues such credit and liquidity) (Regeneration Scrutiny 2009: 1180). A *Creative Clusters Summer School*, organized in Barnsley in 2009, gathered experts in the fields of art, design, culture, technology and other creative industries to share success scenarios that support economic recovery. Consultants from all over the UK and Europe provided four days of "intensive up-skilling" educational workshops to participants from fledging creative industries. These events were held in what were considered to be some of the most successful regional workspaces, including Barnsley's DMC. Less formal and more frequent weekly gatherings of people who work in some of the newly built office and media/design complexes also continue.

In addition to providing educational and skill-sharing venues, such gatherings provide a space to think about new ways to make use of local assets and skill sets that combine the development of creative and culturally-based industries with a more aggressive exploration of new smaller-scale manufacturing opportunities in the arenas of food and drink, engineering, metals and alloy processing, healthcare and environmental technology. Many of these areas of production could benefit from collaboration with the newer high technology CDI sector, and should be seen as an enhancement of the original *Remaking Barnsley* plan.

Much of Barnsley's Renaissance vision was predicated on an ever-expanding stream of outside investment that is now in short supply. The sheer density of programs and new, shared spaces within office blocks that were created in *pre-recession* times to foster inter-enterprise connections, cultural activity, and networking within Barnsley and the larger region, are helping now to support efforts to ride out the current period of austerity. Resource-sharing and the continued creation of production and marketing partnerships, as well as linkages among local and regional enterprises that seek economies of scale among smaller businesses, have the potential to increase market capacity in the UK and abroad. However, on July 1, 2012 the Conservative government abolished regional development agencies such as Yorkshire Forward, in the name of what it called "local empowerment". In the face of this measure, the kind of cooperative skill- and resource-sharing activities that were prevalent prior to 2012, face a serious challenge. Barnsley's Community Plan 2005–2008 recognized that "The borough's prosperity is inextricably linked with that of regional economy", and there are many pre-recession examples of successful collaboration among new creative enterprises and practitioners in and between post-industrial towns in the region that support this goal. The government's promotion of more private sector dominance, along with the encouragement of individual towns and cities to work less cooperatively and more competitively, may foreclose rather than encourage further economic growth. Partnerships within and among smaller cities have the capacity to facilitate the kinds of networking that foster innovation by enabling a pooling of knowledge and resources during times of plentiful and scarce resources. For example, it has been suggested that the Leeds-Sheffield-Manchester triangle, of which Barnsley is a part, could comprise the market equivalent of London if towns and cities collaborate fruitfully (Rivas 2011a). For this reason, many believe that federal changes in the institutional landscape to diminish the importance of regional cooperation are "surrounded by uncertainty on their implementation and consequences, especially regarding … the funding resources for development projects" (Rivas 2011b: 10).

Conclusion

In recalling his childhood memories of the Miner's Strike of 1984, poet Ian McMillan describes how the arts flourished during very hard times. He speaks of his membership in a band called the Circus of Poets and remembers run-ins with the police.

> It almost felt like there was a literary arm to the strike … I was running loads of writing workshops and people would turn up. The thing I remember most is once coming on the bus from Wakefield back into Barnsley … Everyone was talking about [the Strike] … Suddenly this fella stood up .and said, "I've written a poem about it [the Strike] … And the bus went quiet, and he stood and read this poem,

> and there was applause … and then this literary debate happened on the bus! I thought, "Blimey, that is amazing. That is how we can actually take charge of writing when other things have been taken away from us" (McMillan quoted on *BBC South Yorkshire* 2009b).

McMillan believes that the Miners' Strike took more than jobs from Barnsley and its surroundings; it stifled a creative "spirit" that once emerged organically from peoples' working and community lives. It also removed a certain natural aspect of "being human" – the desire to be creative.

Back in the 1930s, the WEA linked labor with learning. It was assumed that all working people had as much of a right to cultivate their creative and critical sides as the wealthy. Many opportunities for life-long learning were offered in a range of fields, including art appreciation, economics and philosophy. The Ashington Group of miners, who became famous in the 1930s for their paintings, and are now memorialized in Lee Hall's recent play, the *Pitmen Painters*, suggest the ubiquity, and yet emphemerality, of opportunities for working people to explore their creative sides. The experiences that miners had in WEA classes provided more than a knowledge base and creative outlet; they supported the right of access to a creative life for every person and place. They illustrated that literal change can come about through the transformation of one object into another (Feaver 2009).

The current recession makes it hard to imagine where new spaces for the creative expression might emerge in Barnsley. Whilst there is great pride attached to a mining identity and the struggles of the past, the aftermath of deindustrialization and the deprivation it wrought leaves a residue of challenges that go beyond unemployment and are difficult to tackle. Barnsley's quest to build a creative economy atop vacant mine shafts exposes the fragility and, in some respects, the incompleteness of its alternative forward-looking vision. Given raised aspirations and a long history of creative responses to adversity, it is unlikely, however, that efforts to regenerate will grind to a halt. The visual images of change that are the result of spatial and economic transformation between 2001–2008 undergird a determination to continue to seek a sustainable future for the town. Barnsley may never become a "Tuscan Village", yet Alsop's "halo", representing aspirations and the desire to retain and build a unique identity upon a foundation of local assets, still inspire the town and region to search for a future that is markedly different from its past. This can only be aided by a continued commitment to addressing the immediate needs of residents and to populating a "living wall". Such a metaphorical (and perhaps real) wall of new cultural and economic activity may continue to encourage the cultivation of active idea- and resource-sharing networks among residents within its bounds. It may also support the animation of city life through the exploration of new small manufacturing opportunities and the incorporation of additional residents and new enterprises from its margins.

References

Anichini, F. (2006), "City and creativity in XIII century Florence", Paper delivered at Center for Medieval and Renaissance Studies, Binghamton University, 2 February.

Barnsley Chronicle. (2006), "Have your say", 1 January.

Barnsley Metropolitan Borough Council. (2002), "Rethinking Barnsley/ Renaissance Barnsley", Report on the Rethinking Barnsley Planning Weekend 9–14 May 2002 (Leeds: Yorkshire Forward).

BBC News. (2005), "Super city of north is unveiled", 24 January. Available at news.bbc.co.uk/2/hi/uk_news/England/4187409.stm [accessed 7 August 2012].

BBC South Yorkshire. (2009a), "Your miners strike stories". Available at http://www.bbc.co.uk/southyorkshire/content/articles/2009/01/28/miners_strike_stories.shtml [accessed 26 June 2009].

BBC South Yorkshire. (2009b), "Ian McMillan's poem for the miners strike". Available at http://www.bbc.co.uk/southyorkshire/content/articles/2009/03/04/ian_mcmillan_miners-strike_poem_feature.shtml [accessed 26 June 2009].

Bruff, G. (2010), "Resilient retail? Supporting the high street through recession and renaissance: learning in Barnsley". *Local Work*, 98 (Manchester: Centre for Local Economic Strategies).

Cachia, R., Ferrari, A., Ala Mutka, K. and Punie, Y. (2010), "Creative learning and innovative teaching: final report on the study on creativity and innovation in education in the EU Member States", (Joint Research Centre) European Commission, Institute for Prospective Technological Studies.

Creative Industries Taskforce. (2008), "Creative Britain – new talents for the new economy", 22 February. Available at webarchive.nationalarchives.gov.uk/+/http:/www.culture.gov.uk/images/publications/CEPFeb2008.pdf) [accessed 7 August 2012].

Davies, J. (2005), "Halo over Barnsley", 28 January. Available at www.johndavies.org/2005/01/halo-over-barnsley.html [accessed 7 August 2012].

DETR. (2000), "Our towns and cities: the future", Report from the Department of Environment, Transport and Regions. Available at www.communities.gov.uk/documents/regeneration/pdf/154869.pdf [accessed 5 August 2012].

EKOS Consulting (UK) Limited. (2006), "Barnsley: emerging policy areas", December, Strategic Economic Assessment South Yorkshire Summary.

Feaver, W. (2009), *Pitmen Painters: The Ashington Group: 1934–1984*.

Gardiner, J. (2011), "Barnsley plans unveiled by CZWG and Holder Mathias", *Building.co.uk* 24 May. Available at www.buildersin-bath.co.uk/2011/05/25/czwg-and-holder-mathias-unveil-70m-barnsley-plan-plan/ [accessed 7 August 2012].

Harrison, A. (2003), "Barnsley College", *The Guardian*, 26 August.

Inman, P. and Wintour, P. (2010), "Cuts will push jobless to 3m", 9 June. Available at http://www.guardian.co.uk/politics/2010/jun/10/spending-cuts-public-sector-staff-thinktank [accessed 25 April 2012].

Insider Media Limited. (2003), "Barnsley: Barnsley's old image for the chop", October. Available at www.insidermedia.com/productsandservices/archive/ybi/2003-10/barnsley [accessed 7 August 2012].

Insider Media Limited. (2010a), "Moving on from Tuscany", April. Available at www.insidermedia.com/insider/Yorkshire/5521-moving-on-from-tuscany/ [accessed 7 August 2012].

Insider Media Limited. (2010b), "Barnsley: Barnsley's old image up for the chop", 8 April. Available at www.insidermedia.com/insider/yorkshire/5221-barnsley/index [accessed 7 August 2012].

Kennedy, M. (2010), "First UK city of culture finalists revealed", *The Guardian*, 24 February. Available at http://www.guardian.co.uk/culture/2010/feb/24/uk-city-of-culture-finalists [accessed 25 April 2012].

McMillan, I. (2009), *BBC South Yorkshire*, 26 June.

Regeneration Scrutiny Commission Report. (2009), Minutes No. 42A, 12 February. Available at edemocracy.barnsley.gov.uk/0xac16000b%200x00564fd1 [accessed 7 August 2012].

Rivas, M. (2011a), "From creative industries to the creative place – refreshing the local development agenda in small and medium-sized towns". February URBACT.

Rivas, M. (2011b), "Barnsley – re-making the Northern England midsized town", URBACT April.

Savage, M. (2009), "Britain's cultural capital … Barnsley?", *The Independent*, 3 October. Available at http://www.independent.co.uk/news/uk/this-britain/britains-cultural-capital--barnsley-1797012.html [accessed 25 April 2012].

Wainwright, M. (2002), "Benvenuto a Barnsley", *The Guardian*, 5 April. Available at www.guardian.co.uk/books/2002/apr/06/books.guardianreview2 [accessed 7 August 2012].

Wainwright, M. (2003), "Starchitect to redesign Barnsley with a laser halo", *The Guardian*, 11 July.

WEA News. (2009), Issue 20, January–March. Available at http://www.wea.org.uk/weanews/ [accessed 2 July 2012].

Websites

www.guardian.co.uk/news/datablog/2011/may/18/ethnic-population-england-wales#data [accessed 17 August 2012].

www.barnsley.co.uk [accessed 10 September 2011].

www.barnsleyandfamily.com [accessed 17 August 2012].

www.barnsleydevelopmentagency.co.uk/key-facts [accessed 5 August 2012].

www.centreforcities.org/centre-for-cities-sets-out-a-new-approach-to-regenerate-englands-cities.html [accessed 17 August 2012].

www.creativebarnsley.co.uk/aims.htm [accessed 2 July 2012].

www.creative-partnerships.com/about-creative-partnerships/ [accessed 17 August 2012].

www.Webmasterworld.com/forum9/8354.htm [accessed 7 August 2012].
www.yorkshire-forward.com/improving-places/urban-areas/barnsley [accessed 16 September 2006].
www.northbarnsley.co.uk [accessed 17 August 2012].
www.linkedin.com/groups/Barnsley-Design-Network-3753373?_mSplash=1 [accessed 17 August 2012].

PART II
Moving Beyond Neoliberal Methodologies in the Study and Practice of Creative Economy Planning

Chapter 7
Creative Revitalization as a Community Affair

Arturo E. Osorio

Mill towns born in the heart of New England frequently foreshadow national business trends and labor force transformations. These communities were often developed and owned as company towns, and grew "organically" as conglomerates of factories where labor mobility, economic change, and cultural and social transformations were recorded in payroll and log books as well as in the town's chronicles and demographics (e.g., Hartford 1990; Devault 1995).

Deindustrialization, starting several decades before the end of the Second World War, became more widespread in the 1970s (Cowie and Heathcott 2003). Global economic restructuring and technological revolutions transformed once proud and prosperous mill towns in New England into a collection of rundown communities immersed in economic recession (Office of Economic Development Mass Government 2001). Almost 40 years after the beginning of their collapse, some of these former mill towns are now experiencing a socioeconomic renaissance nourishing new cultural and creativity *clusters*. One clear example is North Adams, Massachusetts, home of the Massachusetts Museum of Contemporary Art (Mass MoCa) and a variety of art schools, as well as artist's studios and galleries (Oehler, Sheppard and Benjamin 2006).

What happened in these small New England towns between the closing of the last factory and their resurgence as *clusters* of artistry and creativity? How are some communities transitioning from labor-intensive manufacturing into creative and knowledge-based economies? What role do local residents play in this regeneration process? Are they merely passive observers, or are they actively networking, lobbying, and making change possible? Exploring the events and local socioeconomic processes fostering the transformations in one smaller city may contribute some answers to these questions.

The Grassroots Crafting of a Local Renaissance

Easthampton, MA, with a population of around 16,000 in 2009, is an example of a New England community that captures the nationwide socioeconomic processes in its own local history (U.S. Census Bureau 2009). According to the city's historical society, when Easthampton was first settled by European immigrants in 1664, it was very much like the rest of the U.S. at that time; a puritan agricultural community. With the arrival of the industrial revolution in the 1880s, Easthampton became a booming manufacturing-based hub. Immersed in the industrial economy during the Civil War era, it was known as Button Town because of the settlement of large button manufacturing facilities. From the 1940s until the 1960s, when United Elastic Company had its operations there, the town was referred to as the Elastic Maker of the World. From the 1960s until the second-half of the 1990s, when Stanley Home Products and Kellogg Brush Manufacturing based their operations in Easthampton, the community was known as the Home Products Town. Manufacturing was a central feature of town life until the 1970s, when the community began to lose its industry to globalization. Factories first moved to other communities in the U.S. and later abroad. By the end of the 1990s, the large majority of local manufacturers had left town and Easthampton slid into socioeconomic decline. Globalization in this location, as in many other former manufacturing communities within the Frost Belt, translated into cheaper labor abroad and empty factories locally (Kweit and Kweit 1998).

Easthampton residents were not passive spectators during this period, and the economic slump was not the only process underway at that time. Local government and prominent citizens continued to fight to attract new manufacturing operations (The Master Plan Advisory Committee of Easthampton 1987). In parallel to early official actions, grassroots organizations also began to occupy empty factory buildings. The old J.P. Stevens Co. building, located at One Cottage Street, became the first non-manufacturing organization to occupy one of the empty factory buildings in town. The structure was bought in 1976, at the nominal price of $1.00, by Riverside Industries Inc., a local non-profit organization-serving people with disabilities. The purchase paved the way for unexpected change that crafted a local creative class and seeded a local cultural economy.

A Creative Class at Work

Richard Florida defines the "creative class" phenomena as an above average geographical concentration of individuals performing professional activities understood as creative in nature. He identifies this class as a socioeconomic engine for business agglomeration and economic regional development, claiming that its presence is linked to the local cultural economy and associated with healthy and vibrant communities (Florida 2002b; Gibson and Kong 2005; Sommer 2005; Oehler, Sheppard and Benjamin 2006). How these concentrations of individuals come together as a creative class to fuel healthy and vibrant communities has rarely been explored.

Alfred Marshall (1890) was the first to produce a vivid account of the socioeconomic dynamics nowadays commonly linked to the outcomes of a locally resident "creative class" (e.g., Florida 2002b; Sommer 2005; Oehler, Sheppard, and Benjamin 2006). His work describes the phenomenon as the bottom-up crafting of relatively small and prosperous communities that include small cities, towns, and industrial suburbs. As a contemporary observer of the birth of the industrial revolution in England, Marshall portrayed how scientists, artisans, artists, handcrafters and entrepreneurs (i.e., an early "creative class") came together in a geographically contiguous and delimited space to attain individual gains (e.g., knowledge exploitation, access to resources) while sharing communal benefits (e.g., knowledge creation and economic costs) through a local social network.

Marshall's contribution was neither to identify the presence of what later became known as a locally embedded "creative class", nor to propose the role of the creative class in a community. Rather, his work showcases the on-site, socioeconomic conditions that ensure when there is an above average geographical concentration of specialized industries. His work highlights the presence of a clustering of individuals sharing knowledge, and the reduction costs of production, and easy access to resources that result from this sharing among residents. Marshall's work takes for granted a creative and entrepreneurial class, and the emergence of a collaborative network that allows them to take advantage of their clustering.

More than 100 years after Marshall first described the economic side of these vibrant communities, Richard Florida (2002b) associated similar instances of local socioeconomic success with the presence of what he named the "creative class". Using industrial occupational indexes and socioeconomic indicators, Florida (2002b) describes the creative class as an above average geographical concentration of individuals with high educational attainment in creativity-dependant occupations such as arts, crafts, or education professionals. It can be argued that Florida's work on the creative class complements Marshall's original research, as it focuses on the phenomena behind a community's economic success and high concentrations of specialized industries. Florida's construction of the "creative class", based on quantitative data, was quickly incorporated into public policy and regional planning, and the presence of this class became almost immediately synonymous with healthy urban communities. This came after many schemes to attract new manufacturing to post-industrial cities had failed and alternatives to manufacturing were sought to support the development of new local economies.

Florida's work on the creative class, while useful to call attention to the importance of the geographical clustering of creative individuals, has been subject to critique. It has been argued that his operationalization of creativity as an occupation placed the focus on the economic description of a highly dense conglomerate of individuals with high educational attainment within a particular profession, rather than on the assessment of the actual creative practices of individuals (Markusen 2005). To address this issue the same critics offer an updated operationalization of Florida's work. The revised argument centers on the actual performance required to achieve the professional accomplishments listed as creative endeavors, and it includes the attributes and preferences of the performing individuals at this location. This updated operationalization advances the notion that occupation may be used as a proxy for creativity, fostering regional economic development if: (1) it is made "co-equal with industry" (i.e., putting creativity at the same analytical level as any other local industry), (2) it is assumed to be location embedded, (3) it is assessed on a case-by-case basis (i.e., occupation by occupation), and (4) all workers within each occupational category are presumed to make independent and personal location decisions when selecting a place to work and live (e.g., Feser and Bergman 2000; Markusen 2005; Markusen and Schrock 2006). While useful and better fitted for its purpose, this approach still focuses on large communities and uses aggregated data to report comparative indexes. Additionally, the actual practices of local members of the so-called creative class remain unexplored (Gibson and Kong 2005).

This chapter presents an alternative perspective on the creative class. Following Gibson and Kong (2005), I consider the creative class to be a socioeconomic process within a community at large and not a reductive index of activities and occupations within a geographical location (e.g., Feser and Bergman 2000; Florida 2002b; Markusen 2005; Markusen and Schrock 2006). While this project accepts the existence of clusters of creative individuals and their impact on local communities and business, as proposed by Florida and others (e.g., Florida 2002a, 2002b; Gibson and Kong 2005; Oehler and Sommer 2005; Oehler, Sheppard and Benjamin 2006), it does not take them, or their actions, for granted. The occupations of creative individuals are not automatically equated with creativity nor is it presumed that only so-called "creative individuals" undertake certain occupations and live only in large cities, acting in uncoordinated socioeconomic isolation from one another. Instead, I employ a mixed methods approach involving social network analysis and institutional ethnography to explore the way individuals form a creative class that is capable of shaping the socioeconomy of their community.

Framing the Emergence of a Creative Class

Artists and artisans are connected to each other and to other non-artists; they are all around town, yet not always evident. Just like their work, they gradually become a noticeable part of everyday life. In my study of Easthampton, I explore the creative class as an emergent socioeconomic organizing process of individuals within the town. Such a "class" helps to initiate local events that promote, rather than become either the cause or effect of local socioeconomic change.

Combining institutional ethnography and social networks analysis as methods of inquiry, allows a more nuanced approach to the research of social groups. Institutional ethnography is a sociological method of inquiry, first developed by Canadian social theorist and feminist Dorothy E. Smith, to explore the sociology of women (Smith 1987). This method allows for the exploration and contextualization of social relations that structure people's everyday lives as individuals and members of the community (Agar 1996; Denzin and Lincoln 2000; Genzuk 2003; Madison 2005; Aggestam and Keenan 2007). A social networks methodology follows the organizing processes in a community and graphically maps and analyzes the social relationships observed (see Figures 7.1 and 7.2). It facilitates the process of making sense of the observed social processes and documented relationships (Freeman 1996, 2000; Stevenson and Greenberg 2000; Wakita and Tsurumi 2007). The concurrent and complementing use of institutional ethnography with social network methodology, can help to identify and assess the value of the artists and artisans as they become a local creative class. It does this by simultaneously informing their actions as members of the community (institutional ethnography) and as a creative collective (social networks theory). Art and creativity are, in this way, treated as part of the processes of ordinary life.

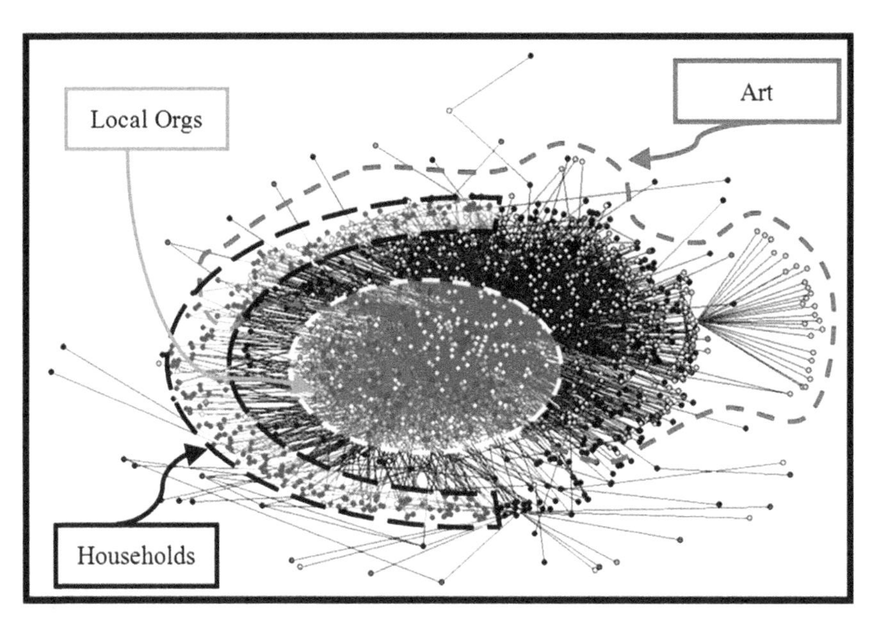

Figure 7.1 Kamada-Kawai analysis of the community's network of organizations

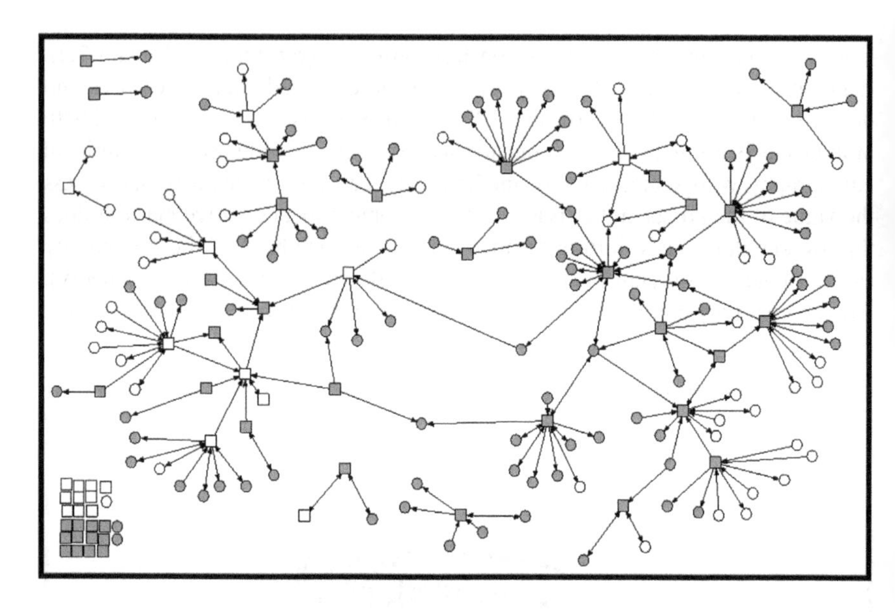

Figure 7.2 Ego maps of questionnaire respondents

Size Does Not Matter – If You Are Creative

In Florida's depiction, the "creative class" is presented as a large urban community dominated by "talent, tolerance and technology" (Florida 2002b). It is argued that local attributes (e.g., lifestyle-related amenities, socially friendly atmosphere) draw, cultivate, and mobilize creative people who are socially compliant model citizens yet incoherent as a group (i.e., they lack a coordinated voice). It is presumed that only large cities are capable of pampering, catering, and hosting these large numbers of professionally creative individuals who are educated and in pursuit of larger incomes (Florida 2002b; Markusen 2005). These professionally creative individuals are also thought to possess the qualities necessary to occupy positions of power, yet they are said to be unable to consciously influence the society they lead, since, as individuals, they are too disparate (often times self-centered) "to be herded together" to become a coherent organization, political party, or the like (Florida 2002b).

This image suggests the potential of a creative class composed of independent individuals to positively affect a local economy if appropriately managed and controlled. In contrast, this chapter presumes the existence of multiple variations of a creative class, each one as unique as the location (i.e., geographically embedded community) where it is constituted. A creative class is presented, not as an aggregate of educated subjects, but as a group of creative individuals purposefully engaged with each other and their neighbors through synergistic relationships that are circumscribed and influenced by the small city where their interactions take place. This alternative perspective accepts the possibility of having as many types of creative people as there are places where they may emerge (Asheim and Hansen 2009). Hence, I argue that large cities are no longer a necessary condition for their cultivation. Small cities (or small communities nested within large urban settings) can become acknowledged as valid hosts and incubators of their own distinct creative class. Furthermore, members of such a class may not need to be highly educated and/or be comprised of wealthy professionals to be able to consciously influence and disrupt community patterns where they live. Likewise, the value of a local creative class may be social as well as economic (Granovetter 1985).

Framed by the ideas above, the story that follows explores how and why an assortment of artists and artisans in the small New England mill town of Easthampton, Massachusetts engaged in recurrent relationships with other individuals in Easthampton between 2005 and 2009. Creative class processes are understood here as a collision of culture and economy that at key moments in time may encourage community transformations through an unexpected ripple effect of local benefits (Gibson and Kong 2005). Artists and artisans in Easthampton present themselves as cultural agents, and their actions in the community are exemplar of these collisions of culture and economy. The relatively small size of this community, along with the mixed-method approaches, allowed me to examine how a geographical agglomeration of organizations and creative individuals unfold over time to promote local creativity that may or may not be sustained.

Once Upon a Town: Seeking and Mapping Friendships and Collaborations within the Local Emerging Artist Community

Visiting downtown Easthampton on a random weekday looking for evidence of artists and artisans can prove difficult. Artists and artisans cannot be distinguished from the rest of their neighbors. Likewise local census records do not provide a true number of local artists and artisans, as many of them either do not live in town or do not earn the bulk of their income as artists or artisans, and are thus not included in the count. Yet, such individuals populate Easthampton. I recognize their faces from local art-related events where they congregate, and from my personal visits to their studios, mostly located in three large buildings across town. They do not wear badges or distinctive clothing – even when performing public art related services (i.e., participating in the local arts walk, working on public installations, doing murals, guiding visitors). If you do not know them, you will miss them altogether. The "open studio sales", the art and crafts sales event where most of the artists and artisans in town come together over the same weekend to open their studios and sell their work to the public in general, only happens twice a year – not often enough to make their faces memorable. Except for these special dates, local artists and artisans live their own normality and seclusion. They are part of local everyday life, yet not always visible – even when their work may be.

Figure 7.3 Local sandwich and coffee shop displaying local art for sale as decoration during the 2006 Windows Project event

Figure 7.4 Local hardware store displaying pencil drawings on its windows during the 2005 Windows Project

While the number of artists and artisans in Easthampton has grown quickly over the last 10 years, many of today's most famous local artists were already here as far back as 15 to 20 years ago, perhaps even longer. Back then, they were not defined as the desirable or successful selves of today, and nobody knew for sure if anyone among them would ever become so. They were disruptive (e.g., constantly questioning the wisdom of the status quo, engaged in non-conventional occupations) local youngsters and/or young "cultural creatives" trying to catch a break while living on a budget and challenging local understandings of normality (Ray and Anderson 2001). Their unconventional (i.e., bohemian) lifestyle was not a model that fit the local mill town mentality.

This early community of artists constituted a small cohesive bohemian sub-group with attributes that set them apart from the rest of the community. Most people around town worked conventional 9-to-5 jobs in the remaining factories and surviving local businesses. They came back home to have supper with the family, watch TV, and go to sleep. Young artists would wake up late in the morning (as often as they could), spend hours walking around looking for inspiration, or putting colors on canvases, or carving furniture. They often closed their days by coming together with other fellow artists and artisans at each others' studios or in common areas at One Cottage Street to discuss town, state, and national socioeconomic issues, critically, until dawn. As young aspiring artists and artisans, their livelihood was not always generated from their creations – at least not at first. The need of income forced many into more conventional occupations around town. By night, some were the grocery baggers, deli counter clerks, city hall employees, jack-of-all-trades repair people, or any number of other temporal occupations. Over time, neighbors befriended and learned to accept them. Yet because of their need for peers to support and encourage their creative side, local artists began to constitute a community and gather by themselves.

As resident bohemians and rebels in an otherwise blue-collar community, these younger selves of the now respectable elders of Easthampton's arts community, seeded the local cultural economy. Their individual presence, however, is not enough to constitute a local creative class. Crafting a creative class and launching a cultural economy is a co-constructive process among artists, and between artists and the community in which they live. The changes in Easthampton sparked by the arrival of more artists and artisans gradually became part of everyday life, as daily interactions across multiple local constituencies shaped new relationships among them and between them and the rest of the community – as illustrated in the socioeconomic map in Figure 7.1. Among these local transformations, perhaps the most important, with the greatest consequence, was the town's involuntarily change from a manufacturing economy into a services and arts and crafts' community.

As time went by, artists and artisans progressively occupied the town's areas zoned for manufacturing. The same regulations that allowed a production line to be set up, allowed a cabinet making operation or a sculpting studio to be installed. Likewise the same toxic disposal regulations and procedures for manufacturing, applied to the use of paints, glues, and other materials required by the artists.

Old factory buildings with solid floors to hold machinery and large windows to let light in were ideal for studios, as natural light was abundant and floors were solid for sculptors and wood workers. Cargo elevators once used to move loads of raw materials, products, and large machinery were perfect to transport large sculptures and canvases.

Even when the local leadership was looking to bring new manufacturing operations into town, the number of locally available properties that could accommodate these businesses was shrinking. The old factory buildings were becoming art studios, artisans' workshops, creative collectives, and art galleries. What started as a temporal solution to deindustrialization became a permanent solution, as global economics continued to lure manufacturing operations abroad. Most of the industries that could be a good fit were already re-located outside the country in places with cheaper labor costs and less stringent manufacturing related regulations. The relocation of local factories in town also resulted in a decline of people declaring their main employment as manufacturing. Likewise, reported levels of educational attainment increased and more people started to work in white-collar jobs, including the service and education industries.

The introduction of new lifestyles in Easthampton became more noticeable around 2004 when the first local arts charrette took place in town. Over time, the local river and ponds, once considered the heart of the mill town, became recreational spaces. Concerns about soil and water pollution from former industrial days found allies among new local artists and artisans, who helped to create cleaning campaigns and community days, as they set an example in the management of their own toxic materials at their studios. Local routes once considered the town's competitive advantage as they moved labor, materials, and products, became noise and safety issues for the new residents. The town was no longer a blue-collar community and, though gradual and incremental, the changes eventually re-shaped local social dynamics.

A board of selectmen, comprised of old time residents who had been actively involved with the local industrial life of the community, governed Easthampton for some time. To the board, artists were transient annoyances. They were not included in development conversations, and went generally unnoticed at the margins. The 1987 master planning process did not include any artists or artisans. In fact, arts and culture were absent from the entire planning agenda. Instead, the master plan reflected the strong voices of the industrial past, as it continued to focus on attracting new manufacturing enterprises to replace the ones that left in the 1970s. It proposed the updating of the empty manufacturing buildings, even though artists and artisans already occupied half of these spaces. This updating never happened, however, and a new plan to construct an industrial park was advanced to attract new factories and jobs. The industrial park did not achieve its goals. It was never fully occupied and most of its tenants left town less than 10 years after moving in.

Oral accounts point fingers in all directions as tensions in the town began to rise. In 1996, the board of selectmen was dissolved and a mayoral system of governance was implemented in hopes of bringing a fresh perspective to addressing the socioeconomic distress. Memories shared by artists, and local written records of the time suggest that, with empty factory buildings and high unemployment, anyone who was not working towards the re-industrialization goal was perceived as working against it. This explains why so many residents continued to view young artists as rebellious "outsiders" even though, by 1996, many of them had been in town for more than 10 or 15 years! Given this background, it is important to ask; what factors did occasion the emergence of a local creative class? The social history of one building may provide some answers.

One Cottage Street: The Social History of Where it All Started

J.P. Stevens Co. closed its factory in Easthampton in 1972, and spent four years trying to peddle their old building at One Cottage Street. The facility was in need of repairs and upgrades. Other manufacturing companies were deterred from buying the building and moving into town by the state of the building, local socioeconomic distress and the demands of globalization. Neither promises to help update the buildings nor lenient town manufacturing regulations were enough to attract new industrial operations.

At the same time, Riverside Industries Inc., a non-profit agency serving people with disabilities, had a pressing need for a new location. Between 1968 and 1972, two separate fires destroyed the facilities of Riverside Industries (formerly known as Occupational & Vocational Developmental Center for the Handicapped – OVDCH), eventually forcing them out of Northampton, MA into the old Brassworks building in Haydenville, MA (Your Riverside Connection, Fall 2003). According to accounts of people working there, the Haydenville location was "*a dreadful place*". The Executive Director of Riverside industries at the time, Ed Pion, drove around the area to find a new location for their activities (Your Riverside Connection, Fall 2003, Interview with Ron Bittel, retired President of Riverside Industries, 2003). The new site would have to fulfill two main requirements: First, it would have to be large enough to accommodate all the current employees and clients under the same roof. Second, it would have to be close enough to its current location so that employees and clients could still work and participate. These two rules were set because nobody wanted to disrupt the organization's social networks. Down the road from their former location in Northampton, and close to its current location in Haydenville, Ed Pion happened to drive by a large, seemingly abandoned building in Easthampton that had potential and, with repairs, fulfilled the two requirements.

Without much to lose and everything to gain, Ed, or "Fast Eddy", as some remember him, looked for the owners to talk them into a deal. Getting an appointment at J.P. Stevens Co. was the first step. After a sales pitch, he convinced the executives that they were losing money by keeping ownership of an old and distressed building, even if zoned and approved for industrial use and located at the heart of New England's crossroads. Ed's argument was strengthened by the lack of buyers for this property, and the building's need of repairs necessitated by J.P. Stevens Co.'s abandonment in 1972. A handshake and a one-dollar bill got Ed the building that was written off from the J.P. Stevens Co. books as a donation to a local non-profit.

With more space than it needed, and a new address to put on the official correspondence, Riverside Industries, Inc. moved its operations to Easthampton. Broken windows, some collapsed ceilings, and rusting pipes, among other pressing issues, welcomed Riverside Industries to their new home. The scope and size of the habilitation plan was limited by the lack of enough capital to do all the required work at once, and the pressing priorities linked to the immediate need to house their operations. In the end, the work included the bare essentials in the areas that were to be immediately occupied. This left a portion of the facility untouched. Riverside's operations and the limited repairs undertaken redefined the property's use. As Riverside Industries' was not a manufacturing enterprise and its focus was occupational therapy for people with disabilities, safety issues prevented the building from hosting any industrial manufacturing activity. In the long run, this realization cemented the town's path away from manufacturing despite the best efforts of local officials to recover industry. This acceptance would prove critical to altering the dynamics of town planning and creating an opening for new social networks.

Figure 7.5 Back view, One Cottage Street, the former JP Stevens building, summer 2004, when Riverside Industries first moved in

Once Riverside Industries, Inc. was relocated, the action plan was to remediate the more than four years of abandonment in the property, set new offices, and start operations. Shortly thereafter, economic mandates as a non-profit, and the abundance of space around them, moved Riverside Industries into the property rental business to raise operational capital. State budget cuts, low donations, and an increasing demand for their services put pressure on the organization's leadership to get new, long term, sources of funding. The scenario for the arrival of the artists and artisans as tenants was thus precipitated by Riverside's ownership of a partially empty and distressed large industrial building, along with safety concerns born from the disabilities of its constituency. Artists and artisans did not need much in terms of services or appearances, and their activities did not represent a risk for the Riverside clientele.

Many individuals and organizations came and went through Cottage Street doors, starting with Warwick Press in 1977. However, the most important and memorable was the One Cottage Street School of Fine Woodworking, known then as Leeds Design Workshop, which moved in to its space in July 1978 (New England School of Architectural Woodworking 2009). The woodwork school, in collaboration with the University of Massachusetts, offered adult, beginning, and intermediate classes in woodworking, eventually including career training in architectural woodworking. The space available for rent at One Cottage Street was ideal.

As time passed and graduations came and went, some locals from the area began to attend classes to launch new careers. Once these new artists and artisans received their diplomas, they started to look around for studio space in the area. The former students required several conditions before they could get settled into their new professional lives. First, they needed a place large enough to fit their operations as cabinet and furniture makers. Second, they needed local ordinances that allowed them to use the required heavy equipment. Third, they were initially local (e.g., from Easthampton and its surrounding towns), which is why they attended this particular technical school, and they wanted a location close to their roots. Finally, because they were just starting a new (and often first) career, they were price-sensitive and in need of cheap space. The graduates also needed access to equipment, advice and a collaborative learning environment just like the one that they had left behind.

One Cottage Street met all their requirements. Riverside Industries, Inc.'s real estate rental venture was still in its early stages and space was abundant and cheap. Deals of temporally reduced rent in exchange for help to clean and shape the building were the perfect conditions to get many started following graduation. The possibility of renting equipment from their former school drafted several alumni into the population at Riverside Industries, Inc. This was, by the account of many, the seeding of the local arts and crafts community in Easthampton. Just as in the SoHo District of New York City, no official plans, agreements, or discussions with town government guided the growth of an arts population (Zukin 1989). The use of the space and development of an arts community at Riverside Industries was organic and suited to meet local needs; it was not a deliberate attempt to copy stereotypical models of cultural investment elsewhere.

At the city level, the small size of the community and relatively low per capita income, combined with a low level of demand for loft living space on the part of Easthampton residents, to attract the interest of people from outside the community. The absence of street level storefronts and living quarters, combined with the dominant presence of Riverside Industries, set its own stamp on the emergence of the arts and crafts community at One Cottage Street. Strong reciprocal links of friendship within the building were fostered by the desire for a collaborative learning environment. This initiated the processes that eventually shifted the demographics of the building into a creative community. By the end of the 1980s, the building was filled with artists and artisans, volunteers and activists, along with people with disabilities – not exactly the mix prescribed by the extant literature on the creative class. Likewise, the town, a mostly white, small former manufacturing community without much entertainment or wealth, also did not fit the image of the ideal home for a creative class. An unusual community fostered an uncommon creative class.

Throughout the transition, Riverside Industries Inc. was happy with its tenants. The artists and artisans, including the woodworkers and cabinetmakers, were good-hearted people, who were not overly conscious of their neighbors with mental and physical disabilities. While the top floors were totally occupied by art studios and workshops, and the domain of the arts and crafts industry, the downstairs contained the therapy, job training, and occupational facilities for people with disabilities seeking to become contributing members of society. Having the acceptance of the upstairs artists became part of the downstairs therapy. Likewise, the hopes and enthusiasm expressed by Riverside clients became an inspiration for the upstairs residents. The collaboration of the artist community eventually extended to the social service clientele and became a model of acceptance around town.

Artists began to befriend downstairs neighbors who had the ability to wander upstairs to learn what was happening there. The relationship was not condescending, but rather based on developing friendships and, eventually, collaborations that began to craft strong reciprocal ties. In 2006, a member of the arts community at One Cottage Street officially exchanged her studio space on the top floor for a classroom on the first floor. She started working for Riverside Industries and was put in charge of the arts rehabilitation center – her pet project as an artist. Soon thereafter, taking advantage of her social capital, she enrolled a roster of trusted colleagues and friends from all over the building, and others from around town, to help her with the newly launched art therapy project. This art therapy project was another unexpected benefit of the relocation of Riverside Industries to the building.

Figure 7.6 Main entrance, One Cottage Street, 2008

The budding relationships between the arts and disability services expanded and began to challenge the definition of who is an artist and what constitutes an artistic or creative endeavour. This had the effect of changing local cognitive maps and encouraging the community to consider anew its evolving identity. The town was no longer strictly blue-collar waiting for the next factory to arrive. People around town started to see themselves as "cultural creative" gradually building a new community.

The Open Studio Sales Choice: Making a Buck or Making a Friend

There are nonetheless other more complicated stories to tell about One Cottage Street. The unplanned layout of the building was not conducive to visits by non-resident artist and customers. It isolated its occupants from the rest of the Easthampton community, and very few, if any, local residents would venture into the building to visit tenants. As the One Cottage Street community grew organically, permanent walls and doors were built to meet personal requirements and the needs of the moment. There was no long-term well thought-out plan. The outcome was a space that supported collaboration, exploration of ideas, and long hours of work at will without interruption. For many residents of the building, artistic creation was their primary livelihood and privacy was a "requirement of the job". Yet, privacy was also a limitation, as outsiders experienced the building's layout as a literal maze. The very same conditions that provided artists with the solitude to create isolated them from the town's social network and, for some, prevented the sale of work.

A solution to the conundrum sprouted relatively early in the life of the arts and crafts community at One Cottage Street. In 1987, about 10 years after the building opened, and about 5 years after it acquired a solid population of networked artists and artisans, a small group of tenants decided to come together and become visible for one weekend a year. They met many times throughout the course of the year, and by the first weekend of December 1987, were ready for the first Easthampton Open Studios holiday sale. This arts and crafts sales event, which from its inception was foreseen as a recurrent affair, built up momentum little by little to a point where, seven years later, in 1994, the group added another sale on the first weekend of June to make it a semi-annual event – as it has remained.

The success of Cottage Street Winter and Summer Open Studios Sale was largely due to the trust-based recurrent exchanges responsible for crafting the social network of its tenants (Granovetter 1985, 1992; Levin and Cross 2004; Smith 2007). These exchanges were born from a common goal and supported by close geographical propinquity, a sense of affinity, and shared responsibility among peers. The common and explicit goal was the coordination of a successful event that would account for a good financial outcome for everyone. From the very beginning, the members of the group – all neighbors at One Cottage Street – knew that a good visitor turnout would only occur if there were a sizable number of arts and crafts offerings from participants. This clearly linked collective participation to individual benefits. Another requirement for success was quality. Everyone in the group was (in)formally pre-screened as an artist or artisan of merit by everyone else in the group. This reinforced the sense of solidarity and crafted new ties across the multiple nodes of the network. A loosely structured governing body meant that decisions were either made by a majority vote or left in the hands of the more knowledgeable in the group, with the corresponding responsibility and accountability towards the rest. Since this process was part of a collaborative venture, efforts were to avoid isolating neighbors at One Cottage. Points of dissent among participants were often solved just by walking down the hall to talk with other members of the group, thus reinforcing network ties and increasing its cohesiveness. The collective understanding of the need for reciprocity reinforced a sense of mission; collective success could only come from individual triumphs and vice versa.

It is worth mentioning that not all artists and artisan residents of One Cottage Street share the same interests or vision. Hence, not all residents participate in the Open Studios initiative. Some of the non-participants claim that this event "does not bring in their market". This is especially the case for those who work primarily under commission. Others do not participate because they do not have the time for it. This can mean many different things, e.g. they cannot work with the group; think of themselves above the quality of the group; or have been hit by a "creativity dry spot" and have nothing to offer. In all, a lack of affinity among artists and artisans can affect the cohesiveness and efficiency of their networking.

Once the Open Studios Annual Sale grew in popularity and became a semi-annual event, the demand for studio space at One Cottage Street tested the limits of the building. By 1992, the former mill factory had no more space to offer. It hosted as many studios as it could accommodate, and although every square foot was already occupied, the phones did not stop ringing. Artists and artisans from the surrounding area kept calling to ask to join the community (Boucher 1998). Cheap rent and opportunities to be in close proximity to those pursuing similar interests were incentives that kept the requests for space and the many "just in case" calls coming.

Eastworks: The Commercialization of Real Estate and Art Production

One Cottage Street is not the only home for the arts and crafts in Easthampton, and not every artist is in town for the same reason. Labeled from the beginning as "carrying the arts a few steps further into the community", Eastworks is another former factory building in town that opened in 1997 as a professionally managed arts and crafts community (Boucher 1998). In contrast to One Cottage Street, the operations at Eastworks did not evolve in a random manner. As a tenant of the Arts & Industry building in Florence, Massachusetts in the early 1990s, Will Bundy, the co-owner of Eastworks and son of McGeorge Bundy,[1] was familiar with the mill re-use model from a commercial perspective. Using this experience, he and his wife, Paula, established Eastworks as a successful commercial venture focused on providing real estate space to individuals and organizations involved in the arts and culture industry. Their model mimicked the "loft living" model in New York City described by Sharon Zukin (1989). It allowed some space for live-in studios as well as easy access to clients in some pre-assigned spaces.

One Cottage Street and Eastworks both accommodate artists and artisans looking to be located in Easthampton, yet the evolution of the communities at these two spaces was entirely different. The One Cottage Street artist and artisan community grew unplanned and was spawned out of the financial needs of a non-profit service organization, Riverside Industries. Space was rented on "first come, first serve" basis whether applicants were artists or not. Eastworks was deliberately set up as a business investment with the specific purpose of providing mixed-use space for a "community of artists, residents, entrepreneurs, non-profits, retail and more" (Eastworks History 2009). The fact that space at One Cottage Street was shared with the rehabilitation services, and tenant operations were not framed as a for-profit venture, but as a financing resource to sponsor Riverside Industries' services, meant that whoever rented the space there would have to be

---

1 McGeorge Bundy was the United States National Security Advisor to Presidents John F. Kennedy and Lyndon Johnson from 1961 through 1966, and president of the Ford Foundation from 1966 through 1979.

comfortable working at, and receiving customers in, a building with a community of people with mental and/or physical disabilities. This shared arrangement eventually proved to be a crucial element in the self-selection of tenants, and the driving networking force in the building (Adamic, Buyukkokten and Adar 2003; Lin 2004; Smith 2007).

Eastworks venture was "professionally handled". Planned and executed from the beginning as a real estate management business, operations were set to be pleasant and efficient. Opened in 1997, Eastworks took advantage of the demand that One Cottage could not address since the latter became full to capacity in 1992. Eastworks' space includes business and home-studios in well-divided and defined spaces. There is also a tenants' screening process to ensure good behavior and create a safe haven for business. This deliberate business model was planned to offer a professional and affordable alternative to artists and artisans who were increasingly priced out of nearby Northampton or reluctant to take part in the gentrification processes happening there.[2] Artists and artisans at Eastworks were sold on the idea of living within a creative community of similar others – an ideal model of homophily and propinquity (Adamic et al. 2003; Levin and Cross 2004; Back, Handcock, Raftery and Tantrum 2007).

Eastworks was successful from the beginning, but its different operations from One Cottage Street produced tensions between both groups as they tried to come together for the Open Studio Sales. By 1998, one year after Eastworks first opened its doors, the semi-annual Open Studios sale at One Cottage Street was in its 11th consecutive year, and so many of the residents at Eastworks saw this event at One Cottage Street as an excellent opportunity to promote their work. As such, the Open Studios sale was perceived by many of the residents at Eastworks as an excellent opportunity to promote their work. Representatives of the Eastworks community approached residents at One Cottage Street and asked for a chance (and the space) to exhibit and sell their work at the semiannual sale. These representatives did not speak on behalf of all of the residents at Eastworks. Several of the artists were not comfortable with the partnership. Rather than become part of the group they tried to capitalize on the already established collective efforts to harvest the benefits of the already scheduled Open Studios events by setting up

2 Northampton, Massachusetts is a city located 10 minutes down the road from Easthampton. The economic and cultural changes of this city's downtown have resulted in the relatively recent gentrification of the community. Northampton has become a place where old and new come together. It is simultaneously the kind of community that people love to visit for a good time and a nice, quiet city in which to live. Because of its local cultural accomplishments and collection of nearby institutions of higher education (e.g., Smith College) this city has become a destination point. Northampton thrives with numerous restaurants and shops that reside in buildings centuries old (Aiello 2004). In 1989, median family income in Northampton was $39,808, with 79 families earning incomes of $150,000 or more. Ten years later, in 1999, the median family income in this city had increased to $56,844 and 358 families passed the $150,000 mark (U.S. Census Bureau 2009).

independent open studio events on the same dates as the One Cottage Street Open Studios Sale, and posted signs on the road reading "more art ahead". While this was an easy tactic to attract foot traffic and bolster sales of their work, their actions took advantage of the Open Studios collective efforts without contributing to it.

The "perpetrators" were soon labeled as "free loaders" and their practices damaged their local social capital (Granovetter 1985, 1991; Granovetter and Swedberg 1992). In contrast, those who tried the hard way to work out a collaboration, managed to eventually succeed, even when it meant just a one-on-one partnership. Collaborators were always welcome by One Cottage Street occupants. Anyone approaching the community asking for space was invited in, as long as the tacit rules (e.g., be a team player, have quality work) and explicit rules (e.g., pay fees, be sponsored by a regular participant) of the event were followed. This openness on the part of the artists and artisans working at One Cottage Street continued even after a formal collaboration between the two groups fell apart a year after it began due to growing conflicts associated with the execution of the open studios week – the same event that brought them together in the first place. As time passed, participation in the Semiannual Sale by artists and artisans who were not members of the One Cottage Street community became the norm. The presence of town-wide events to promote the collaboration of the arts community then served as a bridging process that brought together different artists and artisans around town.

The surge of art related activities supported by the regular presence of artists and artisans, prompted the town's mayor to comment on "the possibility that Easthampton might become better known for its association with the arts" (Boucher 1998). Yet this comment paid only lip service to the arts, as it appeared in an article promoting the arts revival in Easthampton. In the same year, local arts were altogether absent from the Decennial Master Plan Meeting, even though this community continued to grow. One reason for this absence was that the agendas of the artists and artisans at One Cottage Street and Eastworks were often at odds. A case in point was the understanding of the role of the arts in the community. At Eastworks the arts were presented as a professional endeavor that required a business-like mentality. At One Cottage Street, the arts were viewed as a means to reflect on society and the artist was a social commentator. These contrasting ideologies reflected the way that each group of artists approached the open studios events and governed their relationship within each building. Eastworks' business orientation focused on efficiency, while One Cottage Street's non-profit status meant that many artists focused on collaboration and creative exploration over straight sales. Year after year, the Open Studio sales have nonetheless expanded. More people around town are involved and more artists and artisans open their doors to the public in general. The organizing is still "organic" though the coordination of efforts happens among the constituencies of each building, and not as a collaborative effort of each group coming together through their representatives. Efforts are often duplicated or ignored as a result.

All of the art-related organizing around Easthampton and the emerging businesses, such as the new art galleries, a framing shop, and a high-end design and decoration store, signal a transformation. They suggest that local artists and artisans have begun to act within the town as a group with common interests and goals. Art-related events and public art exhibits have grown in number and call attention to the artists and their activities. Local government has gradually come to recognize how artists impact the local economy, and how they draw people to town to spend money. As a consequence, bridges are being built among the varied constituencies in Easthampton, and this is helping to craft a new cultural economy and local identity. Local authorities are beginning to take pride in the presence of artists and artisans. This was especially evident at the 2005–2006 local art charrettes, and later at the 2009 Easthampton Master Plan meeting.

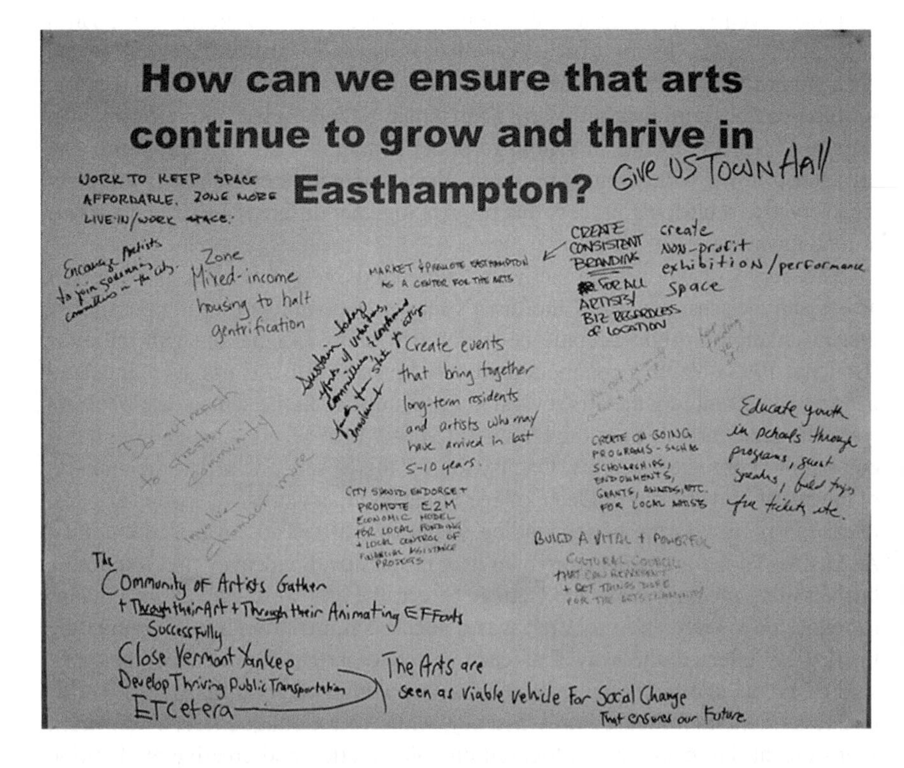

Figure 7.7 "Wish List" written by participants in the charette event, 2005. Among the items was the stated need to have ownership, control or access to the city's old Town Hall ("Give us Town Hall" statement in the upper right hand corner). Almost five years later this came through the collaboration of city officials, local business, the local Art's Council and multiple artists around town

The goal of the former was to start a citywide conversation to clarify the role of the artists and artisans in the context of the socioeconomic renaissance of Easthampton. The subsequent Master Planning event represented the first time in the history of the town that the arts and culture were included in the planning processes designed to chart the future of socioeconomic development.

The Creative Class: The Grassroots Unfolding of a Cultural Economy

From a theoretical perspective, this chapter is about the tools and methods needed to follow the local unfolding of a "creative class". Such approaches illuminate the disruption of old social patterns and the co-constructing processes that unfold when residents (re)organize their environments to meet new challenges and opportunities. In contrast to conventional perspectives that assume the a priori existence of the creative class, this work articulates the creative class as the bottom-up collective outcome of a particular set of local socioeconomic dynamics that become institutionalized over time. Conventional approaches presume the creative class as something to be found and measured. This work presents the phenomena as something that unfolds and must be studied from the actors' point of view.

Generally accepted views pose the creative class as a picture-perfect result of the concurrence in time and space of three elements: creative individuals, "good" organizations, and "good" environments. This project argues that the creative class is a contingent and continuously evolving process, whose emergence at any one point in time may or may not be sustainable, regardless of the existent organizations and environmental conditions at the time of their concurrence. The case of Easthampton and other smaller post-industrial cities also suggest that organizations and environmental conditions are not distinct entities that exist a priori. Rather, they co-construct each other through socioeconomic processes in particular spaces over time. This challenges the current underlying premises about what kinds of places attract a creative class, who can be a member of this class, and what role the creative class' plays in the community.

The self-organizing of artists and artisans and local social networks are never the result of straightforward, inevitable or necessarily positive processes. Relationships among community members run in every possible direction and are subject to power struggles and ideological battles, as was the case with the open studio event. Furthermore, even in small localities such as Easthampton, members of the creative class are not homogeneous, sharing a single network or identity. They are a collection of unique individuals, who may share some interests some of the time, and disagree at other times. Their multiple interests and power struggles are intertwined in their everyday life dynamics, and only their enacting of a balance between their individual goals and their communal interests ensures their sustainability as a creative class network in town. These ideas suggest that creativity is not only a lonely professional exercise undertaken by individuals trained to be creative but also a communal endeavor achieved through interpersonal dynamics. Hence geographical agglomerations of creative individuals do not constitute a creative class unless they are linked to each other and enact their individual creativity in the context of the collective. Finally, this creative class to be of local value should frame their actions in the benefit of the hosting community.

The U.S. census data, while useful to show regional and national tendencies, is too broad to study the concrete experiences of smaller post-industrial communities such as Easthampton. The U.S. Census data handling protocol puts a lower limit on the size of the census track to be reported at 25,000 residents when dealing with detailed data such as occupations. This limit on the reporting size, set to respect the privacy of the respondents, blinds the researcher to happenings at the ground level. The findings in this study about the evolution of the creative economy in Easthampton would not have been possible without the use of a ground level mixed-methods approach, and the complementary tools of institutional ethnography and social network analysis. These methodologies allowed the documentation of a grassroots organizing process enacted by local artists and artisans.

Future Research

This chapter provides one example of a small community revisiting its identity and setting itself on a path to socioeconomic regeneration that incorporates local arts and crafts, not as the ultimate economic goal, but as one social means to foster and channel creativity as an engine of local renaissance. Change in Easthampton came as a transformative progression that first and foremost altered the community's sense of self and internal planning processes. City officials did not guide this process; rather, it was the result of a long-term, unplanned, grassroots movement initiated by local residents. Creativity became the local currency, and artists and artisans became one vehicle for generating local income linked to the town's changing identity. This socioeconomic transformation may be seen as an example of the unexpected outcomes that have been consistently and prescriptively cited in the extant creative class literature (Gibson and Graham 1992; Florida 2002b; Markusen 2005; DeNatale and Wassall 2006; Donegan, Drucker, Goldstein, Lowe and Malizia 2008). But is this truly a casual unexpected outcome or a recurrent pattern in cities of similar size? Are these processes replicable in other small communities or purely local and internal to Easthampton? How far can self-help go towards regeneration within small post-industrial cities, and when do cultural economies need to be "jump started" through the infusion of government and other outside investment? These are some of the questions that require further research. The answers will help us to better understand the socioeconomic dynamics and potential redevelopment trajectories of the varied responses to deindustrialization that involve creative work in non-global cities.

References

Adamic, L.A., Büyükkökten, O. and Adar, E. (2003), "A social network caught in the web", *First Monday*, 8:6. Available at http://www.firstmonday.org/issues/issue8_6/adamic/index.html [accessed 30 July 2007].

Agar, M. (1996), *Professional Stranger: An Informal Introduction to Ethnography*, 2nd Edition (London, UK: Academic Press).

Aggestam, M. and Keenan, J. (2007), "Contraversations constructing conflicts: lessons from a town-gown controversy", *Business and Society* 46:4, 429–56.

Aiello, B.L. (2004), "The changing tastes of a community: gentrification and the taste hierarchy of Northampton", *Annual Meeting of the American Sociological Association* (Hilton San Francisco and Renaissance Park 55 Hotel, San Francisco, CA, 26 May 2009).

Asheim, B. and Hansen, H.K. (2009), "Knowledge bases, talents, and contexts: on the usefulness of the creative class approach in Sweden", *Economic Geography* 85:4.

Boucher, C. (1998), "Address for the arts Easthampton becomes home for growing number of artists", *Daily Hampshire Gazette* (Northampton, MA) – Wednesday, 19 August 1998, 17. Record Number: 125F79EE3972D538.

Cowie, J. and Heathcott, J. (eds) (2003), *Beyond the Ruins: The Meanings of Deindustrialization* (Ithaca, NY: Cornell University Press).

DeNatale, D. and Wassall, G.H. (2006), *Creative Economy Research in New England: A Reexamination* (Boston, MA: New England Foundation for the Arts).

_______ (2007), *The Creative Economy: A New Definition* (Boston, MA: New England Foundation for the Arts).

Denzin, N.K. and Lincoln, Y.S. (2000), *The Discipline and Practice of Qualitative Research*. In N.K. Denzin and Y.S. Lincoln (eds), *The Handbook of Qualitative Research*, 2nd Edition (Thousand Oaks, CA: Sage Publications), 1–28.

Devault, I.A. (1995), *Sons and Daughters of Labor: Class and Clerical Work in Turn-of-the-Century Pittsburgh* (Cornell University Press).

Donegan, M., Drucker, J., Goldstein, H., Lowe, N. and Malizia, E. (2008), "Which indicators explain metropolitan economic performance best? Traditional or creative class", *Journal of the American Planning Association* 74:2, 180–95.

Easthampton Massachusetts: A Historical Website for a Historic Milltown (last updated 14 November 2010). Available at http://www.historiceasthampton.com/Home.html (home page) [accessed 1 August 2009].

Eastworks History (first published online 2006). Available at http://eastworks.com/history.shtml [accessed 20 November 2009].

Feser, E.J. and Bergman, E.M. (2000), "National industry cluster templates: a framework for applied regional cluster analysis", *Regional Studies* 34:1, 1–19.

Florida, R. (2002a), "Bohemia and economic geography", *Journal of Economic Geography* 2:1, 55–71.

_______ (2002b), *The Rise of the Creative Class*, 2004 Paperback edition (New York, NY: Basic Books).

Freeman, L.C. (1996), "Cliques, galois lattices, and the structure of human social groups", *Social Networks* 18, 173–87.

_______ (2000), "Visualizing social networks", *American Statistical Association 1999.* Proceedings of the Section on Statistical Graphics, 47–54.

Genzuk, M. (ed.) (2003), *A Synthesis of Ethnographic Research* (Los Angeles, CA: Center for Multilingual, Multicultural Research, Rossier School of Education, University of Southern California).

Gibson, C. and Kong, L. (2005), "Cultural economy: a critical review", *Progress in Human Geography* 29:5, 541–61.

Gibson, K. and Graham, J. (1992), "Rethinking class in industrial geography: creating a space for an alternative politics of class", *Economic Geography* 68:2, 109–27.

Granovetter, M. (1985), "Economic action and social structure: the problem of embeddedness", *American Journal of Sociology* 91:3, 481–510.

_______ (1991), "The social construction of economic institutions", in A. Etzioni and P.R. Lawrence (eds), *Socio-Economics: Toward a New Synthesis* (Armonk, NY: M.E. Sharpe), 75–81.

_______ (1992), *Economic Action and Social Structure: The Problem of Embeddedness*, in M. Granovetter and R. Swedberg (eds), *The Sociology of Economic Life* (Boulder, CO: Westview Press), 53–81.

Granovetter, M. and Swedberg, R. (eds) (1992), *The Sociology of Economic Life* (Boulder, CO: Westview Press).

Handcock, M.S., Raftery, A.E. and Tantrum, J.M. (2007), "Model-based clustering for social networks", *Journal of the Royal Statistical Society* 170:2, 301–54.

Hartford, W.F. (1990), *Working People of Holyoke: Class and Ethnicity in a Massachusetts Mill Town, 1850–1960* (New Brunswick, NJ: Rutgers University Press).

Kweit, R.W. and Kweit, M.G. (1998), *People and Politics in Urban America*, 2nd Edition (Abingdon: Routledge).

Levin, D.Z. and Cross, R. (2004), "The strength of weak ties you can trust: the mediating role of trust in effective knowledge transfer", *Management Science* 50:11, 1477–90.

Lin, N. (2004), *Social Capital* (New York, NY: Routledge).

Madison, D.S. (2005), *Critical Ethnography: Method, Ethics, and Performance* (London: Sage Publications).

Markusen, A. (2005), "Urban development and the politics of a creative class: evidence from a study of artists", *Environment and Planning* 38:10, 1921–40.

Markusen, A. and Schrock, G. (2006), "The artistic dividend: urban artistic specialisation and economic development implications", *Urban Studies* 43:10, 1661–86.

Marshall, A. (1890), *Principles of Economics* (New York, NY: Macmillan and Co).

New England School of Architectural Woodworking, Available at http://www.nesaw.com/about-us/School_History.html [accessed 1 August 2009].

NGA Report. (2002), *A Governor's Guide to Cluster-Based Economic Development* (National Governors Association (NGA)).

Oehler, K., Sheppard, S. and Benjamin, B. (2006), *Mill Town, Factory Town, Cultural Economic Engine: North Adams in Context* (North Adams, MA: Center for Creative Community Development).

Office of Economic Development Mass Government. (2001), *Toward a New Prosperity* (Boston, MA: Commonwealth of Massachusetts).

Osorio, A.E. (2003), Interview with Ron Bittel, retired President of Riverside Industries.

Ray, P.H. and Anderson, S.H. (2001) *The Cultural Creatives: How 50 Million People Are Changing the World* (New York, NY: Three Rivers Press).

Smith, D.E. (1987), "A sociology for women". Chapter Two in *The Everyday World as Problematic* (Boston: Northeastern University Press).

Smith, M. (2007), *Implicit Affinity Networks* (Provo, UT: Brigham Young University).

Sommer, D. (2005), "Art and accountability", *Review: Literature and Arts of the Americas* 38–2:71, 261–76.

Stevenson, W.B. and Greenberg, D. (2000), "Agency and social networks: strategies of action in a social structure of position, opposition, and opportunity", *Administrative Science Quarterly* 45:4, 651–78.

The Master Plan Advisory Committee of Easthampton. (1987), *The Easthampton Master Plan Update* (Easthampton, MA: Land Use, Incorporated).

U.S. Census Bureau, (2009), Easthampton, Massachusetts 2009 Census Data. Available at http://www.census.gov/ [accessed 1 June 2010].

Wakita, K. and Tsurumi, T. (2007), *Finding Community Structure in Megascale Social Networks.* Available at http://arxiv.org/PS_cache/cs/pdf/0702/0702048v1.pdf [accessed 31 May 2009].

Your Riverside Connection. (Fall 2003), The thirty-fifth anniversary edition of a periodic newsletter.

Zukin, S. (1989), *Loft Living: Culture and Capital in Urban Change* (New Brunswick, NJ: Rutgers University Press).

Chapter 8
Creative Communities and Everyone Else: Towards a Dialogic Understanding of Creativity after Liverpool's Capital of Culture Year

Susan Fitzpatrick

Introduction

The city of Liverpool, has been subject to shifts in its economic fortunes from the Industrial Revolution through the post war Keynesian municipal state, to the present entrepreneurialization of urban governance (Harvey 1989). As with other post-industrial urban centres in North America and Western Europe, the contemporary network of governance in Liverpool aspired to accommodate what Richard Florida terms the "Creative Class" (2002). Liverpool City Council and its partners sought to re-position the city as a center for culture, knowledge and creativity.

The most explicit performance of this latest experiment with city identity is the European Capital of Culture, a year-long event hosted by Liverpool in 2008. The official program included a large number of performances, exhibitions and other cultural interventions. It also gave a form and focus to conversations about the nature of public life in the city. This chapter is based on research that I conducted in Liverpool on the community involvement dimension of the Capital of Culture between 2007 and 2009. On the one hand, I will argue that the event gave rise to productive conversations about governance, democracy, uneven development, and cultural policy, amongst those acting within the city's existing spaces of creative activity. This includes the groups, organizations, venues, and communities of interest whose activities constitute Liverpool's existing creative infrastructure. On the other hand, the official event's strategic direction imposed limits on the possibilities of creative action by burdening the idea of "being creative" with the responsibility of bringing about a rational and communicative public sphere peopled by individuals rich in specific kinds of social and cultural capital. I will argue further that in the discursive construction of this imagined public sphere, the official event failed to take account of the existing cultural fabric of the city, and instead constructed a series of identity-based categories into which existing activities should proceed. In other words, the official event brought with it an environment of expectation into which people and activities should be ordered. In considering the potentially contradictory effects the event has wrought, this chapter contextualizes qualitative interview data from those working within these "existing spaces of creativity" and those who worked within the temporary branch-organization of Liverpool City Council, the Liverpool Culture Company, whose remit was to "deliver the official programme of 2008" (Liverpool City Council 2003).

The groups I describe as constituting existing spaces of creativity, and who I contacted in the course of data collection, are:

1. Yellowhouse – A charitable organization that provides young people with the equipment to develop film and theater projects;
2. The Royal Standard – An art collective who run a multi-purpose project space;
3. Tenantspin – A community media project which curates and broadcasts web based video;
4. A Writers Group for people living with Multiple Sclerosis (MS) and their carers [this group were asked by the Liverpool Culture Company to participate in the Community strand of events that formed part of the official program of activities for the Capital of Culture year];
5. The Picket – A music venue and music academy which provides young people with the courses, equipment and spaces to develop skills in music production and performances;
6. Fazakerley Federation – A Community Centre serving the Fazakerley area of Liverpool, which is situated on the northern periphery of the city. The centre is run independently of local government funding;
7. A club for the over 50s – This member-funded social group meet for weekly afternoon socials in their local community hall [this group were also approached by the Liverpool Culture Company to participate in the community strand of the Capital of Culture Year activities].

To further explore some of the consequences of these complex effects of the Capital of Culture event, this chapter reflects on the work of the socio-linguist Mikhail Bakhtin and his development of the idea of dialogic communication (2006). Dialogism acknowledges the possibility of public life as consisting of many voices, formal and informal, which are relational and responsive, subject to difference, contradiction, and mis-understanding, as well as to rational responses. The responses of the different interest groups who participated in the research are analysed through their dialogic character. The chapter closes by looking at where this perspective takes us in terms of the understanding the communicative ethics of large civic celebrations that are often the centrepieces of a creativity economy agenda.

Liverpool in Context

Liverpool grew in size and wealth as a port city because of its strategic importance to the transatlantic slave trade. The city continued into the late Victorian era as an important node in the network of global trade, as it was a "principal exit point for British manufactures bound for the crucial specific markets of India and Pacific Asia, as well as a major warehousing and transit port for imported produce from 'the East' generally" (Webster 2008: 44). Liverpool experienced extensive damage during the Blitz of WW2, and in its post-war years, has seen a spectacular economic and population decline. This was exacerbated by the partial closure of the city's Docks in the early 1970s, a move which sought to make way for the increasing mechanization of docking and haulage.

In subsequent decades, the city struggled to develop alternatives to the specifically colonial mode of production that dominated the city's economy for two centuries (Wilks-Heeg 2003). Many of the city's residents moved to new towns, such as Skelmersdale, Runcorn, and Warrington, in areas neighboring Liverpool in West Lancashire and Cheshire. In 1931, the population of the city was 846,000; in 2001, it stood at 439,000, and then declined to an estimated 435,500 by 2007 (Liverpool City Council 2010).Aiming to bring Fordist style economic stability to the local economy (a decisive move away from the non-nationalized, casualized dockside and maritime industries), central government sought to re-locate large scale motor manufacture to Liverpool . The Standard Triumph Motor Company shifted some of its operations from the established home of UK motor manufacturing, the West Midlands, to a plant in Speke in the South of Liverpool, in 1959. The plant was subject to chronic under-investment in tools and equipment, and management inefficiency (Beynon 1978). The Management of the plant never committed the Liverpool operation to full productivity. The factory at Speke in Liverpool was capable of producing 100,000 motors per year, but in its brief history, the maximum production of the plant was 30,000 (Marren 2009). The plant closed in 1978, the first of several plant closures to further dent Liverpool's dwindling employment base. This reflects the city's problematic relationship both with central Government-directed policy experimentation and private industry who were not ready to commit to investing in the city. From the 1980s onwards, the focus turned to the question of how the service sector would form the basis of the local economy.

In the past 20 years, development and regeneration have focused on the city center, rather than addressing the deficiency of the city's transport infrastructure, which has resulted in the lowest income areas of the city (to the North of the center) being poorly connected to employment hubs, including the airport, and surrounding business and retail parks in the south. Liverpool City Council, and its regeneration partner, Liverpool Vision, pursued a policy of defining the centre in terms of leisure, tourism and retail opportunities. In addition to hosting the European Capital of Culture, a second event occurred in 2008 which reflects the desire to re-imagine Liverpool as a thriving neoliberal city. The opening of the

Liverpool One retail development represented the privatization of the public realm on an unprecedented scale within the UK. The previously public land on which it was built was leased to the developer Grosvenor for 250 years. This process of transferring the land meant replacing traditional rights of way with "public realm arrangements" policed by private security guards known as "quartermasters" or "sheriffs" (Minton 2006). This transfer received much the same treatment in the local press as the news that the Capital of Culture was to be held in the city. There was a sense of pride, entitlement, even relief (Echo 2003, 2008). Abounding in the narratives around these two events is what Coleman has termed "urban patriotism" (2008). These occurrences were rendered eventful through discursive constructions suggesting that the events symbolized a long overdue, and much deserved, shift in the competitiveness, fortune and spirit of the city.

The remainder of this chapter focuses on the community involvement strand of events within the official program, called Creative Communities. This program initiated a discourse of civic involvement during the build up and throughout the host year. Narratives of neighborliness and civic responsibility were used by artists to enter into communities and facilitate interventions with residents in the shape of poems, plays, videos, animations, songs, drawings, paintings and other artefacts (Four Corners 2008). From 2007 to 2010, I undertook in-depth interviews with a total of 31 key actors in Liverpool, who represent a range of perspectives. All were involved in designing, delivering, or participating in either the Creative Communities program, or other forms of regeneration or community arts activity in the city.

I was interested in the extent to which the official program initiated a critical dialogue in the city around the ethics of representation and power. Official Capital of Culture discourse continually represents the event as an unprecedented moment of change, a decisive turn away from municipal managerial style of local government, towards an entrepreneurial style of governance in which networks of power enmesh state and private sector interests. This fails to acknowledge the nature of the critical dialogue that emerged amongst existing groups of interest in the city. The failure led to a crisis of legitimation for the event, as many of my informants discuss the event's inability to engage with existing groups, organizations and communities of interest who make up the city's existing spaces of creativity.

The chapter presents data gathered from key actors based in a range of sites of cultural production in Liverpool, and frames my argument that the event was interpreted and produced according to a confounding array of contexts and contingencies. This array constitutes the dialogic aspect of the Capital of Culture year and builds on Anderson and Holden's insightful work on Liverpool's Capital of Culture, in which they present the event as multiple "dis-jointed assemblages":

> There is not a single initial event and subsequent "interpretations" of it. There are and will be other ways in which the event happens (e.g. the event as disappointment, the event as the continuation of the same) … (Anderson and Holden 2008: 155–6).

Capital of Culture cannot be reduced to any one interpretative context, or indeed dichotomized between two oppositional contexts. A public event of this nature is, and remains multiple, although this multiplicity was marginalized by those who designed the official program for 2008.

This chapter critiques assumptions bound up with culture-led regeneration by arguing that the public sphere is a becoming and generative space of discourse. This approach acts as a counterweight to the breadth of impact-related literature on the large scale, urban, cultural event. Such research is oriented according to the logic of the event itself by confining analysis to answering how the event achieved the aims it set for itself (Impacts August 2010). Research into event-led regeneration might start by asking more fundamental and relevant questions such as "why is a temporary event being framed as a way to address deep seated social inequalities in a locality?", or "how is the event produced (as a meaningful occurrence) by different communities of interests in the city?".

The European Capital of Culture

The inaugural "European City of Culture" was held in Athens in 1985. At the time, it was the sum of various efforts within the European Union to "enhance the profile of European construction at the level of popular consciousness" (Psychogiopoulou 2008: 13). Alongside the wariness of attempts to foster a common European identity, was a desire to preserve what the Hague Summit of Heads of Government, in 1969, termed Europe's "development, culture and progress" (European Union 1985). The discourse of the event at that time tended to position Europe as "exceptional" in its heritage and culture (Gold and Gold 2000). The event was intended to "improve communication amongst the artists and intelligentsia of Europe" (Mecouri in Gold and Gold 2000: 221). In its first five years, the European Capital of Culture took place in the major historic European cities such as Athens, Florence and Berlin. The event at this stage was small in scale and concentrated on domestic audiences; it commanded little from the EU in terms of funding (Gold and Gold 2008). The rationale of the European City of Culture shifted when Glasgow became host city in 1990. The event was re-branded as a vehicle for failing post-industrial urban economies to intensively reshape their image to attract more capital from tourism, retail and cultural institution-based outlets. The European Union sought fresh impetus into the notion of a common European identity in the 1990s, driven by the economic desire to mitigate the effects of Fordist rationalization. The City of Culture became a key strategic event in the success of a European Union economy, and the discourse of the cultural identity within the EU shifted to a celebration of diversities, a "culture of cultures" (Psychogiopolou 2008).

Following 1990, when the event was held in Glasgow, a pattern emerged in which cities with small- to medium-sized populations, such as Cork 2005, and Patras in Greece, 2006, played host to the European Capital of Culture. These cities have broadly followed the rationality put forward in the extensive research document published by the European Commission in 2004. This study suggests the event's chief objectives are to raise the international profile of the host city, attract visitors, and enhance the pride and self-confidence of those residing in the city (Palmer Rae 2004). These objectives, though generated by research into the effects on cities that hosted the event in the past, have become directives for what the event ought to do when hosted by cities in the future. They provide guidance about how to cultivate an environment of aspiration in host cities, and anticipate certain conditions for a public sphere within which a successful event can occur.

Capital of Culture advocacy research tends to focus on the effect of intensified and positive media representations of the city, and how these changed the way residents feel about the city (Palmer Rae 2004; Garcia 2005; Impacts August 2010). This maintains the construction of the Capital of Culture as an external effect, which the resident population is subject to, but cannot shape or produce. It is therefore pertinent to ask how the event is rendered meaningful within existing spaces of creativity in the city. What are the modes of communication between the official event and these existing spaces? How do existing creative spaces such as independent art collectives, writers groups, community centres and music venues come to produce the event?

Dialogic Communication

Dialogism, a nuanced socio-linguistic concept developed by the Bakhtin circle in the late 1920s,[1] can add to our understanding of how these events are rendered meaningful within existing spaces of creativity. Dialogism allows us to analyze the "communication situation" as we encounter it in our everyday lived experience. It suggests that:

> all language and existence is relational and responsive. Each speech or communicative act is dialogical through its responsiveness to other past, present and future speech acts (Holloway and Kneale 2000: 1).

Without wishing to place too strong an emphasis on the idea of dialogism as the communication act between two people, Hirschkop describes the context of the

1 For background of the Bakhtin circle see Holloway and Kneale (2000); Holquist in Bakhtin (1986: ix–xxii).

concept in a way that begins to open out its wider importance to language and the social world:

> It refers to what other writers would call the intersubjective quality of all meaning: the fact that it is always found in the space between expression and understanding and that this space – the "inter" separating subjects – is not a limitation but the very condition of meaningful utterance (Hirschkop 1999: 4–5).

The illuminating concept of Dialogism, as presented by the Socio-linguist Mikhail Bakhtin (2006), allows an enquiry into the Capital of Culture that enables us to consider the event as something that arrives into a pre-existing (and multiple) public space that will not allow itself to be re-constituted into a re-formed blank space onto which ideal modes of interaction can be inscribed. The question of "who speaks" and "under what conditions … these voices emerge" is crucial to understanding the power relations that inhere within "flagship" civic celebration events. The supposition that "you cannot force everyone to speak the standard or official language, […] But the very existence of this language, and its relative position in political and cultural life, alters what is possible for every other kind of language" is a useful place to begin (Hirschkop 1999: 256).

In charting the historical development of the novel as an art form, Bakhtin argues that a centripetal force is apparent in language, that seeks to perpetually overcome the heteroglossia (many voices) in order to get closer to a unitary language, one that insures the "maximum of mutual understanding in all spheres of ideological life" (2006: 271). I use these insights to understand how the official language in the Capital of Culture policy and strategy documents seeks to unify and centralize, whilst the existing heteroglossia, that is, the existing groups in the city undertaking a variety of creative endeavors, make their own demands on language simultaneously with the attempts by the official event to control and unify public life. Dialogism offers a potential to understand how we might extract meaning from this confounding process of many voices making demands upon the discursive field of creativity.[2] It presents a useful context in which to unpack the complexities that account for the "chains of meaning" involved in Liverpool '08. Bakhtin's dialogism attempts a "phenomenology of practical doing" rather than pursuing an "all-encompassing explanatory system" or theorizing social action as a process which conforms to linguistic precepts offered by official discourse

2 Somewhat pre-empting the project of post-structuralism, Bakhtin saw, in the crowded and competing multiple voices of subjects in everyday life, a fertile ground to develop ways of thinking about the social world. Bakhtin's pursuit of "the dialogic" in his account of aspects of European literature calls into the question the structuralist ethos of treating language as a closed formal system, in the sense of giving "linguistic structure pride of place among the facts of language" (Saussure 1983: 10) at the expense of other, more chaotic, unruly, elements of linguistic experience (Hirschkop 1999).

(Gardiner 2004: 32). It offers ways of forming new knowledge about culture-led regeneration and encourages the development of alternative perspectives on creativity in everyday life. These perspectives can, in turn, address the inequalities that arise out of power relations implicit in many existing culture-led regeneration strategies.

The following section explores the official language of Liverpool's European Capital of Culture in 2008 to see how it altered what was possible for the existing, resident, on-going and mutable groups of interest that formed their own "spaces of creative discourse" prior to the 2008 event. The term "spaces of discourse" describes spaces of articulation that have formed as a result of people with similar interests, needs and desires coalescing around specific practices. The Capital of Culture official event in Liverpool highlights these existing spaces of discourse, by arranging artist interventions which interceded in the habitual activities of the groups in order to promote the event as one based on community participation. These interventions formed a strand of the official Capital of Culture program and were collectively named "Creative Communities" in official Liverpool '08 literature.

The Capital of Culture and the Re-imagining of Liverpool

The European Capital of Culture was an unambiguous vehicle for Liverpool's re-positioning as a competitive city, able to make the claim that its economy was moving away from stagnation and dependency to that of a self-sustaining city based on economies of leisure, retail and knowledge. This mirrors Markusen's notion of the politics of decline (quoted in Parkinson 1990), where failing economies look to external government agencies for resources because of internal failure to govern adequately. Since Fordist rationalization of the late 1960s, Liverpool has been a test bed for every centrally allocated urban policy experiment, including Community Development Project, an Inner Area Study, Urban Development Corporation (see Parkinson 1990), Enterprise Zone, a City Challenge Programme, and most recently the European Union structural fund Objective One[3] (Couch 2003; Jones and Wilks-Heeg 2004).

Historically, the social structure of the city has suffered from a series of divisions along the lines of class, sectarianism and race. Successive administrations' entrenched leadership style prevented the development of effective alliances between leaders and community groups, and failed to address the social divisions of the city (Mangen 2004). Another significant narrative in the story of Liverpool's post-war civic leadership, was its pursuit of municipal socialism at the height

3 Objective One funding is allocated by the European Union. It is the highest priority designation for European aid and is targeted at areas where prosperity, measured in Gross Domestic Product (GDP) per head of population, was 75% or less of the European average (Objective One 2012).

of Thatcherism in the 1980s. Between 1983 and 1987, the Trotskyist faction within Labour called Militant Tendency ran the City's Labour Council. As with Manchester, Sheffield, and London, city leaders in Liverpool engaged in open confrontation with national government (and in Liverpool's case, the national Labour Party leadership) over capital and revenue spending limits, changes to public taxation, the abolition of metropolitan government, the privatization of local government services and the break up of local authority housing and education (Parkinson 1990).

The logic of "crisis displacement" played a part in how Liverpool was understood by the rest of England after the city's intensive period of economic decline following the world recession of the early 1970s. Habermas' development of the term (1976) suggests a process by which a crisis originating from within the economy (such as market failure and the flight of capital) is transferred into the political realm of the state (see Jones and Ward 2002). In the case of Liverpool, the city itself was interpreted as the problem, not the crisis tendencies of global capitalism. This is apparent in the othering scenario that the city has been subject to within the UK. Media representations from the 1970s onwards tended to perpetuate an image of Liverpool as a spatial intensification of the sentimental, uneducated and idle underclass by focusing on narratives of unemployment, rioting, and football hooliganism (Boland 2010). As a city in economic decline within a neoliberal climate of competitive cities and regions, the Capital of Culture official discourse sought to draw a line underneath the othering process by insisting that Liverpool could take its place amongst other northern cities as a centre for knowledge, retail and creativity.

The process of cultural "othering" that Liverpool experienced is complicated by the suggestion that, historically, residents of the city sought to distance themselves from the mores of the rest of the country. Historian John Belchem has theorized this as a "proverbial exceptionalism":

> a "difference" which extends far beyond historiographical discourse. Liverpool's apartness, indeed, is crucial to its identity. Although repudiated by some as an external imposition, an unmerited stigma, Liverpudlian "otherness" has been upheld (and inflated) in self-referential myth, a "Merseypride" that has shown considerable ingenuity in adjusting to the city's changing fortunes (Belchem 2000: xi).

A similar language appears throughout the official Capital of Culture documents in order to position the resident in the role of amicable host of the event. Whilst official Capital of Culture discourse does not refer explicitly to any given historical episode of the city's recent past, it strategically incorporated this "otherness" in the process of regulating Liverpool identity as a singular and finalized brand:

> Liverpool is not a chocolate box city. It is unconventional, pioneering, unruly, unpredictable … In the late 20th Century, Liverpool had to draw on its enormous capacity for resilience and re-invention … this has been a response which has been distinctively Liverpudlian … combative, comic, determined, laced with a healthy cynicism. It is a distinctive brand, which attracts worldwide recognition (Liverpool City Council 2003: 101).

These exclamations tacitly regulate the subjectivity of the Liverpool resident acting the role of the "Scouser"[4] on the "stage" of Capital of Culture. The numerous social and political problems that the city has encountered over the past 40 years are woven into a narrative that places Capital of Culture as the long awaited turning point, which the city deserves, and where the historic problems of civic leadership are glossed over:

> Cities in the future will be differentiated not by their physical environment but by the quality of experience they offer. Liverpool is releasing its latent energies, moving completely away from old style city governance to a new model where creativity is at the core of innovative regeneration. Ours is a creative citywide agenda. A liberating agenda, empowering the people of the city and helping to unleash their creative potential. *Liverpool is Changing* (Liverpool Culture Company ((LCuC)) 2006: 11, their emphasis).

The perspective of an urban elite "delivering" culture to groups, which they homogenize into assets or stakeholders, is the central subject of contest amongst many of the actors interviewed for this research. Since 2003, official discourse surrounding Capital of Culture has sought to formalize and regulate its position as deliverer of culture to the city by establishing an explicit link between creativity and the raising of social capital amongst groups deemed in most need of help.

Creativity to the Rescue: The Capital of Culture Participant

Liverpool City Council set up and funded a separate company to prepare the city's bid for the 2008 Capital of Culture. In 2003, its remit changed to delivering the aims of the official culture program. The Liverpool Culture Company was established as a private limited company in 2000 and dissolved in 2009. The City Council controlled the budget for Capital of Culture, but internal confusion over the roles of the two bodies resulted in the City Council seeking independent legal advice and produced a memorandum of understanding between the two organizations (Mayor for Liverpool 2010). This continues the pattern of an unclear distinction between administrative and political arms within the City Council, which has characterized it since the 1970s (Parkinson 1990).

4 The informal collective pronoun for those hailing from Liverpool.

Throughout its nine-year existence, the Liverpool Culture Company Executive Management and its Board of Directors was subject to regular changes in structure and personnel. An enduring part of the structure was the Creative Communities team. Their remit was to increase participation in the official Capital of Culture program amongst residents of the city (DTZ Debenham Tie Leung 2005: ii). The report entitled "Building the Case for Creative Communities", authored by the consultancy firm DTZ Debenham Tie Leung Pieda Consulting Ltd, clearly identifies target demographic groups in receipt of state welfare, for participation in the successful Capital of Culture:

> Three interrelated "groups" can be characterised as symptomatic of non-engagement and exclusion:
>
> Local concentrations of Incapacity Benefit recipients,
> Local concentrations of Income Support recipients, and
> NEET – 16 to 24 year olds Not in Education Employment or Training.
>
> (DTZ Pieda Consulting 2005: vi)

Through its discursive regime of "building positive social capital" (DTZ 2005 section 3: 16) Creative Communities promoted the idea of individuals as self sufficient, not reliant on the state, and preferably employed. The DTZ report does not suggest that participation in Creative Communities' program will lead to paid employment, which might leave the program vulnerable to the suggestion that it failed. Rather, it produces and maintains the importance of the capability of the individual to manage his or her own affairs within an idealized communicative civic sphere:

> … *at their best* creative activities have the following characteristics: people take part, actively – They are not passive recipients (DTZ 2005 Section 3: 11 their emphasis).
>
> arts activities … are particularly suited to an adult education approach in which participants make decisions and choices and take responsibility for their actions and outcomes (DTZ 2005 section 6: 32).
>
> Cultural activity in urban regeneration has been evaluated in case study form demonstrating that they act as a motor for individual and community development bringing important benefits such as … developing self confidence, enhancing organisational capacity, supporting independence (DTZ 2005 section 6: 33).

Care is taken to build a case for establishing conditions to support a public sphere where everyone is able to thrive without the welfare state. In addition, no reference is made to the job market nor any of the conditions which shape it. There is a determined and narrow focus on the provision of cultural capital to the individual.

The project of establishing conditions for the emergence of the responsible, active, independent citizen, directly links to the neoliberal desire for less government, where lack of jobs and other infrastructure become the responsibility of the individual. This aim of encouraging creative expression in people to bring about a change in their expectations of the job of local government is clearly signposted in an interview with a Neighbourhood Manager within the City Council:

> C17: But the culture program itself is more than that it's about reaching into the communities, it's about getting people to use culture to address particular issues in their lives ... The other issue that relates to that which is that it has a legacy of the public sector doing everything, so the Council, or the "Corpy" as its known locally, did everything, from transport, social services, everything .... I'm talking about residents, residents will say "oh, the Corpy will do that for me, the Corpy does it, you know, why aren't you doing that for me?", you know "why aren't you making that decision for me" you know, quite frankly, you're an individual, you're capable of making your own decisions, and you're capable of being independent, but quite frankly, there are communities in this City, who have a reputation for being particularly labour intensive in terms of the support they need, and Council Members, because this is the ways its been...we haven't got the resources, the Council's resources are finite, we can't contentedly live that way, we need to raise people's expectations of themselves, and their confidence levels, and with some capacity building, and some confidence building, gradually release the apron strings (C17 Liverpool City Council Neighbourhood Manager).

Care is taken to avoid any explicit link between participating in Creative Communities activities and finding employment. Instead, the language of "empowerment" and of "finding voice" on a particular issue becomes the achievable end result of Creative Communities intervention. The relationship between job creation and doing creative activity remains ambiguous. The lack of any direct link between the two is obfuscated by the emphasis placed on the emancipatory event of self-expression, after which, as a socially responsible, active citizen, the resident is then assumed to be able to pursue employment without further recourse to state intervention.

Another pervasive trope of the official discourse of the Capital of Culture is an insistence upon the unifying potential of the official culture program. The language of participation, integration, increasing involvement, cohesion, engagement and "mainstreaming" of the arts into a "wider range of policy contexts" (Liverpool Culture Company/ DTZ 2005: 14) suggests an evangelistic belief that a unified public sphere is first of all achievable, and that the Creative Communities program will deliver that vision:

> The programme objectives can be summarised in a single aim, that is: to bring more people into circulation through the arts, applying creative activities more widely than was previously the case and linking them more clearly to the regeneration process. As such the objectives are strongly aligned to the development of human and social capital through positive engagement […] (DTZ 2005)

> it's a well-known fact that Liverpool has lots of diverse groups, but there's been issues about representation of those groups on various things. And to give an example there will be the World in One City next year so it's very much about insuring that BME [black and minority ethnic] groups, gay and lesbian groups, and deaf and disabled groups play a full active part in next year [2008] (Creativity Diversity Manager Liverpool Culture Company).

In the above quotes, an image of the Habermasian civic sphere begins to emerge, in which consensus is eventually possible by presupposing "that people engage in discussion under conditions that neutralize all motives except that of co-operatively seeking truth" (Habermas in Young 1987: 69). Truth, in the particular setting of Liverpool's Capital of Culture, is aligned with the rightness of neoliberal forms of governance, yet neoliberal governance as a meta-narrative of the Capital of Culture is not acknowledged as a guiding principle. Instead, it is obscured beneath layers of art-as-empowerment discourse. Official Capital of Culture literature summons into being a world in which every utterance can have a single meaning, understood in the same way by all speakers. This is a model of communication in which consensus is assured, yet which veers away from the more embodied and rhetorical devices of the communicative act.

Official forms of mediating the pursuit of creative expression in the Creative Communities initiative, remain implicitly committed to a model of the public sphere as rational, universal, monological, and capable of impartial reason. This is accomplished through a multi-layered and highly nuanced process of communicating the notion of creativity to the public, which actually utilizes a number of discourses of heterogeneity. There are problems with this model. Firstly it involves an unambiguous insistence that creative expression is capable of facilitating contact between the public and local government, as the following quote from a council employee involved in delivering the official program of '08 illustrates:

> lets get real, we haven't gone "oh by the way here is your local jobs, education, training advice", we haven't done that, but what we've done is just got more people involved in activities, so that they feel more involved in their own area. And what you'd like to think is that that will lead on to them getting involved

> in other things. It mightn't work, some people might go back to their house and shut the door mightn't they, but we've given opportunities, we've opened up opportunities for them, indirectly and directly (Neighbourhood Manager Liverpool City Council).

Second is that the model ignores difference, and the multiple understandings of a particular speech act or action. Missing from existing critiques of top down civic celebration events is an engagement with other, more informal orders of communication, which inhere within existing spaces of creativity in the city and which produce other interpretations of the event.

The View from within Liverpool's Already-existing Spaces of Creativity

The view from within the groups I spoke to was that the Capital of Culture was not a "year zero" moment, more that it coincided, and was incidental to the affairs of each group. Furthermore, the agenda set by the official program does not capture the political questions and concerns being asked within Liverpool's existing creative spaces. What follows is an exploration of the antagonism that arises when an official version of creativity encounters the multitude of interests which find expression in the city's existing creative spaces.

My research focussed on the opinions of those who organized and participated in spaces of creative discourse, and who had varying degrees of contact with the official Capital of Culture program during 2008. Two examples include a writers group made up of people living with Multiple Sclerosis, and an over 50s social group. Both were approached by the Creative Communities initiative to participate in a showcase exhibition called Four Corners, which began in 2006, and formed part of the community strand of activities for Capital of Culture year. This was managed first by Liverpool Culture Company, and since 2009 by Culture Liverpool. Each year, five different arts organizations are commissioned by the City Council to work in different geographical areas of the city. Residents in these areas work with an artist to produce installations, videos, and events that are consolidated into an annual exhibition, by the Four Corners Artistic Director. The sum of the work is shown at The "Four Corners" exhibition, which in 2008, was held for four days at the Bluecoat Art Centre in central Liverpool. The exhibition presented the social history of Liverpool residents as a conspicuous concern of the official program of events. The work which feeds into the Four Corners project can be read as way of mediating council policy to residents because its geographical and thematic boundaries were pre-arranged by the City Council. The thematic content of the project revolved around the question "what makes a good neighbour?". The project was also funded and administered by the City Council, who commissioned arts organizations to do the "outreach" work.

The question "what makes a good neighbor?" seems tangential to concerns that participants of the project raised in interviews. Their concerns focus on a desire for greater transparency from the Council in regard to its decision to opt for investment in the city centre and the consequent dis-investment of outlying areas. The question that may capture the agenda of those quoted below might be "What makes uneven urban development persist?", rather than "what makes a good neighbour":

> you see the way things are, here now, its more charity shops, right, go to town, you'll see everything, all the fancy shops, everything like that. But the way things are going now, how can people afford it? […] I'd love to see regeneration up here as well. And what you've got here, this club […] Some kids would appreciate it, if this was a gigantic, bigger club, the kids would love it […] Part of it is political. All the European funding and all that. If it works, great, but everything seems to be down towards town, the new buildings and all that (C25 Over 5s Social Group member).

> there used to be nobody living [in city centre] so I actually think that there are quite attractive flats and everything is good, but that's only in the centre of Liverpool. I very much believe that we shouldn't have big shops overtaking. the regeneration thing, […] if you can't keep your local shops going, your local activities, you can't keep the community going (C27 Over 50s Social Group member).

Concerns about the policy of prioritizing the development of the city centre over the peripheries is also voiced in discussions about the extent of the Capital of Culture's geographical reach.

> I've not heard just in this street, even mention the capital of culture, so to me, if they haven't mentioned it because it hasn't had an effect on them, one way or the other (C30 MS Writers group member).

> you're looking at an awful lot of people up in Fazakerley, there's a lot of elderly people as well, that haven't got the funding to actually go into the city centre, or to take the children down, there's one off's, but couldn't attend a lot of the activities that were centred down in the city centre, so I think there was a lot of feedback there, to actually say, it's not being brought out into the communities (C28 Co-Manager Fazarkerley Federation).

The Four Corners project unfolded somewhere between public and private space. There was a public exhibition held in the Summer of '08 showing the sum of various collaborations between artists and community groups in the shape of sculpture, poems, songs and films. Yet the political rationale for the project, and the assumptions made about who "the community" is, which appear in the DTZ report, were not offered up as the subject of public discussion. The emphasis which is placed on art as a tool to "give voice" to the community by these official events disguises the extent to which the idea of "community" within policy discourse is the sum of a set of pre-constructed identity-based categories into which participants are expected to fit.

The official program of events of 2008 attempts a "formal systematization and functional coherence", which is one of the hallmarks, according to Foucault, of the process of subjugating other knowledges (Foucault 2004):

> I am referring to a whole series of knowledges that have been disqualified as nonconceptual knowledges, as insufficiently elaborated knowledges: naïve knowledges, hierarchically inferior knowledges, knowledges that are below the required level of erudition and scientificity … what I would call, if you like, what people know (and this is by no means the same thing as common knowledge or common-sense but, on the contrary, a particular knowledge, a knowledge that is local, regional or differential, incapable of unanimity and which derives it's power solely from the fact that it is different from all the knowledges that surround it (Foucault 2004: 7–8).

This process constructs a universe of meaning that includes a rationale for the use of creative activity, the intended recipients of that activity, its means of delivery, intended and actual outcomes, and finally, extended praise for the success of the project (LCuC 2008, 2009).

The discursive construction of arts-based creative activity in Creative Communities literature has a specific schema that subjugates the kind of knowledges referred to above. This is because Creative Communities requires creativity to fit into a rational and neoliberal order within the illusion of a unified civic sphere. The antagonism arises when this order encounters the dynamic flow of already existing organizations and activities by residents. In order to illustrate the inherent difficulty of fixing the idea of being creative to a set of technocratic objectives imposed by policy, it may be useful, at this juncture to turn to descriptions given by these residents of how and why they have set up and participated in existing creative activities.

What are some of the motivations for people to establish or become participants in the range of creative spaces that came before the Capital of Culture rationale? Below, one of the founders of the Picket Venue and Music academy describes its beginnings.

> You're talking about the 1980s. It was really a response to mass unemployment in Britain, in particularly the North West, and my aim was, with colleagues and the Merseyside unemployment centre, was to attract young people through music, away from negative stuff, unemployment, heroine, into creative stuff, […] My response was instinctive […] what do I understand, and what do I enjoy […] Punk said pick up a guitar, […] talk about your experiences, your angers, your unhappiness, or your situation, and express yourself, so I followed that ethos, and that's why I think its important to people to have that in their lives,

> to have some opportunity to be creative and to speak about their experience (C2 Director of The Picket music venue and academy).

The manner in which C2 identifies the potential of creative activity to form a diversion for young people from the negative effects of mass unemployment and poverty has some overlaps with the current official discourse that arises out of an active citizenship agenda. Yet there are important distinctions. For example, the motivation to establish The Picket was not led by policy-led regeneration events such as the Capital of Culture. Rather it was a response to the social effects of the unemployment which resulted from Fordist rationalization in Liverpool at the time. People who had shared experiences with those who used the organization's facilities, led it, and so it was not structured as a temporary, top-down project.

> Been involved in the community for 20, 21 years plus, I got involved because my son, he's got learning difficulties so I got involved in the school, which sort of spilled out into the community. Got involved in setting up the tenants and residents group [...] I was doing some stuff up in the school, with the children, set up a parents club, [...] 'cos children with special needs, they don't have the front gate contact with the friends, with the parents, so got involved in there [...] and eight years ago, I became full-time here, in the community centre (C28 Co-Manager of Fazakerley Federation).

> most of the community centres round here have been started by mothers that have wanted activities for the young people, or elderly groups (C27 Over 50s Social group member).

Both responses reflect on how the contingent and particular necessities of everyday life establish the desire and need for a space for interest groups to coalesce. With C2, an absence of facilities in the local area forges a space of support and conviviality with others who share similar life experience. Another informant describes the circumstances in which he entered into the MS writers groups:

> I finished work and I was looking for a focus for my creative ability, somewhere to express that and I looked on local writers websites, then just by chance I got a newsletter from the local MS Society and it was just a little line that said MS Writers group, Kensington [...] they were an independent group of people with MS, and their carers, [...] who were very focused on creative writing (C29 MS Writers group member and Four Corners participant).

Again, participation in a local space of discourse is rationalized with reference to the highly particular and unique set of circumstances that C29 experienced. In contrast, the Four Corners project in which the MS writers group eventually participated in the Capital of Culture is described in the exhibition catalogue document as a project interested in "capturing memories, aspirations and supporting community cohesion" (Four Corners 2008). A difficulty arises here when official discourse applies its generalized concepts of creativity's place within an idealized civic sphere, to a situation in which groups arise out of a multiplicity of contingent factors that do not necessarily fit with the pre-ordained image of the communicative and rational public conjured by official event discourse.

In similar fashion, the excerpt below reflects on how an over 50s club began as a response to the absence of facilities in an area with large numbers of elderly residents.

> We need this place to function for other people, 'cos there is no other thing in the area, [...] We don't get funded, [...] the idea of it is to drop in, and I think it's vital for this area, because there is a lot of old people here, and the ones that do come, it's two hours out, and it gets them out of the house (C26 Over 50's group organiser and Four Corners participant).

These creative environments respond to a need and provide a support network at a local level. They do not foreground their activity by consciously stating that they are "spaces of creativity". In contrast, official discourse constructs the Capital of Culture as "year zero" for the city, a starting point that fails to acknowledge the conditions of emergence of the event, and in the process, negates the esteem in which the already established spaces of creativity and conviviality are held by many Liverpool residents.

Some of those who sustain these already established spaces agreed to participate in the Capital of Culture. Rather than leading to additional promotion or funding for their organizations, the official event instead spoke on their behalf, suggesting that the experience of "being creative" was a novel and emancipatory process (Four Corners 2008). Official discourse around '08 tended to dwell extensively on the spiritual benefits of the "arrival" of culture into community settings. These stage-managed "arrivals" occurred with an accompanying emphasis on spectacle, in the form of Four Corners projects. One such project, The Gathering, turned a major road in South Liverpool into a market and street festival for a day. In another event called "Be My Guest", residents from around the city were asked to talk about personal memories and experiences, which were "collected in an intimate one person multi-media performance that toured around South Liverpool" (Four Corners 2008: 25). In these staged events, The Capital of Culture positions creativity as something that has just arrived in people's lives, and from that position, proceeds to construct official program activities as social interventions. In the legacy literature attached to the Capital of Culture year, there is an absence of the question "who is speaking". For some interviewees, the significance of '08 is this very absence of the political.

In the following section, I analyse interviewees' reflections on their own pre-'08 spaces of creativity. Particular attention is directed to the articulation of the contentious relationship between state and society, and the functional role creative expression was given in mediating council policy during the Capital of Culture.

Theorizing a Contested Creativity

The writers group for people living with MS were approached by the Creative Communities initiative in 2008 to participate in a short film, which was screened as part of the Capital of Culture official program. The film was called "Communities on the Edge", and sought to bring the perspectives of a number of different groups living and working in the city to bear on the kind of regeneration projects that were rapidly re-shaping the city centre, an example being the retail development Liverpool One. The writers group wrote poems and scripted monologues that were then edited into the film by the independent media company brought in to direct and edit the film. There was a consensus amongst the group members I spoke to that their contribution to the film was minimized in the editing process. The reflection below is from a member of the MS Writers group, who re-counts her experience of producing work for the Creative Communities exhibition Four Corners:

> SF: do you think the Four Corners was a success?
>
> C30: Four Corners project? Yeah, for the people who ran it, who got the funding for it, I just feel that, yeah, it's got their names on papers, it's got their names on films and hopefully from there, and hopefully for them, they'll do better things. But for us no, we were *allowed.* We were given sort of the, what do you call it, the rostrum to stand on and have our say, it's not going to do anything for us personally (C30 MS Writers Group member and Four Corners Participant).

Another member of the group describes the difference between their embedded position in the community centre where they hold their meetings and the marginality they experienced as participants within the schema of the Four Corners project:

> There is an energy we provide in the community centre itself. We are sort of regarded as, you know, "the MS writers are here", they give us our own space, they give us our own room, and they see us as part of the community, [...] So I see our group and others who you know about as a particularly powerful socio-cultural phenomenon which maybe the Culture Company, the '08 people were not aware of. And I think it's those sort of groups they need to bring on board [...] I felt with the film that was done with us, there was a missed opportunity there, we weren't harnessed, we were sidelined [...] but you think, ok, we know they're playing the political game, but next time? We'll do it on our terms (C29 MS Writers Group member and Four Corners Participant).

These objections point us to the process in which a community art intervention such as Four Corners project has its' own clearly demarcated boundaries, final aims and objectives, while participants position their abilities and contributions outside of these narrowly defined boundaries. The experiences of the MS Writers

Group cannot be satisfactorily "captured" by their participation in the project, yet the policy context of the Capital of Culture insists on Four Corners as an emancipatory "voice-giving" experience for participants.

The discursive construction of creativity and community in Liverpool's Capital of Culture's official program perceives the public sphere according to the logic of accommodation. Official discourse hones a space that facilitates the accommodation of diverse groups, and from that, builds a narrative of improvement in the civic life of Liverpool. The data from interviewees expresses a confounding array of examples of how people produced their own meaning of the event. In their diversity, they express the impossibility of knowing the event through a single interpretive framework. Often the official literature constructs the event as unproblematic and one dimensional, as seen in the formula participation in artistic intervention = emancipated participant. There is no consideration of the complexity of the artistic intervention, in terms of, for example its partiality, its politics, its awkwardness, or its failure. The consideration of the inherent complexities of groups of people working together, becomes a critical retort to the conventional interpretation of the public sphere constituted by '08, and the subsequent cultural strategy developed for the city (Liverpool First 2008). To refer to Anderson and Holden's point quoted earlier, the Capital of Culture "happens" repeatedly over time, it may re-enter our consciousness as disappointment, or as a moment of intensity, or as a continuation of the same. The Capital of Culture is not a stable and fixed entity in time. The ways people have and will continue to respond to it are dynamic and contain an inherent multiplicity. Cultural policy discourse seeks to arrest this dynamism by constructing a series of technocratic objectives which interpret the event for us, and in the process, suggest what the public sphere ought to be and who should act within it.

Academic debates taking place around the subject of Capital of Culture in the UK often fore-ground economic inequality in UK host cities of Glasgow and Liverpool in order to eventually pose such questions as "Whose Liverpool is being celebrated? Whose story is dominating – and which story is being marginalised" (Mooney 2004: 338). Rather than continue to make demands for a more inclusive event, one that is more representative or relevant to a wider range of subject positions, further work needs to be done in the area of challenging attempts to formalize the public sphere. What or who benefits from these large scale spectacle led events which pre-suppose the nature of the identities that make up the public sphere, and furthermore, pre-suppose the conditions in which these groups ought to speak?

Conclusions

People coalesce into spaces of creativity in multiple, contingent and unexpected ways, according to needs, desires and interests that are not accounted for by the idealized civic sphere summoned in the official discourse of Liverpool's Capital of Culture. The Bakhtin Circle challenged the linguistic certainties of forms of official discourse through a theorization of language as heterogeneous, multi-faceted, open ended, dynamic and animated by the contingencies and innovation of everyday speech. Such a theorization provides a useful interpretive framework in which to develop a critique of event led regeneration. In the course of this chapter, I have argued that the official discourse of event led regeneration constructs the voiceless citizen, who is given voice via a creative encounter facilitated by the official event. This idealized citizen is then made to be a conspicuous beneficiary of event led regeneration in the form of the Capital of Culture. Outside of these discursive constructions however, is the complexity and multi-faceted languages that animate everyday life, and indeed, the spaces of creative activity which pre-exist this particular event.

This chapter argues that both within Capital of Culture's early policy documents, and the delivery of activities, exhibitions and other events, the official discourse remain faithful to the singular notion that an event of this kind can initiate a new communicative and rational civic sphere, which will inevitably aid in the process of Liverpool becoming a viable neoliberal city, able to engage in inter-urban competition for economic resources. In the process, the communities which constitute Liverpool's existing spaces of creativity have been re-described by the official event discourse to fit in within this ideological re-imagining.

Recalling Anderson and Holden, there are ways that the event continues to happen. It re-appears in people's consciousness as a frame of reference for the lack of democracy in Liverpool's civic leadership. Informants in this research have suggested that the event's policy and promotional discourse sought to speak on their behalf; they critically discuss their experience and refer to how they were implicated in the new Liverpool. Informants point to the lack of accountability, and the policy of uneven development in which economic investment is perpetually focused on the city centre. It is through these narratives that the event is rendered meaningful to participants, over and above any promises made by the official event that the provision of forums of creative expression can "give voice" or achieve any sense of emancipatory fulfilment within the city's population. They became "the city's greatest assets" or "survivors" or as "combative, comic, determined, laced with a healthy cynicism", and so on. These constructions produce the neat and finalized meanings found in the event's promotional discourse, and reflect Bakhtin's notion of unified discourse. The authors of Liverpool's official version of creativity may have intended their construction to be read as "a real force overcoming heteroglossia, setting defined borders ... guaranteeing a certain maximum of mutual understanding and crystallizing in the real ... 'correct language'" (Bakhtin 2006: 270). What this chapter suggests however, is that amid these attempts to unify, are spaces of discourse which complicate the certainties of official discourse with their own socially situated, everyday experience.

I have sought to delineate a parallel narrative of the existing spaces of creativity in the city as sites of enjoyment, support, conviviality and moments of transformation. Can we therefore conceive of the Capital of Culture in its official pomp as the alternative site of creative expression, creating a new and different scene? Its appeal to a mass re-imagining of the City, which cherry picks from history, fails to reflect on its own agenda and attempts to delineate a unitary discursive field of creativity, which can only ever have limited appeal to existing groups of interest in Liverpool, who continue to create their own spaces of discourse.

This chapter also reflects on the structural and moral problems with technocracies delivering cultural activities to communities. Technocratic delivery of cultural activity in Liverpool, tied as it is with the wider social and economic desires of City Council policy, has led to unproductive and uninteresting frontiers of creativity, bounded by discourses of citizenship and employability. Yet the existing spaces of creative activity in the city, created and maintained by those who act *within* them, are providing the physical, and cognitive space for people to experiment, learn, reflect, and create in ways that undermine claims these groups are given voice through the technocratic channels of event led regeneration.

Neoliberal forms of urban development establish event narratives as normative contexts in which "the people" can interact with the decisions that shape urban public space. The question of "why an event" is therefore a useful starting point. Public space is a complex site of on-going, and competing claims for legitimacy and expression. The existing spaces of creative activity I have spoken to in this research acknowledge this complexity, and work in, around and through it, often with the singular aim of surviving financially. The vehicle of Capital of Culture seems a tokenistic way to support the city's creative infrastructure. An ideological shift is required which moves away from the temporary civic boosterism of event led regeneration to a city which works on an arms length basis to provide accessible, flexible and affordable spaces for individuals and groups to create and maintain projects, sites, scenes and activities and conversations that are already occurring.

Tully's useful concept of the public as a "strange multiplicity" of voices that make demands on public life (1994), suggests that the logic of accommodation (*into* the public sphere from a perceived external starting point) is unhelpful. A future direction might be to understand that public space is the site onto which new formations, groups and political subjects inscribe and are inscribed. This is a public space where those speaking on behalf of an official rendition of public space constitute just one voice amongst many others.

Bibliography

Anderson, B. and Holden, A. (2008), "Affective urbanism and the event of hope", *Space and Culture* 11:2, 142–59.

Asen, R. (2000), "Seeking the 'counter' in counterpublics", *Communication Theory* 10:4, 424–46.

Bakhtin, M. (1986), *Speech Genres and Other Late Essays*. Edited by C. Emerson and M. Holquist (Austin, TX: University of Texas Press).
Bakhtin, M. (2006), *The Dialogic Imagination Four Essays by M.M Bakhtin*. Edited by M. Holquist (Austin, TX: University of Texas Press).
Belchem, J. (2000), *Merseypride: Essays in Liverpool Exceptionalism* (Liverpool, UK: Liverpool University Press).
Beynon, H. (1978), "What happened at Speke?" 6-612 Branch Trade and General Workers Union, Liverpool.
Boland, P. (2010), "'Capital of Culture – you must be having a laugh!' Challenging the official rhetoric of Liverpool as the 2008 European Capital of Culture", *Social and Cultural Geography* 11:7, 627–45.
Coleman, R. (2008), *Forging of an Urban Patriotism: Illusions of People and Place*. Keynote paper at Capital, Culture, Power: Criminalisation and Resistance Conference. 2nd to 4th July (Liverpool: University of Liverpool and John Moores University).
Couch, C. (2003), *City of Change and Challenge: Urban Planning and Regeneration in Liverpool* (Aldershot: Ashgate).
DTZ Pieda Consulting. (2005), Liverpool City Council. Building the Case for Creative Communities – Final Report.
Echo. (2003), *Liverpool Echo*. "We did it". 15 June.
Echo. (2008), *Liverpool Echo*. "Liverpool One: huge shopping development opens". 29 May.
European Union. (1985), *European Union* Official Journal, C153 (22 June 1985), 2.
Florida, R. (2002), *The Rise of the Creative Class – And How its Transforming Work Leisure and Everyday Life* (New York: Basic Books).
Foucault, M. (2004), *Society Must be Defended: Lectures at the College de France, 1975–1976* (London: Penguin).
Four Corners. (2008), Exhibition Catalogue. Bluecoat Art Gallery (Liverpool City Council).
Four Corners. (2008a), *"Communities on the Edge"* DVD (Liverpool City Council, Everyman Playhouse Theatre, River Media).
Garcia, B. (2005), "Deconstructing the City of Culture: the long term legacies of Glasgow 1990", *Urban Studies* 42:5/6, 841–68.
Gardiner, M. (2004), "Wild publics and grotesque symposiums: Habermas and Bakhtin on dialogue, everyday life and the public sphere", *The Sociological Review* 52: S1 28–48.
Gold, M. and Gold, J. (2000), *Cities of Culture: Staging International Festivals and the Urban Agenda 1851–2000* (Aldershot: Ashgate).
Gold, M. and Gold, J. (2008), "Culture and the City. History in Focus". *Institute of Historical Research*, Issue 13: The City. Available at http://www.history.ac.uk/ihr/Focus/City/articles/gold.html#t20 [accessed 5 November 2011].
Habermas, J. (1976), *Legitimation Crisis* (London: Heinemann).

Harvey, D. (1989), "From managerialism to entrepreneurialism: the transformation in Urban governance in late capitalism", *Geografiska Annaler Series B Human Geography* 1:71, 3–17.

Hirschkop, K. (1999) *Mikhail Bakhtin. An Aesthetic for Democracy* (Oxford: Oxford University Press).

Holloway, J. and Kneale, J. (2000), "Mikhail Bakhtin: dialogics of space", in M. Crang and N. Thrift (eds), *Thinking Space* (London: Routledge), 71–88.

Impacts '08. (2010), *Creating an Impact: Liverpool's Experience as European Capital of Culture*. By Garcia, B., Melville, R. and Cox, T. (Liverpool: Liverpool City Council, University of Liverpool and John Moores University).

Jones, M. and Ward, K. (2002), "Excavating the logic of British urban policy", In N. Brenner and N. Theodore (eds), *Spaces of Neoliberalism* (London: Blackwell), 126–47.

Jones, P. and Wilks-Heeg, S. (2004), Capitalising Culture: Liverpool 2008. *Local Economy* 19:4, 341–60.

Liverpool City Council. (2003), Liverpool '08 Bid Executive Summary Report.

Liverpool City Council. (2010), Consultation Portal. Available at http://liverpool-consult.limehouse.co.uk/portal/planning/csrpo_consultation/csrpo?pointId=1245921856105 [accessed 5 November 2011].

Liverpool Culture Company. (2006), *The Art of Inclusion Liverpool's Creative Communities* (Liverpool: Liverpool City Council).

Liverpool Culture Company. (2008), *Inter Cultural Capital Liverpool 2008: Sharing Experiences, Sharing Culture.*

Liverpool Culture Company. (2009), *Liverpool '08 European Capital of Culture: Impacts of a Year Like No Other*.

Liverpool First. (2008), Liverpool First Draft Liverpool Cultural Strategy.

Mackenzie, I. (unpublished manuscript) "From identity to event: the changing nature of the public sphere".

MacKenzie, I. and Fitzpatrick, S. (Forthcoming), *Public Art Festivals, Significance Movements and the Constitution of the Public Sphere.*

Mangen, S. (2004), *Social Exclusion and Inner City Europe: Regulating Urban Regeneration* (Basingstoke: Palgrave Macmillan).

Marren, B. (2009), *The Closure of the TR7 Factory in Speke, Merseyside 1978: The Shape of Things to Come?* Paper presented at the University of Liverpool School of Geography's "Developing Theoretical Approaches in Labour Geography Conference", 11th June 2009.

Mayor for Liverpool (2010), www.amayorforliverpool.org [accessed 7 July 2010].

Minton, A. (2006), *What Kind of World are We Building? The Privatisation of Public Space*. Report written for the Royal Institute of Chartered Surveyors. Available at www.annaminton.com [accessed 31 January 2012].

Mooney, G. (2004), "Cultural policy as urban transformation? Critical reflections on Glasgow, European City of Culture 1990", *Local Economy* 19:4, 327–40.

Objective One. (2012), http://www.objectiveone.com/O1htm/01-whatis/faq.htm [accessed 10 January 2012].

Palmer Rae Associates. (2004), *European Cities and Capitals of Culture. A Study Prepared for the European Commission. Part One*. Available at http://ec.europa.eu/culture/pdf/doc654_en.pdf [accessed 31 January 2012].

Parkinson, M. (1990), "Leadership and regeneration in Liverpool: confusion, confrontation or coalition?", in M. Parkinson and D. Judd (eds), *Leadership and Urban Regeneration* (London: Sage), 241–57.

Psychogiopoulou, E. (2008), *The Integration of Cultural Considerations into EU Law and Policies* (The Netherlands: Martin Nijhoff).

Saussure, F. (1983), *Course in General Linguistics*, edited by C. Bally and A. Sechehaye (London: Duckworth).

Waterman, S. (1998), "Carnivals for elites? The cultural politics of arts festivals", *Progress in Human Geography* 22:1, 54–74.

Webster, A. (2008), *The Empire in One City? Liverpool's Inconvenient Imperial Past*, edited by S. Haggerty, A. Webster and N.J. White (Manchester: Manchester University Press).

Wilks-Heeg, S. (2003), "From world city to pariah city? Liverpool and the global economy", in R. Munck (ed.), *Re-inventing the City? Liverpool in Comparative Perspectives* (Liverpool: Liverpool University Press), 36–52.

Young, I.M. (1987), "Impartiality and the civic public", in S. Benhabib and D. Cornell (eds), *Feminism as Critique* (Minneapolis, MN: University of Minnesota Press), 57–76.

Chapter 9
Active Ageing: Creative Interventions in Urban Regeneration

Sophie Handler

Located on the eastern fringes of London, Newham is easily identified as an area of deprivation, a place frequently cited in indexes of social exclusion and deficit (Newham is consistently ranked among the top 11 deprived areas of UK).[1] Historically, though, this is an area that has carried an altogether different inflection. For almost two centuries Newham was an area defined by the strength of its industrial manufacturing base (in chemical works, in printing, in marine engineering, in food processing). In the recurring narratives of local reminiscence, Newham appears, in the inverse image of post-industrial deprivation, as a potent, industrial powerhouse: boasting one of the largest dockyards in the world, the host region for household name industries (Tate and Lyle), a place able to define itself as an integral part of the country's "southern industrial belt" (Sainsbury 1986: 52).

The last 50 years, though, has seen many of those industries disappear with the closure of the docks (in 1981) and the collapse of local industry, and with it the vitality of that local identity has largely collapsed too. Mergers, relocations, mechanization and industry failure have all contributed to the radical alteration of Newham's local landscape. The move from a manufacturing base to a service/entrepreneurial economy has involved real socio-economic and physical impact, including severe job losses, a period with unemployment rates above 20%, and the borough's physical fabric substantially altered, with over 1,000 acres of former industrial land left open for subsequent re-development.[2]

1 "The 2004 Index of Multiple Deprivation shows that Newham has very high levels of deprivation, ranking the borough 11th nationally (of 354 local areas, where 1st has highest deprivation) and 4th among the London boroughs." London Borough of Newham (2007), *The State of the Borough: An Economic, Social and Environmental Profile of Newham* (London: Local Futures), 7.

2 London Borough of Newham (2007), *Local Implementation Plan 2005/06 to 2010/11* (London: Forward Planning and Transportation and Physical Regeneration and Development), 1.

Coming to terms with this post-industrial legacy has involved the compensatory introduction of an ongoing series of regeneration programmes designed to actively revitalize the area. Over the last two decades Newham has received a substantial amount of regeneration funding from central government to lift it out of post-industrial decline. In 1993, the borough was awarded £37 million to encourage inward investment via the Stratford City Challenge program. Six separate Single Regeneration Budget rounds, from 1994 onwards, introduced central government funding into the area with match funding supplied by local authority, private sector and voluntary organizations. The Neighbourhood Renewal Fund (NRF) was introduced in 2001 to tackle long-term, multiple deprivation. The New Deal for Communities (NDC) scheme, in turn, was introduced to tackle social exclusion specifically (Newham was one of 39 "deprived" boroughs to be awarded NDC funding in the UK).[3] Infrastructural, commercial and local enterprise projects, in addition to commitments to local housing development[4] have all attempted to drive forward the process of borough-wide regeneration and renewal. That regenerative ambition has most recently, and most visibly, been marked out by the legacy promise of the 2012 Olympics (Newham as one of the five Olympic host boroughs).

These funding rounds have ranged in their forms of support over the years: from general attempts to target youth offending rates and improve educational attainment in schools (as a way of tackling social exclusion) to more direct attempts to attract inward investment into the borough through improved transport links. Allied cultural strategies, meanwhile, have evolved alongside these more high profile initiatives in an attempt to draw the local community into a "coherent" process of regeneration. Arts programming initiatives, for instance, have increasingly become a standardized way of incorporating local inhabitants into that borough-wide process of regeneration. In 2007, a group of 12–21-year-olds were invited, as part of an arts project, to respond to local regeneration through a self-curated exhibition raising "regeneration awareness" and inviting responses to the physical fabric of local change (Regeneration and Renewal 2007). More tangentially, the work of organizations like East London Dance, have developed work in conjunction with "landmark" regeneration developments in an attempt to paint a localized picture of cultural renewal in progress. A dance commission facilitated by East London Dance in partnership with local amateur dancers was used to support the inauguration of a new retail development in Stratford

---

3 In 1997, key stakeholder partners within the borough agree on a common vision and regeneration strategy for the borough that marks this shift from industrial to enterprise economy and a shift to service and a small business economy "by 2010, Newham will be a major business location and a place where people choose to live and work". Ibid., p. 2. See also, <www.newham.gov.uk/informationforbusinesses/externalfundingopportunities/gloassary.htm> [accessed 12 August 2011].

4 Newham anticipates building 50,000 new homes by 2020. See, London Borough of Newham (2007), *Focus on Newham* (London: LBN).

with images from that commission used as the prominent visual emblems of community cohesion (placed strategically onto the advertising hoardings of this new development under construction).[5]

For elderly witnesses of radical post-industrial transition, the experience of regeneration can easily be felt as an unsettling process. For those residents with long-standing relationships to place there is often a marked sense of disconnect from an environment that has become so radically altered, particularly with such a visible loss of local industrial landmarks. As these lifelong residents of Newham have grown up over the last 60, 70, 80-odd years, through the war, the decline of local industry, and the restructuring of the economic and demographic landscape, that sense of disconnect and alienation can often be quite strongly felt and can manifest as nostalgia for the past that is, in turn, coupled with a deep sense of loss.[6] Personal investment in place is, arguably, more charged, more keenly felt the longer you live in a place. But the assumption of elderly alienation and withdrawal from change, is entangled in more complicated stereotypings of an older generation where it is all too easy to portray "the elderly" as passive subjects of change, as attached to the retrospective image of place. Stereotypes like these, however, deny the possibility for a more present and prospective identification with place.

Ageing Facilities is an informal creative/research initiative that has been attempting to unsettle this retrospective and "passive" stereotyping of older age through a cultural practice of small-scale urban interventions. Developed in Newham, in partnership with local "elders'-only" groups, these interventions try to build on and amplify those creative relationships to transitional urban environments that already exist among its older population, and in this way make space for a marginalized older population in other ways.

5 The current retail component of the Stratford City regeneration program (a "landmark" retail development by global corporation Westfield) is currently being supported by a parallel series of community arts projects that includes a recent dance commission with local dance organization East London Dance. For more on this see <http://uk.westfield.com/stratfordcity/community/east-london-dance> [accessed 12 August 2011].

6 Many of these elderly participants were involved directly in the old industries of Newham, as: dockyard workers, a mint seller in Rathbone Market (now currently in the process of being rebuilt), as workers on the printing floor of Paragon a major printing works (now moved and merged). For more on this see the oral history work of Kelly, C. (2005), *There Go the Ships: East End Lives in Words and Pictures* (CK Editions: Maidstone, Kent). This oral history includes many of those elderly participants who have been involved in this ongoing programme of interventions.

From a choreographed dance class action for 50-plus-year-olds (temporarily re-appropriating, for one night only, the space of a closed-off park after dark), to *An Alternative Street Furniture Guide* (for "pensioners" only) this creative practice[7] becomes a way of temporarily reconfiguring routinized relationships to urban space while also unsettling those generational norms that govern and limit the use of urban space.

These interventions have been "acted out" in Newham over the course of the last six years, with a consciousness that many of those involved in these interventions have been long-standing residents and witnesses of radical urban change over the last 50 years. For many of those elderly participants involved, the fundamental shift in Newham's identity, the whole process of post-industrial restructuring, has occurred over the course of their lifetime (many once employed in what are the now redundant printing and food processing industries).[8] To operate here within a site of post-industrial change, is to allow these interventions to start, in their own small way, to position an older generation more centrally within the informal planning and re-imagination of Newham's future and to challenge the implicit youth-focused discourse of cultural economy theories and practice.[9]

Active Ageing

There is a growing level of interest around ageing issues and a perceived need to engage with older people that is fuelled in large part by demographic forecasting: the knowledge that by 2050 a third of Europe's population will be over 60 years old (Dittmann-Kohli, Westerhof and Bond 2007: 1), and that declining fertility rates and the lengthening out of life expectancy are creating "unprecedented" demographic trends (United Nations 2002: xxviii). In the next 10 years, the proportion of over 65-year-olds as a percentage of total population is expected to almost double from the mid-1970s whilst the number of the "oldest old" –

7 See <www.ageing-facilities.net> [accessed 21 February 2009]. These ongoing programs of interventions under the *Ageing Facilities* banner have been funded by the RIBA Modern Architecture and Town Planning Trust and the RIBA/ICE McAslan Bursary 2006 and 2008.

8 For a potted history of Newham as it has changed radically over the last 150 years, as its status shifts from former industrial heartland to post-industrial "gateway" to London see *The Newham Story: A Short History of the London Borough of Newham*, available to download via <www.newham.gov.uk/nr/rdonlyres/52364e5a…/newhamstory.pdf> [accessed 12 August 2011]. Also, Pewsey, S. (2001), *Newham: Past and Present: The Changing Face of the Area and Its People* (Swindon, UK: Sutton Publishing) and Sainsbury, F. (1986), *West Ham: 1886–1986* (London: Plaistow Press).

9 It is worth noting in this context that Florida's theory of the creative class and its implementation focuses primarily on a younger demographic. See Florida, R. (2004), *The Rise of the Creative Class: And How It's Transforming Work, Leisure, Community and Everyday Life* (Basic Books: New York).

those 80 years and over – is projected to increase at an even faster rate (Dittmann-Kohli, Westerhof and Bond 2007: 2). Concepts like "active ageing" championed by the World Health Organization (WHO)[10] are the global policy responses to this heightened sense of urgency around ageing issues: the need to actively engage with and manage what is often formulated as the "problem" of a rapidly rising older population with its associated political and economic costs.[11]

There is a curious process of distancing though at work even within this heightened preoccupation with a rapidly ageing population – a process of distancing that runs parallel to an attentive problematization of an ageing population, and involves the marginalization of an older generation more generally. The fear of ageing and its associated morbidity, as deep-seated cultural anxiety is, arguably, more easily protected against via a basic defense mechanism of distancing and denial (Woodward 1991: 19). Within spatial practices, within the context of urban design and planning, that marginalization of an older generation manifests itself in the lack of critical and exploratory work around ageing issues. Most work on ageing, conducted within a specifically urban context, tends not to move beyond a baseline interest in physical needs, often limited to detailed questions of accessibility – the streetscape's pitching of pavements, for instance[12] – or to questions of physical health framed under the watchwords of "global health" and "wellbeing".[13] In the context of regeneration and urban renewal, there is, perhaps, an even more noticeable reluctance to engage with the elderly.

Recent research that has started to examine the impact of regeneration policy on the elderly repeatedly highlights the way in which the elderly are all too often locked out of regeneration programming, from policy formulation level to

10 The official World Health Organization definition of active ageing is: "the process of optimizing opportunities for health, participation and security in order to enhance quality of life as people age." See, World Health Organization, (2002), *Active Ageing: A Policy Framework* (A contribution of the World Health Organization to the Second United Nations World Assembly on Ageing, Madrid, Spain: April), 12.

11 For the "problematization" of the elderly – where research tends to map out the perceived "problems" of old age, see Johnson, M., "That was Your Life: a Biographical Approach to Later Life" in V. Carver and P. Liddiard (eds) (1978), *An Ageing Population; A Reader and Sourcebook* (Milton Keynes: Open University Press), 99.

12 See for instance the work of the Salford SURFACE Inclusive Design Research Centre – a research center currently committed to detailed auditing of pavements in terms of accessibility (which includes, for instance, close monitoring of the impact of blister hazard warning paving on elderly passers-by).

13 These global health watchwords, are in turn, picked up by politicians and re-framed as useful political slogans. Most recently, the coinage of a new term "wellderly" (a word-merging of *wellbeing* and *elderly*), makes the policy point that there is now, increasingly a desirable class of fit/active/autonomous elderly individuals for whom the term wellbeing is particularly apt – for whom compulsory retirement may no longer be a feasible policy option. See "Harman Wants an End to Compulsory Retirement Age", *The Guardian*, 12 January, 2010.

consultation on the ground.[14] There is, arguably, a prevailing sense in which older people are more easily cast as passive recipients (as opposed to active participants) in regeneration and change, even though transitional, post-industrial environments (those undergoing regeneration and renewal) are experienced more acutely as sites of transition by those lifelong elderly witnesses of urban change.[15]

Recent attempts to involve older people more actively as direct participants in urban renewal have tended to carry their own limitations as positive attempts at elderly inclusion have been constrained, in turn, by the limiting terms in which the notion of elderly participation is understood. The Centre for Urban and Community Research, for instance, has devised a toolkit to encourage and enable the direct involvement of older adults within the process of urban renewal.[16] But there is, arguably, a limited sense of participation and agency at play within this toolkit structure where participation is limited to involvement within *existing* local authority planning procedures. Elderly engagement, in a context like this, is restricted to movement within existing local authority processes: learning how to navigate the working mechanics of planning procedures (with all its in-built expectations as to what might be, reasonably, changed within an urban context). The possibility to shape and lay claim to a transitional urban environment – to even question what urban change might mean – is not so readily facilitated within formalized processes like these.

Informally, however, below the radar of planning authority and intent, there is perhaps more room for maneuver. Here, on-the-ground, through small-scale, agile, interventionist practices, through informal cultural practices, elderly groups and individuals can in small ways (not even necessarily consciously) take part in processes that are able to reshape and renew otherwise routinized relationships to place.[17]

---

14 See the work of undertaken by Age Concern, the Economic and Social Research Council's "Growing Older Programme" in A. Smith, T. Scharf, C. Phillipson, A.E. Smith and P. Kingston (2002), *Growing Older in Socially Deprived Areas* (London: Help the Aged); Phillipson, C. and Scharf, T. (2004) *The Impact of Government Policy on the Social Exclusion of Older People; A Review of the Literature* (London: Social Exclusion Unit, Office of the Deputy Prime Minister); and, most recently, Smith, A.E. (2009), *Ageing in Urban Neighbourhoods; Place Attachment and Social Exclusion* (Bristol: Policy Press).

15 For a recent study that addresses this exclusion of the elderly from processes of urban regeneration, see Smith, A.E. (2009), *Ageing in Urban Neighbourhoods; Place Attachment and Social Exclusion* (Bristol: Policy Press).

16 For a detailed report of these toolkits see, Rooke, A. and Wuerfel, G. (2007), Mobilizing Knowledge – Solving the Interaction Gap Between Older People, Planners, Experts and General Citizens within the Thames Gateway (London: Centre for Urban and Community Research, University of London).

17 It should be noted that these small-scale direct actions are not enacted as consciousness-raising exercises on the spatial politics of elderly marginalisation. The focus of each intervention is constructed, for the moment, for the surreal pleasures of the intervention itself – of dancing, or sitting out-of-place, and for that playful possibility of

With an Emphasis on Youth

Paradoxically, these elderly-focused interventions started out in response to a project geared towards youth: working on a feasibility study with muf architecture/art[18] in Newham with a remit to identify small pockets of leftover public space for use *with an emphasis on youth* (muf architecture/art 2004). Commissioned by the Stratford Development Partnership (SDP) as part of Newham's borough-wide regeneration programming, there was a sense that the brief – to find youth-oriented uses for leftover open spaces – sat in direct correlation to the demographics of the borough, then one of the youngest in the UK. The strategic ambitions for borough-wide urban renewal aligned, in this way, easily with its majority constituency – its younger demographic – with 41% of the borough then youthfully aged under 25.

From conversations back then at muf about the logic of this project we began to turn over the assumptions of this youth-focused brief: the hidden politics of placing a specifically generational emphasis in regeneration programming, the logic of planning out an urban design brief that is so explicitly youth-specific. Thinking around these issues led to the evolution of *Ageing Facilities* as a separate practice (outside of muf) with an overt agenda: to redress the generational imbalance in urban planning and design *with an emphasis on the elderly*. In a layering over of generational provision within a borough where the emphasis on youth provision has been so dominant – these interventions work now with an unapologetic elderly bias in a deliberate demographic re-balancing act.

In the last decade or so, a series of themed funding rounds operating under New Labour's social inclusion agenda, have attempted to frame urban regeneration within the UK explicitly around the interests of marginalized groups.[19] But that social inclusion agenda, like the Stratford Development Partnership brief, with its explicit emphasis on youth, seems to carry a more deliberate focus on the needs of a younger generation (children, adolescents, younger adults of working age), generations with economic prospects, as society's upcoming labor force (Smith 2009: 1). Here, the premise for youth inclusion is articulated in terms of prospective value, as younger people are seen as necessary to securing the "future health of society" as "future citizens".[20] This premise has the curious

distorting and amplifying existing activities elsewhere. (The critical polemic and poetics of the intervention is to a degree kept apart.)

18 muf architecture/art is a collaborative practice of artists and architects that has been devising innovative, socially engaged projects within the public realm since 1996.

19 See, for instance the following reports published by the Social Exclusion Unit (1998), *Bringing Britain Together: A National Strategy for Neighbourhood Renewal* (London: SEU); the Social Exclusion Unit (2001a), *A New Commitment to Neighbourhood Renewal: National Strategy Action Plan* (London: SEU).

20 Fitzpatrick, S., Hastings, A. and Kintrea, K. (1998), *Including Young People in Urban Regeneration. A Lot to Learn?* (Bristol: Policy Press), vi and 23. These youth-funded

distancing effect of pushing the debate about regeneration in terms of generating value further into the future.

As a term, "regeneration" does not naturally align with the commonplace stereotyped image of ageing as a process of naturalized *degeneration.* The "inevitable" physical, mental degenerations of the ageing body – as cultural trope – mirrors more closely the symbolic *counter*-discourse of regeneration: the picture of a post-industrial landscape as a degenerating scene of physical deterioration and loss that regeneration works consciously to resist.[21] In the embedded symbolism of *regeneration,* meanwhile, there is an obverse, almost literal kind of re-generation at play: a process of revival and rejuvenation, *with an emphasis on youth.* This is the hidden, symbolic dimension of regeneration that is not ordinarily addressed even though it is confirmed in the practice of social inclusion with policies that manifest an inbuilt generational bias towards youth inclusion.[22] The attentive focus on a younger demographic in regeneration programming is arguably one way of replaying that implicit metaphoric alignment of regeneration with rejuvenation, literally.

To focus, however, on youthful regeneration, brings unintended consequences as it turns attention away from the possible incorporation of an older generation into mainstream regeneration programming. In the calculation of regenerative value it is all too easy to lock out a generation seen to be beyond relevant, "productive" working age (there is, after all no age beyond old age to regenerate productively on into the future).[23] But a regenerative emphasis on *future* citizenship comes at a cost – on multiple levels.

---

programmes often take place on the basis of training up the citizenry of tomorrow on the principle of "us today, you tomorrow" (and the unspoken assumption, that there comes a point where a sector of the population is relegated to the past).

21 For standards stereotypes of the eldery see, O'Reilly, E. (1997), *Decoding the Cultural Stereotypes about Ageing* (New York: Garland Publishing). For recent commentary on the more general failure to incorporate older people in regeneration programming see, Smith, A.E. (2009), *Ageing in Urban Neighbourhoods; Place Attachment and Social Exclusion* (Bristol: Policy Press).

22 More ordinarily, understandings of "regeneration" revolve around a spiritual terminology of "revival" and "rebirth". See, Tallon, A. (2010), *Urban Regeneration in the UK* (Abingdon, Oxon: Routledge), 5.

23 For a general analysis of how the socio-structural marginalisation of the elderly corresponds to the strategic devaluation of those lying outside of the mainstream labor force (i.e. the retired elderly), see Hockey, J. and James, A. (1993), *Growing Up and Growing Old: Ageing and Dependency in the Life Course* (London: SAGE). In a culture where economic value is tied to values of independent adulthood, those outside of the labor force, including those beyond retirement age, are all too easily rendered redundant, marginalized while the presumed economic value of the retired population is often overlooked, denying the genuine and ongoing "mainstream" economic contributions of a retired elderly population working as volunteers.

With regeneration efforts focusing on a younger demographic there is, invariably, a knock-on effect on the real costs of elderly welfare in terms of resource allocation – diverting funds in real terms from elderly interests and local authority care.[24] Similarly, so long as that priority emphasis on youth persists, the possibility for meaningful *inter*generational programming is precluded (limiting the possibility to think creatively across generations beyond the tokenistic use of the "intergenerational" as a one-off, public gesture of social cohesion).[25] Within creative cultural programming, meanwhile, where the focus is, again, still largely youth-oriented, that marginalization of older adults is played out all over again. As Jenny Hockey and Allison James have pointed out, there is a deeply-ingrained cultural tendency to see creativity and creative potential as coded for youth, thus blocking older adults from access to original, creative enterprise.[26] It is that same reluctance to identify creativity in older age that allows articulations of a "creative class" to ignore the elderly and place a generational emphasis almost exclusively on the creative capital of a younger generation (Florida 2004: 294). With its focus on a younger demographic Richard Florida's age-limited conceptualization of a creative class only serves to reinforce the familiar stereotype of creative impoverishment in older age.

But if creative strategies for urban change are more easily aligned towards a younger demographic, it begs the question: how does a changing environment incorporate those long-standing, elderly residents of a place as it undergoes change? As a place like Newham reinvents itself from its heavy, post-industrial past and its lost legacy of manufacturing (the fallout from the dockyard closures of the 1970s and the borough's major employer), does it also ignore its older generation, as it starts to forget and reconfigure the old fixtures and features of its landscape? The paradox in the discourse of social inclusion is its costs of exclusion on-the-ground, as ostensibly inclusive policies of urban renewal actually end up perpetuating a different, generational kind of social exclusion. Alternative cultural strategies, however, offer a way of combating this exclusionary dynamic to produce more

---

24 In the current climate of cost cutting local councils are having to make cost-cutting decisions where the impact on elderly services is proving to be very real. This can, as MP Emily Thornberry has pointed out, mean the difference between carers now only being able to get an older person up after midday as opposed first thing in the morning. See Radio 4, *Any Questions*, 14 January 2011.

25 A case in point is the "Open Your Eyes" project run by local organization Fundamental. Here, a regeneration program started out as an intergenerational project where the future of the borough was envisioned through collaborative film animation between a primary school and a local elders' group (but the project ends up in fact as an exclusively youth-focused creative enterprise).

26 For the contrasting way in which creativity is valued and actively nurtured in children but often overlooked in older age, see Hockey, J. and James, J. (1993), *Growing Up and Growing Old: Ageing and Dependency in the Lifecourse* (London: SAGE), 93. The notion of childhood creativity – in particular, the creative power of play in childhood – can be traced back to the work of the developmental psychologist Jean Piaget in the 1920s.

participatory planning processes instead and, arguably, a more equitable set of outcomes.

Regenerating Newham

For the past eight years, the Newham Education Business Partnership (NEBP) has operated a dedicated regeneration tour for local schools and business – a scenic tour of the major regeneration sites and schemes of the borough: the Barrier Park, "London's first major public parkland project for over 50 years", the Olympic Viewing Point for the 2012 Olympic park; Green Street ("an exciting new shopping district"), and the Cultural Quarter in Stratford, "a new cultural centre for East London".[27] This borough-wide version of planning tourism is the promotional arm of Newham's ongoing regeneration strategy, as the borough continues through a £15 billion investment programme to adapt to its post-industrial legacy and promote its "young, energetic and increasingly skilled" working-age population.[28]

Through tours like the NEBP regeneration tour, Newham has been able to recast itself as, variously, a sports and entrepreneurial city (as a host borough for London 2012), "a place where people choose to live, work and stay" (the borough's recurring slogans), and as a vital cultural center on the margins of London (with a newly commissioned performing arts center in Stratford City Centre). Local cultural organizations, such as "Fundamental" have been operating with a similar ambition to promote "regeneration awareness" across Newham setting up an architecture forum (the *Architecture Crew*) for 13–19-years-olds. Here, working through artists' collaborations, film-making, and the production of newspaper supplements, the *Architecture Crew* provides a creative platform for its young members to voice their own views on local regeneration schemes.[29] But even in this context, the creative cultivation of heightened "regeneration awareness" can look more like a simple retracing of official regeneration tour-route steps – i.e. exploring, rather than engaging with, and critiquing, the primary sites of regeneration and renewal.

Meanwhile, off the official regeneration tour-route, there is this more ordinary culture-infused activity that continues from week to week as generic cultural

27 For more information about the NEBP Regeneration Tour and its sites, visit <http://wwwnebp.org.uk/regen/> [accessed 12 August 2011].

28 For a sense of the range of regeneration projects underway across Newham now, see The London Borough of Newham's dedicated "Regeneration Project" webpage where details of its regeneration programmes sit under the heading "Europe's biggest regeneration project" <http://wwwnewham.gov.uk/InformationforBusinesses/RegenerationProjects/RegenerationProjects.htm> [accessed 12 August 2011].

29 For more information about the Architecture Crew visit the Fundamental Architectural Inclusion website <http://fundamental.uk.net/> [accessed 12 August 2011].

activity, on-the-ground: the Monday morning dance class for elders, a Thursday lunch club in Custom House for pensioners with a "Rem" (reminiscence) session afterwards, and a housebound readers' network with a weekly distribution of books via the local mobile library. This generic cultural activity operates under the radar of mainstream cultural regeneration agendas, a kind of "active ageing" in the generic sense of the term.

While a borough, like Newham, projects into 2012 and beyond by explicitly and visibly building and imaging its future, here, at the level of the local pensioner group, that aspirational dynamic is not so readily apparent. Even within the relatively short span of time that these elderly-focused interventions have been operating in Newham (over the last six years), it is impossible not to notice the state of uncertainty under which these groups operate. They exist under the constant threat of diminished budgets, and are subject to the wavering support of its local authority (the 50-plus dance class in East Ham through the Autumn of Spring of 2007–8 meets each week with an opening caveat that the class may not be able to meet up next week – owing to possible funding cuts). In a way, these clubs and groups operate below a threshold of visibility and certainty, where prospective planning is not always feasible. But the modest persisting routine of these ongoing cultural activities, in spite of their precarity, also offers a comfortable space out of which interventions can start pushing and testing out possibilities for expanded elderly use of urban space.

A Practice of Minimal Intervention

In an inverse image of top-down strategies of regenerative urban change and renewal, *Ageing Facilities*, has been operating its practice of small-scale interventions from the bottom-up, from the ground level of local pensioner groups, clubs and networks. By extending the reach of these groups outwards into the public landscape of the street, each intervention is designed to expand the field of elderly activity, quite literally, gently reconfiguring the reach of existing cultural routines (dancing indoors by day, turned into dancing outdoors by night).

There is a basic groundedness to the mechanics of these "elderly interventions" that start out from within the existing scene of elderly routine: a housebound readers' network, a weekly dance class and a lunch plus reminiscence club. It is that ordinary cultural activity (of dancing, reading, reminiscing) that is gently reworked and extended, through each intervention, into the surrounding urban environment. The weekly routine of an elders'-only dance class, in this way, is gently reconfigured into the choreographed appropriation of a closed-off park by twilight – a small extension of existing "elderly practice". On a practical level, it is the grass-roots mechanics of this interventionist practice that enables a more informal, easy reclamation of urban space – outside of formal planning and design processes.

It is also what enables their autonomous repetition as sustainable acts of intervention that start and end from within the existing cultural scene of weekly elderly routine. But there is also an element of poetic provocation at work here too, beyond the pragmatic mechanics of practice, as each intervention starts to function as a kind of conversational prop: to galvanize and provoke public dialogue and debate around: the place of older people within the urban environment; as creative inhabitants of urban space; as informal planners of their own environment – even within those urban environments undergoing large scale processes of master planned regeneration and change.

As an informal practice of small-scale interventions, *Ageing Facilities* borrows from an existing spatial typology of intervention used by creative practitioners (designers, artists, activists, urban curators) to generate alternative models of urban inhabitation and critique, that operate beyond the parameters of conventional, top-down models of urban planning and change.[30] There is a valuable spatial repertoire of grounded principles and movements to be found in these alternative urban interventions, where the coupling of social critique and direct action is based on a methodology that is fundamentally grass roots, direct, small in scale, and flexible. These interventions work in ways that are often playful, but constructive (in a Situationist vein), even as each intervention operates carefully and deliberately within a minimal economy of means, below the threshold of visibility.[31] There is a degree of spatial ingenuity and agility enabled in a practice of minimal intervention that allows for a more carefree sidestepping of top-down planning strategies, and reconfigures, for a moment at least, generic relationships to urban space.

---

30 See, for instance, the work of artists' collective SYN – who use the unauthorized insertions of "furniture" into the urban environment – direct action interventions – as a way of testing out and critiquing the relative porosity – the openness – of public space. For a selection of interventionist practices see the compilation catalogue, Thompson, N. and Sholette, G. (eds) (2004), *The Interventionists: Users' Manual for the Creative Disruption of Everyday Life* (Cambridge, Mass.: MIT Press).

31 For a starting definition of "urban intervention" as a critical urban practice see City Mine(d) "Urban Interventions in Economic, Political and Cultural Citizenship", PS^2 (2007), *Space Shuttle: Six Projects of Urban Creativity* (Belfast: PS^2), 60–63.

Civil Twilight: An intervention based on Spatial Proposition #9
"A Last Dance in the Park: An Illicit Tea Dance in the Park After-Hours"[32]

Date: 10 June 2008.
Duration: 90 mins.
Location: MUGA, Barking Road Recreation Ground, East Ham.

Figure 9.1 **"A Last Dance in the Park: An Illicit Tea Dance in the Park After Hours"**

32 A spatial proposition from The Fluid Pavement entitled "A Last Dance in the Park". Proposition # 9 reads, in full, as follows: "A Last Dance in the Park is a one-off, unauthorized event: a tea dance in the park after dark. The event is both a historic re-enactment of a lost tradition – dancing in the park on a Saturday afternoon – as well as an illicit flashmob act of unsanctioned congregation in the park after closing time. With a dress-code of fluorescent-enhanced period (minimum) 1950s dress, 'A Last Dance in the Park' combines irreverent misbehaviour with old-style conformity to tradition. For one night only the deadening weight of preservation and tradition is provided with a bit of light relief, as historic re-enactment becomes an opportunity for pensioners to lay claim to the public realm en masse by night." From Handler, S. (2007), *The Fluid Pavement* (London: RIBA Modern Architecture and Town Planning Trust), 120–21.

For one night only (under the guise of an *Ageing Facilities* intervention) an "elders-only" dance class breaks its usual Monday morning routine to go dancing in the Multi-Use Games Area (MUGA) of its local park after dark. Three times the size of the church hall (where the class usually meets each week, to dance indoors) the open air MUGA is, in many ways, the inverse image of the class' own, more limited, literally contained reality. Built as an explicitly youth-designated space the MUGA was constructed – through NRF funding – under the banner of local area renewal. For 90 minutes, though, on the tenth of June the terms of that youth-focused provision is temporarily upset, with the open air MUGA momentarily "reclaimed" by the "elders-only" dance class (a temporary panacea for generational exclusion of sorts).

Tea dancing is, in many ways, a standard form of regular, "elderly" activity. It is a popular pastime and cultural given for an older generation, where tea dance classes exist as part of the basic repertoire of local authority provision for its pensioner communities.[33] In Newham, that contemporary practice of tea dancing now carries a particular inflection as it corresponds to the memory of an older, different kind of dancing routine: of dancing outdoors in the park, on a Saturday afternoon in the old Beckton and Central Parks.[34] Nowadays, that tradition of dancing outdoors is a practice that is largely lost and confined to memory. Dance routines have become spatially more limited, placed in designated community centres and other "public" community provision contexts, the confines of a church hall, for instance, and other sheltered substitutes for the more expansive setting of the public park. (What was once a visible and audible public practice of dancing has now been distinctly contained.)

Unavoidably perhaps, a whole older generation's use of urban space ends up having to adapt to the changing face of the borough over time, as it adjusts itself, physically, to accommodate a younger generation (as priority generation for advancing local economic renewal). The decade-old MUGA at the end of the road is, in a sense, a visible marker of that generational adjustment, as it stands within the public space of the local park, subtracting from its more general, cross-generational "public space". Meanwhile, the old practice of dancing outdoors has turned, literally, inwards, indoors (here, into a church hall). This is a generational adjustment of a different sort, as the parks themselves (like those bordering the MUGA) have, increasingly, over time turned from spaces of elderly occupation

---

33 The London Borough of Newham advertises and part-subsidizes three separate classes across the borough. For a demonstration of the popularity of tea dancing as pensioner activity see the recent mapping exercise conducted at Goldsmiths University by Helen Thomas which demonstrates the lengths to which over-60-year-olds will go to seek out these kinds of dance events <http://dance.gold.ac.uk/activities.html> [accessed 14 June 2009].

34 The Newham Archives in Stratford Library contain a number of dance posters from 1948 advertising this old-time routine of dancing outdoors in the park. One announces, "Summer Season Dancing in Central Park Every Thursday and Saturday" for an entry fee of three pennies.

into spaces of real anxiety and fear. The elderly are often seen avoiding parks like these because of a perceived fear of crime, the "unsettling" presence of congregating youngsters, and the felt need to avoid all open spaces as it starts to get dark (from 4pm in winter) (Neuberger 2008: 118–20).

Cultural practices invariably shift over time, through a lifecourse. They can narrow down and be narrowed spatially, in concrete, real terms. The formerly expanded field of an informal park-based dance practice becomes limited to the confines of a church hall. But that narrowing down is also the chance to offer up through these interventions a propositional what-if? What if a remembered practice becomes a pretext for shifting routinized relationships to space? What if the raw material of existing routine were turned into the basic steps for reconfiguring relationships to space beyond a familiar time and place? What if a set of weekly-danced steps at one end of the road (the tango, cha cha cha, the salsa, the foxtrot …) were turned, for one night only, into an alternative dance routine at its other – a choreographed exercise in "generational re-alignment" – before returning to class the following week?

Liminal zones, like the suspended time zone of the closed-off park after dark, tolerate the inversion of norms and routines.[35] The twilight hour permits the construction of an alternative time and space: testing the boundaries of routine cultural activity; challenging the set-piece generational configurations of urban space; indulging in the small thrills of semi-illicit transgression (moving into a closed-off public park after dark). There is the possibility here of defying those self-imposed temporal routines that can so often self-restrict any kind of movement outdoors after dark and makes possible what is simply the ordinary pleasure of being, for a while, temporarily, out of place.

As a one-off experiment, Civil Twilight (as a literal form of "active ageing") helps "prove" in a fairly ordinary way that the MUGA is as good a place to dance in as dancing indoors in a church hall (not least because the MUGA is at least three times its size). This is the pragmatic dimension of the intervention, as the act of intervention offers up unexpected benefits and functional implications for the class when it returns to dance in the church hall as usual the following Monday morning, broaching the question of the intervention's own "legacy". Does the intervention remain as a one-off action or is it repeatable in some way – as an ongoing series of sustainable urban acts? Is there a desire even (among those involved) to repeat this intervention again? To take part in what might be seen to be a self-consciously elderly, self-segregating act? Or, is there a more inter-generational version of this action to be acted out at some point, perhaps in a different, multi-generational configuration?

From the outset, the intention of the intervention was always to equip the dance class with its own set of facilities, to facilitate the acting out of the intervention

35 For the potential of liminal zones to operate as powerful sites for the inversion of accepted social norms, see Turner, V.W. (1975), *Dramas, Fields and Metaphors: Symbolization in Human Society* (Ithaca, NY: Cornell University Press), 255–6.

itself: 1) a lightweight trolley for serving tea, 2) a set of disposable chairs to be used outdoors, with 3) knee rugs for warmth and, 4) a set of external electrical sockets – for the music player to plug into the power supply of a nearby roadside feeder pillar. Now permanently housed back indoors in the church hall, this mobile kit of tea dance parts is there to practically enable the repetition of this intervention by the class, at its own will. This is the sustainable dimension of the otherwise temporary intervention: a promise to move beyond the limits of urban interventionist practice as a one-off gestural statement and to enable its re-enactment (in any number of different configurations).

Resistant Sitting: An intervention based on Spatial Proposition #4 "The Adaptation of Low Walls by Pavements into Temporary Seats"[36]

Date: Summer 2009.
Duration: Ongoing.
Location: Roadside Abacus® bollards (borough-wide).

Doreen Massey describes ageing as a "closing-in", an increasingly curtailed existence that is seen to be the inevitable by-product of the ageing process: restricted mobility, confinement indoors to more sedentary environments of care homes, sheltered housing complexes, age-segregated spaces and community provision scenarios for "elders" only (Massey 1986: 75). But the notion of an inevitable closing in older age, a narrowing down to a more limited field of existence is, in its lived reality, more complicated. As Massey herself starts to imply, the process of closing in will almost invariably start to induce small compensatory mechanisms – physical and psychological mechanisms – that manage to narrow down into conditions of mounting dependency. (Massey observes her father adopting the compensatory technique of walking with his eyes tied to the ground, to avoid tripping on broken paving stones (Massey 1986: 75).)

From 2005 onwards, as part of an early *Ageing Facilities* initiative, that notion of resisting an "inevitable" closing in emerges in a different context – through a conversational intervention of sorts. Here, the usual reminiscence format of a bi-weekly pensioner lunch-plus-reminiscence club in Custom House, is temporarily

36 A spatial proposition from *The Fluid Pavement.* Proposition #4 reads, in full, as follows: "Hollowed and padded out with detachable made-to-measure cushions made to fit into these hollows, the Accommodating Low Wall, provides a much-needed seat for those in need of a rest when there is no obvious sitting spot nearby. The detachable cushions, which are to be removed and fixed into these hollowed-out walls according to need, are of a standard size and will fit any hollowed-out wall in the borough. The cushions will be available from a central distribution point in the borough. They are free of charge on display of a freedom pass." Handler, S. (2007), *The Fluid Pavement* (London: RIBA Modern Architecture and Town Planning Trust), 110–11.

reconfigured intermittently over the course of four years. Conversations within the club were, in this way, guided to digress gently but deliberately away from the usual reminiscence rules of the game. From talking about the way things were, and the often nostalgia-infused rhetoric of the way things have changed since, conversation was turned instead to telling stories about surroundings lived-in now. Over time, these conversations started to reveal a whole repertoire of tactics and techniques used between the home and streetscape as a way of adapting to, and actively adapting in turn, a surrounding environment as it starts to be experienced in a different way in older age. A radiator in the front hall gets re-appropriated, for instance, as an informal walking stick. A generic roadside bollard is turned into an impromptu public seat when there is nowhere else to sit. (The lack of adequate street furniture provision – surfaces over and again, in the club, as a recurring complaint.)

In a way, what these conversations reveal, above all else, is the way in which the urban environment surrounding the Rem club itself, is already measured and mapped out clearly in the mind. There is a whole typological inventory here of "good enough" places to sit when there are not enough standard seats around: the appropriated wall of sittable height, low enough to sit on; a series of rightly paced bollards along a stretch of road that function as a post-operative walking aid. This kind of mental mapping of the urban environment, holding a re-appropriated streetscape in the head, reveals the mundane, day-to-day reality of how people start, in their own way, to operate as informal planners of their own environment. As everyday tactics, however, acted out on-the-ground they reveal a certain creative flexibility too: appropriating the borderline features of a street (low walls, bollards), turning in-between spaces into accommodating rest stops (temporary "public" personal-use seats), resisting the absences of formal street furniture provision where so often, in older age, the urban environment can be experienced as a series of subtractions (a regularly used bus stop seat, for instance, on Upton Lane arbitrarily removed).

There is a basic principle of hospitality that underpins public street furniture provision – the generosity in furnishing a street at all. But that generosity is all too often undercut by arbitrary subtractions (the arbitrarily removed bus stop seat, for instance) and its allied logic of the design coded streetscape that determines the extent and limit in the use of public space at all. This is the embedded logic of street furniture design where the pitching of a bus stop seat will be designed to a deliberately sloped incline to curtail the length of a sitter's stay (Norman: 2000). All of these modifications have a real impact, a genuine knock-on effect on degrees of comfort and accessibility.

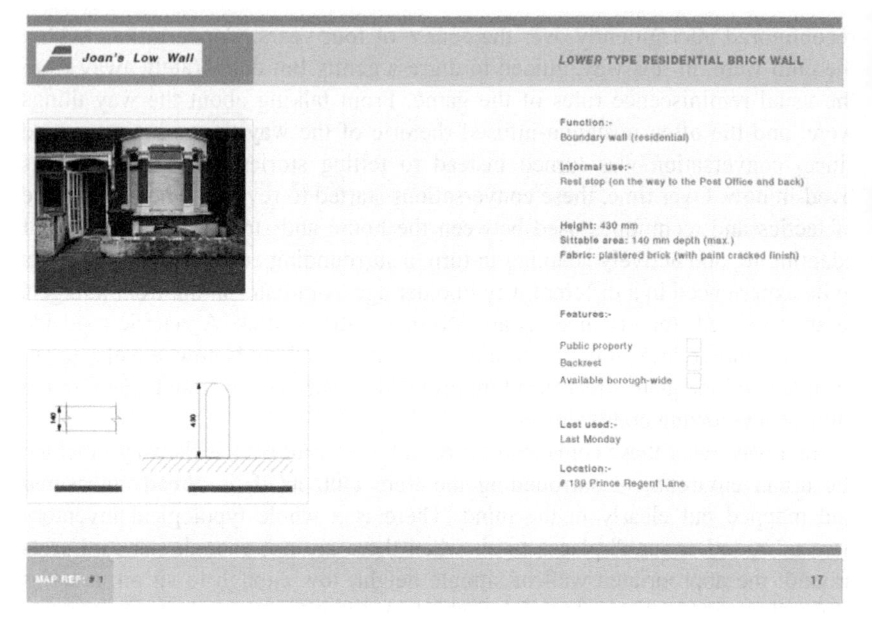

Figure 9.2 Alternative street furniture guide

This happens at both the baseline of physical needs and beyond, where the related psychology of how accommodating an urban environment even makes you feel, comes with its concomitant emotional costs determining and limiting time spent outdoors – if at all.[37] To sit, then, deliberately, out of place (on a bollard, or a wall) is, arguably, already in its own way an unconscious form of resistance,[38] a latent form of urban critique that is, in a sense, a direct intervention of its own sort (like the choreographed intervention of Civil Twilight – there dancing, only here sitting deliberately out of place).

37 For the related psychology of physical falls in older age and its impact – in terms of emotional cost on older people's identity (not just on a baseline of physical mobility), see Davenhill, R., "Developments in Psychoanalytic Thinking and in Therapuetic Attitudes and Services" in Davenhill, R. (ed.) (2007), *Looking into Later Life: A Psychoanalytic Approach to Depression and Dementia in Old Age* (London: Karnac), 24.

38 For other, examples of elderly resistance that presents itself as obstructive/playful mis-behaviour (in care home settings principally) see Hockey, J. and James, A. (1993), *Growing Up and Growing Old: Ageing and Dependency in the Lifecourse* (London: SAGE), 172–3. Here, a range of behavioural strategies (including techniques of stubbornness) are deployed to communicate, challenge and subvert the oftentimes frustrating states of dependency experienced by the elderly in carer-cared for relations. (Here, resistance is expressed through seemingly innocuous jokes and willful acts of stubbornness.)

Figure 9.3 Resistant sitting: the reversible cushion

It is in that same spirit of "creative resistance" that *An Alternative Street Furniture Guide for Pensioners* (produced by *Ageing Facilities* in 2009) starts to formalize and give value to these informal acts of sitting out of place, as it maps out this landscape of "elderly-appropriated spaces" – spaces that would otherwise be held hidden, invisible in the head. In a formal parody of standard local authority street furniture codes, this alternative street furniture guide presents a comprehensive catalogue of low walls, bollards, the ordinary features of the street that are misused as temporary seats. Giving value to these alternative sitting spots, the guide expands the restrictive design code discourse of conventional guides while offering up what is also a serious spatial proposition for planners: the parasitic appropriation of existing features of the street (see Figure 9.2).

The guide also comes with a portable cushion attached: an artist-commissioned cushion made-to-measure the generic Abacus® roadside bollard (for added comfort, when sitting on the bollard's cast iron seat) (see Figure 9.3). Designed by artist Verity-Jane Keefe, with a reversible, two-sided fair and wet-weather face, for all-weather use – including a minimum of 30mm padding – the cushion exists as the poetic accompaniment to the alternative street furniture guide. A playful formalization of sitting out of place, constructed in collaboration with elderly users, the cushion legitimates symbolically, as much as it tries to physically facilitate, that "resistant" reclamation of urban space in older age. (The prototype first generation version of the bollard-cushion is currently being adapted for more robust application as a series of personalized multiples, for more widespread, borough-wide use.)

Playfully resisting the generic solution, these personalized, bollard-bespoke cushions are being designed to remain literally in the hands of elderly use as personalized objects (free to use or dispose of at will). They are designed to facilitate more confident (less tentative) acts of sitting out of place, and to encourage the ad hoc retrofitting of public space as an individual, autonomous practice that transforms the generic streetscape into a more personalized space. The cushion, in this sense, represents a small-scale gesture of personal investment to a place – amplifying ownership of and attachment to the most mundane of places.

Reality Orientation

As witnesses of urban change over a life course, an older generation often maintains a long-standing attachment to place that persists even amidst the radical change of a regenerating landscape.[39]As a theoretical concept, that sense of place attachment can easily be read as attachment to place borne out of an over-attachment to the past – a nostalgia of sorts, where attachment to place links into personal memory of that place (G.D. Rowles' notion of "autobiographical insidedness").[40] In Newham, the changing, degenerating face of the post-industrial landscape could easily be lamented alongside the disappearance of printing works and old markets – sites of personal history, places that carry personal associations, private memories, places of work retired from now. The nostalgia of personal history fails to keep up with the evolving shape of an urban landscape undergoing change in an inevitable process of alienation. In a growing disconnect from the surrounding environment, the only connection with a place is with the place as it once was in the past. In this stereotyped construction of older age as rooted in the past, change from the familiar is met with an anxiety and/or hostility, and the assumption is made that the older you get, the more alienated you feel from the change that surrounds you.

---

39 "I've never wanted to leave here." Still, after over 60 years of living within Newham, this is the familiar refrain of real attachment to a place that has changed a great deal within those 60 years. Kelly, C. (2005), *There Go the Ships: East End Lives in Words and Pictures* (Maidstone, Kent: CK Editions), 21.

40 For an overview of this notion of place attachment and autobiographical insidedness see Peace, S., Wahl, H.W., Mollenkopf, H. and Oswald F., "Environment and Ageing" in Bond J., Peace, S. and Dittmann-Kohli, F. (eds) (2007), *Ageing in Society* (London: SAGE), 215. See also Rowles, G.D. (1978), *Prisoners of Space* (Colorado: Westview Press); Rowles, G.D., "Geographical Dimensions of Social Support in Rural Appalachia", in G.D. Rowles and R.J. Ohta (eds) (1983), *Aging and Milieu: Environmental Perspectives on Growing Old* (New York: Academic Press).

Heritage-based regeneration strategies are the outward signs of this retrospective, nostalgia-infused notion of attachment to place as it was. The past, here, is valued as a compensatory, but still empty, restoration of what was (compensating for the losses of urban change). In 2008, the re-landscaping of Newham's Central Park proceeded, with Heritage Lottery funding, in line with its former 1898-footprint, with regeneration and renewal conforming to the "restorative" value of what once was. Regeneration strategies like these are all, in a sense, nostalgic responses to that retrospective notion of place attachment. They mirror the working mechanics of the reminiscence club: valuing the elderly in terms of their past. What was as against what is, or might be, becomes an easier way to relate to an older generation (for a generation to relate to itself).

But place attachment could be thought of differently, as a theoretically more grounded concept, more circumspect and rooted in the here and now. In this instance, attachment to place might carry a degree of pleasure and promise in what already exists, and in what might exist, providing prospects for the elderly as much as for an upcoming generation.

***

In elderly care, there is a term "reality orientation", that is now unfashionable. It refers to a technique used to orient a person mentally back into the present when reminiscence presents as cyclical, circular behavior (Woods 1996: 583). The injunction of reality orientation pulls a reminiscing individual back into the here and now – not the what-was – and it re-orients that person into the surrounding present. It is an outmoded technique that applies equally to the urban interventions described above, as each intervention works to actively shift retrospective modes of engagement with the elderly into the here and now. Even as these interventions emerge, paradoxically, out of scenes of reminiscence, scenes of nostalgia (the Rem club setting, the remembered routine of dancing in parks), the interventions themselves function as small-scale exercises in urban reality orientation. They cultivate real-time and prospective modes of engaging with a surrounding place, making space, literally, for the elderly appropriation of urban space.

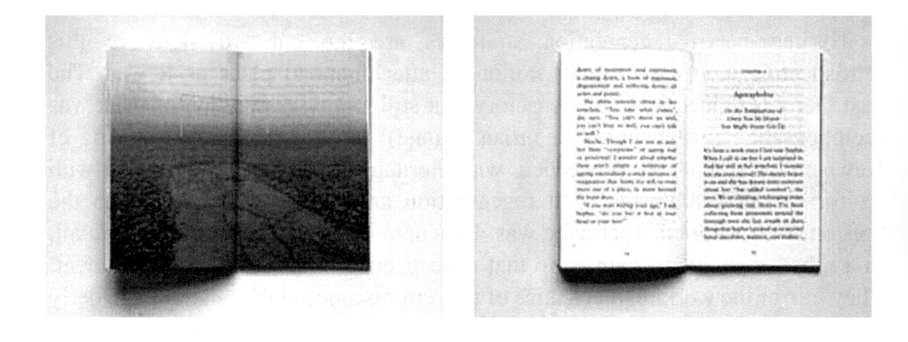

Figure 9.4 ***The Fluid Pavement*** **– extract**

The Fluid Pavement and the Housebound Readers Network

When *Ageing Facilities* began its work on ageing six years ago (in response to muf's feasibility study with its emphasis on youth), its practice started out, quite simply, as a straightforward mapping exercise: trying to give voice and space to a minority elderly population in a fast-regenerating borough. To document and map out people's changing relationships to Newham in older age, invariably involved a process of moving between radically different settings: from the confines of a geriatric ward to a pensioner lunch club to care homes and sheltered housing complexes (for the more isolated housebound elderly).

In practice, the breadth of that mapping process was only made possible through the informal cultural network of Newham's housebound library service (a weekly mobile library service run by the local authority, distributing books to its readership of more isolated, housebound pensioners).[41] The followed-out trail of that library service became, for *Ageing Facilities*, the starting point for a whole series of borough-wide conversations about the spatiality of ageing set within the real context of a marginalized, harder-to reach elderly demographic.

In a spirit of methodological circularity (where the methodology of a research process responds to the landscape out of which it emerges), that research and mapping exercise ended up being resolved, quite deliberately, as a semi-fictional story, with the story itself returned to circulate back in the same housebound readers network out of which it first emerged –an urban intervention of its own sort. Available in Large Print, that story – *The Fluid Pavement* – reads as a playful research travelogue: "a semi-fictional journey through regenerating Newham,

41 The library service is generally seen to be a vital social and not just a cultural lifeline for the elderly specifically (particularly within rural communities). Current calls, led by Somerset library campaigner Steve Ross, for a national inquiry into the closure of public library services are frequently articulated in terms of the important value library services hold for the elderly. See BBC Radio 4, *You and Yours*, 13 January 2011.

investigating the spatiality of ageing from Plaistow to Canning Town, and featuring such notable places as the shelves in Poundland, Edie's porch and that double avenue of trees in Beckton Park (gone missing)" (Handler 2007). Written as a readable account of ageing and growing old within a landscape of regeneration and change, the story is meant to be digested at a readerly pace, written in a tempo and tone that reflects and responds directly to the local readership network out of which it first emerged.

Circulating now around the borough on a mobile bus, *The Fluid Pavement* tries to resist the more commonplace fate of mapping exercises that all-too-easily get filed away. As a mobile, spatial artefact in its own right, the book offers up a useful model for the kind of urban interventionist practice that generates its own space for critical reflection. In its own way, *The Fluid Pavement* carves out a real, material place for mirroring elderly expression, for airing hidden feelings, thoughts and taboos that would otherwise remain housebound, tied to a care home, or a hospital bed as private thoughts. The storyline of *The Fluid Pavement* gently amplifies as it circulates the quiet cultural economy of this housebound readers' network – a space for reading, reflection and beyond (into flights of the imagination).

In the end, the novel ends with a blueprint of alternative fantasies: a series of 10 spatial propositions, "*from the sublime to the absurd*" (Handler 2007: 104–23) that offer the reader different ways of laying claim to public space in older age. These 10 propositions present a whole set of spatial possibilities and fantasies that include: fluidified pavement for recovered ease of movement (*Spatial Proposition #5)*; a set of pensioner-twinned trees to enable a sense of "age-equivalent space" (*Spatial Proposition #8)*; the accommodating low wall to make sitting outdoors that bit more comfortable (*Spatial Proposition #4)*. Dreamt up by its psychogeographic author these ten spatial propositions exist as the creative extensions of actual scenes and desires recorded within the storyline itself. They are offered up to its elderly readership as concluding fantasies that point to other possible ways of appropriating and thinking about a given space. Since the publication of *The Fluid Pavement*, these propositions have gone on to serve as the catalyst for those interventions subsequently initiated by *Ageing Facilities*. On the 21st March 2007, during the public launch of *The Fluid Pavement* members of the housebound readers' network voted for their top three interventions (while following out the route of the novel's storyline on the back of the mobile library bus). Their votes, in order of preference, were as follows: 1) the accommodating low wall (with its portable cushion), 2) a dance in the park after dark, and 3) a set of pensioner-twinned trees (a fantasy proposition to plant mature trees across the borough that are twinned to match the age of its pensioners). Variations of the first of these interventions (the dance in the park and the portable "public" cushion) have now been realized (under the *Ageing Facilities* platform).[42]

---

42 That portable public cushion (also known as the "Clutter Cushion" is currently on display at the Canadian Center for Architecture, Montreal as part of its exhibition "Imperfect Health: the Medicalization of Architecture" November 2011–April 2012.

There is a curious paradox in the semi-fictional scene of the novel becoming the literal, "real-world" pretext for initiating interventions on-the-ground as an ongoing programme of elderly-specific projects. Each project, realized out of that fictional fantasy-world, has become the basis for reconfiguring relationships to a changing urban environment, directly. Arguably, fictional space provides a kind of poetic breathing room that tolerates, provokes and facilitates the reader in imagining other ways of inhabiting regenerated public space in older age. It actively engages with people's shifting relationships to their surrounding environment in the face of wholesale urban regeneration and change. "Preparing the ground" in this way, generating fictions and fantasies, is a first step in using creative methods to shift the physical and rhetorical spatial dynamic of regeneration strategies that are typically more easily tailored to attend to the desires and needs a younger (adult) demographic.

Like the interventions that have followed since, the element of fantasy in *The Fluid Pavement* as semi-fictional story offers real room to think about and enact interventions that enable Other ways of imagining, inhabiting and relating to environments-in-transition creatively, informally, playfully – outside of formal planning processes. This follows the "playful-constructive" mantra of the Situationist urban interventionists of a generation ago (Sadler 1999: 77). But it is in the grounded poetics of these interventions (where fantasy propositions are realized at the mundane level of "elderly routine") that the everyday scene of generic cultural activity is gently extended. The concrete, socio-political ambitions of each intervention become a mechanism for literally opening out the field of elderly activity somewhere along the line between fantasy and generic routine. "Active ageing" starts to take on a more playful inflection here, as each of these interventions moves beyond a baseline of physiognomic needs, and the more familiar functionalist response to ageing as "problem-solving" activity, revealing instead, along the way, the hidden assets of elderly routine. In the end, the constructive play of each intervention is only ever facilitating what already exists – i.e., the poetic extension of existing elderly activity.

Postscript

It should be noted that these *Ageing Facilities* interventions try to operate discreetly, as far as possible, as non-spectacle in order to avoid the reduction of either "pensioner participants", or of the creative practice of the intervention itself, to an aesthetic spectacle for public consumption. Choreographed outside of local authority public relations teams, each intervention functions with two guiding principles in mind: to enable the embodied critique of otherwise routinized relationships to a given place through the performative act of the intervention itself, and to initiate a process of critical reflection in relation to these interventions, including thinking pragmatically about the public policy implications of each.

As each intervention starts to feed back into a broader discursive circuit of thinking around ageing – through website, publications, conferences – the last act of each intervention (beyond the "participative" moment of the intervention itself) is always a turning towards its dissemination into an expanded field of reflection and critique. The intervention, in a way, is only ever the starting point for thinking more broadly about the spatial politics of creative forms of "urban regeneration" and its (hidden) relationship to ageing.

Bibliography

Carver, V. and Liddiard, P. (eds) (1978), *An Ageing Population; A Reader and Sourcebook* (Milton Keynes: Open University Press).

City Mine(d) (2007), "Urban interventions in economic, political and cultural citizenship", PS², *Space Shuttle: Six Projects of Urban Creativity* (Belfast: PS²).

Davenhill, R. (ed.) (2007), *Looking into Later Life: A Psychoanalytic Approach to Depression and Dementia in Old Age* (London: Karnac).

Fitzpatrick, S., Hastings, A. and Kintrea, K. (1998), *Including Young People in Urban Regeneration. A Lot to Learn?* (Bristol: Policy Press).

Florida, R. (2004), *The Rise of the Creative Class: And How It's Transforming Work, Leisure, Community and Everyday Life* (New York: Basic Books).

Handler, S. (2007), *The Fluid Pavement* (London: RIBA Modern Architecture and Town Planning Trust).

Hockey, J. and James, A. (1993), *Growing Up and Growing Old: Ageing and Dependency in the Life Course* (London: SAGE).

Kelly, C. (2005), *There Go the Ships: East End Lives in Words and Pictures* (Maidstone, Kent: CK Editions).

London Borough of Newham (2007), *The State of the Borough: An Economic, Social and Envrionmental Profile of Newham* (London: Local Futures).

Massey, D. "My mother lives now in a nursing home", in I. Borden, J. Kerr and J. Rendell (eds) (1996), *Strangely Familiar: Narratives of Architecture and the City* (London: Routledge).

muf architecture/art (2004), *Small Open Spaces that are Not Parks* (London: SDP).

Neuberger, J. (2008), *Not Dead Yet: A Manifesto For Old Age* (London: Harper Collins).

Norman, N. (2000), *The Contemporary Picturesque* (London: Bookworks).

O'Reilly, E. (1997), *Decoding the Cultural Stereotypes about Ageing* (New York: Garland Publishing).

Peace, S., Wahl, H.W., Mollenkopf, H. and Oswald F. (2007), "Environment and ageing" in J. Bond, S. Peace and F. Dittmann-Kohli (eds), *Ageing in Society* (London: SAGE).

Pewsey, S. (2001), *Newham: Past and Present: The Changing Face of the Area and Its People* (Swindon, UK: Sutton Publishing).

Phillipson, C. and Scharf, T. (2004), *The Impact of Government Policy on the Social Exclusion of Older People; A Review of the Literature* (London: Social Exclusion Unit, Office of the Deputy Prime Minister).
Rooke, A. and Wuerfel, G. (2007), *Mobilizing Knowledge – Solving the Interaction Gap Between Older People, Planners, Experts and General Citizens within the Thames Gateway* (London: Centre for Urban and Community Research, University of London).
Rowles, G.D. and Ohta, R.J. (eds) (1983), *Aging and Milieu: Environmental Perspectives on Growing Old* (New York: Academic Press).
Sadler, S. (1999), *The Situationist City* (Cambridge, Mass.: MIT Press).
Sainsbury, F. (1986), *West Ham: 1886–1986* (London: Plaistow Press).
Smith, A., Scharf, T., Phillipson, C., Smith, A.E. and Kingston, P. (2002), *Growing Older in Socially Deprived Areas* (London: Help the Aged).
Smith, A.E. (2009), *Ageing in Urban Neighbourhoods; Place Attachment and Social Exclusion* (Bristol: Policy Press).
Social Exclusion Unit. (1998), *Bringing Britain Together: A National Strategy for Neighbourhood Renewal* (London: SEU).
Social Exclusion Unit. (2001a), *A New Commitment to Neighbourhood Renewal: National Strategy Action Plan* (London: SEU).
Tallon, A. (2010), *Urban Regeneration in the UK* (Abingdon, Oxon: Routledge).
Thompson, N. and Sholette, G. (eds) (2004), *The Interventionists: Users' Manual for the Creative Disruption of Everyday Life* (Cambridge, Mass.: MIT Press).
Turner, V.W. (1975), *Dramas, Fields and Metaphors: Symbolization in Human Society* (Ithaca, NY: Cornell University Press).
United Nations. (2002), *World Population Ageing: 1950–2050 [executive summary]* (New York: UN Department of Economic and Social Affairs, Population Division).
Woods, R.T. (1996), *Handbook of the Clinical Psychology of Ageing* (Chichester: John Wiley and Sons).
Woodward, K. (1991), *Aging and Its Discontents. Freud and Other Fictions* (Bloomington and Indianapolis: Indiana University Press).
World Health Organization. (2002), *Active Ageing: A Policy Framework* (A contribution of the World Health Organization to the Second United Nations World Assembly on Ageing, Madrid, Spain: April).

Chapter 10
Access as the "Gorgeous Norm": Creating Spaces that Inspire Ownership through Occupation

A conversation between Liza Fior and Myrna Margulies Breitbart

Prelude

In March 2011, I had the opportunity to meet briefly with Liza Fior, one of the founding architects of muf architecture/art, a London-based firm that was established in 1995. muf is noted for innovative research methodologies and a commitment to engaging communities in the improvement of public spaces and their access to them.

I was introduced to muf through a mutual acquaintance, Jane Rendell. My curiosity about the firm derived from a (then) limited knowledge of their gender politics and their use of the design process to co-create inspired environments with occupants. This includes temporary alterations of space as opposed to building construction alone. Known more for their detailed site-specific studies of individual spaces and temporary occupations of space than starchitect buildings, their work was formally recognized when the firm was awarded the prestigious 2008 European Prize for Urban Public Space for a new town square they helped to create in Barking, East London.

The intersection of muf's work with a book on the post-industrial creative economy derives, in large measure, from the out-of-the-box, highly effective methods the firm employs to induce the participation of users of space in the rethinking and regeneration of their neighborhoods. A part of this process involves the design and use of spatial artifacts to suggest the possibility of new social relationships and to promote a more critical stance on changes taking place in the local environment. muf's ability to subvert and redirect already-established planning agendas in favor of animating creative ideas is now legendary. This effectively combines with the clever, often humorous and artful, means they use to provoke larger aspirations for change and disrupt tried and true notions, such as the private ownership of urban space. The repertoire of critical spatial practices that muf employs opens up spaces (even at the tiniest of scales) and occasions the emergence of multiple new and unexpected possibilities in the urban regeneration process. Adding to their appeal is their track record of inviting many diverse participants into a complex analysis of environments that privilege residents' personal knowledge, often in the interest of altering public policy.

On the plane over to the UK, I perused examples of muf's projects and composed several detailed questions that I expected would frame my interview with Liza Fior. What I failed to anticipate (and should have) was far from the structured interview I was used to. With the recorder between us, Liza and I neglected to follow any linear thread that might have produced a lucid transcript. Instead she took me on a mesmerizing journey that moved backwards and forwards through richly visual examples of projects that could not be entered into recorded sound. The depth of my interest in muf expanded with each example, and I wanted the conversation to extend far beyond our allotted time. A read-through of our transcribed conversation and subsequent communications reveals a series of themes that can be gleaned from the shared associations and individual digressions that resulted from an interviewee too interested in the experiences of the interviewer, and an interviewer who brought many overlapping references to the table, including memories of creative environmental interventions discussed at an anarchist conference in Venice in 1984.

What follows is a sampling of the themes we touched upon in our conversations, as well as a description of the methodologies, and built and unbuilt proposals, that have been generated through muf's efforts to work with residents and neighborhoods to create economically sustainable creative activity in post-industrial settings. While part of the London conurbation, many of the neighborhoods that are described below have strong identities of their own and share the characteristics of smaller de-industrialized cities.

"Start off gently and take advantage of polarized, anger-producing situations. Take the right risk at the right moment. Make visible what really is going on by using cultural asset mapping (with broad inclusive definitions of culture) to publicize and highlight what is there and what is missing."

Project Example: Making Space in Dalston (2008–2012): Value What's There

Dalston, East London is a neighborhood that operates somewhat in isolation from other parts of the city. The area is changing rapidly as large-scale developments meet a densely occupied area with a high index of poverty and unemployment. There is a history of social activism yet Dalston is the only area in London to see off neighboring riots through standing firm in support of Turkish businesses. New coffee shops seem to open on a monthly basis and the neighborhood maintains a variety of small independent businesses, a rich ethnic diversity and many well-loved cultural organizations.

Mark Stern and Susan Seifert refer to "natural" cultural districts as areas that, among other things, are often characterized by a "prof/pov" demographic: a higher than average proportion of the population with strong post-graduate degrees and a higher than average proportion of people with few educational qualifications at all (Stern and Seifert 2007). Dalston fits this profile. When muf began work to establish an alternative master plan that "values what's there" it sought to maintain this diverse yet fragile demographic balance. The idea was to give value to the dynamic created by the new professional and artistic class without giving it *too much* value and causing the displacement of existing residents and their creative pursuits.

When muf entered the community Dalston was already polarized and there was a palpable degree of anger due in part to the rush of developers to demolish buildings that contained thriving small businesses and community organizations that defined the area's identity. The demolition of one building in particular, the Dalston Theatre, symbolized for residents the borough's disinterest in the area's existing assets. muf responded by saying, "Right, rather than everything being so polarized, it's time to pause". A new approach to participative research aimed at unearthing possibilities for more shared public spaces was initiated. Working with J & L Gibbons landscape architects, muf commenced the project *Making Space* as a "gentle" way of intervening into a controversial regeneration process, transforming it into one that actively engages residents, improves their quality of life and nurtures indigenous creativity and diversity. Two approaches were employed simultaneously: one involved making visible and celebrating existing social, cultural and physical assets; and the other generated design strategies and a programme of cultural production to "nurture the possible" in ways that would benefit both residents and visitors. There was also a conscious effort to involve the voluntary sector and international artists in the animation of shared public and semi-public spaces. muf believed that Dalston did not have to be branded as a "cultural district" since it *already was one*. They felt nonetheless that more could be done to use local arts to strengthen social networks and to derive more economic benefit for long-term residents from existing cultural activity.

Mapping is the primary tool that muf employs to develop a conversation and to implicate disparate parties as stakeholders. To this end, a catalog of cultural and arts-related events in Dalston was compiled from a local blog and flyers found all around the neighborhood. This "close looking" culminated in a map that revealed an amazingly high level of activity, and gave value to fragile ecologies, through a creative visualization of space that represented Dalston as *it is* and as it *could/should be*. The map became an invitation to organizations to come up with ideas about how to animate neighborhood space and further develop its potential. This process began with a first meeting when the maps were presented back to those represented on it. Participants were asked to evaluate them both for accuracy and for what they suggested. The comments were varied, and included "who is this for, the estate agents?" A self-selected, shifting number of the 20–50 attendees, who continued to come to meetings, constituted a stakeholder group to whom muf and J & L Gibbons presented the project monthly as co-authors and critical friends. The high risk was the insistence of including a representative from the local campaigning organization Open Dalston as part of the client steering group –

the first time they had sat on that side of the table. Open Dalston began campaigning against the development plans and associated demolition, and continued to engage with other development plans for the area. The mapping and proposals (76 projects in total) that had been initially provoked as a critique of a top down master-plan, were officially embraced as the first phase of change, although in a neighborhood with great scarcity of open spaces and development sites, few projects were neutral. Later in this chapter two of those projects – the Dalston Barn and Gilette Square – are discussed in more detail.

Open Dalston maintained their autonomy, organizing an exhibition called "What's Best for Dalston?". Residents were invited to address key issues embedded in the Master Plan commissioned by the Council, and were able to influence it. Finally, one of those representatives from Open Dalston became custodian of the Eastern Curve garden, the largest of the six of the 76 projects that were built.

Figure 10.1 East Curve Garden

Use Culture to Reframe What is Already in a Place. Shift Redevelopment Plans to Focus on Augmenting Existing Assets. "It's Got to Be Gorgeous!"

> Supporting creativity is not about single public art commissions, but rather making it easy for all the arts, including music, visual arts, film and performance to be part of the public realm (Liza Fior).

The process employed in *Making Space* in Dalston was as much about reframing and renaming what was already there as adding new design elements. Existing development plans required demolition of several buildings in order that a plan of sufficient scale "stacked up". One clear intention of muf was to shift the role of public realm projects from a focus on creating more of a gateway and "wayfinding" for visitors (which assumes the lost-ness of the incomer) to documenting and preserving the positive activities already present in the neighborhood. A place that is valued does not need signposting. The resulting map asked the question, "Does value always have to be displaced? Can value be shared?"

muf employed the Dewey decimal system pertaining to the humanities and culture to categorize creative spaces that already exist. Residents eventually added more activities to the cultural asset map, including churches, open spaces, heritage buildings and more cultural venues and businesses aimed at enhancing the local and visitor economy. muf then used this data to propose 76 small projects that ranged in cost from £5,000,000 to £5,000.

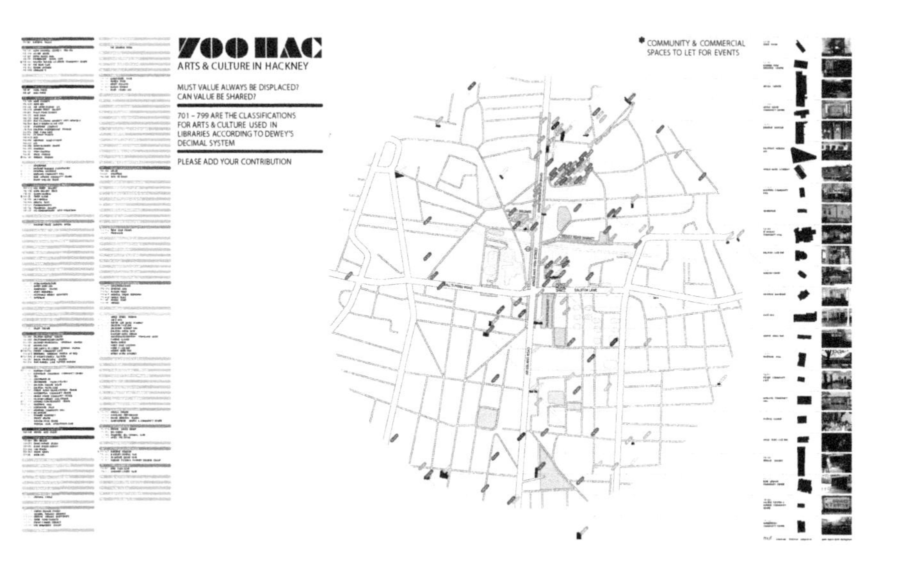

Figure 10.2 Dalston map: "Does value have to be displaced?"

Project Example:
"Tailgating as Municipal Housekeeping" in Pittsburgh, Pennsylvania

The process of data collections and analysis used in *Making Space* in Dalston drew on muf's experience of working in the North American context, and the opportunity this presented to question the relationship between a cultural institution and its context. Pittsburgh, Pennsylvania, once a classic "shrinking city" that experienced a massive loss of economic opportunity, and the out-migration of more than 20% of its population, has since worked to build a more diversified economic base with high-tech industry, health care, culture and tourism.

One neighborhood has remained outside this regeneration scenario, with the exception of eight days a year when the Pittsburgh Steelers play a home football game in their North Side stadium, drawing over 60,000 non-residents back into town and filling vacant lots with barbeques and music. The North Side is separated from the rest of the city by a massive highway and is designated as the site of a new casino. muf was one of four teams selected to come up with creative ideas to spark a more permanent regeneration for the North Side initiated by Jane Werner, director of the Children's Museum, which also houses a Head Start nursery and provides free admission to the museum to the Head Start children and their families. The museum is one member of a 10-institution coalition of local cultural organizations that calls itself a "Charm Bracelet" and is promoting the regeneration of the North Side as a "family district". The design teams were invited to Pittsburgh for three days and then returned with ideas four months later.

When Liza Fior entered this neighborhood she immediately picked up on the irony in first "white flight", and then white flight bringing the suburbs back to town with all their barbeques. This notion, interwoven with the municipal housekeeping movement, sparked the idea for the principle of "Tailgating as Municipal Housekeeping". Jocelyn Horner, student at the London School of Economics in London, introduced Liza Fior to the history of Municipal Housekeeping, a movement of women's clubs and settlement houses in the 1890s that saw women as the domestic caretakers of not only their households, but their cities as well.

The Pittsburgh Steelers happened to be playing the day before the three-day organized program. Luckily Fior came a day early in order to walk the North Side in advance of the official tour. She came across the tailgating phenomenon, which occurs when mainly white suburbanites return for the day from the suburbs. muf devised the concept of "tailgating as municipal housekeeping" by working from the premise of bringing out the resources of the museum and amplifying them to the scale of the city. The idea is that each of the "charms" (as these cultural institutions in the North Side were named) opens their "gates" and operates as "municipal housekeepers" using the "properties" of tailgating. These include the temporary occupation of vacant sites to demonstrate what is "possible"; the addition of missing amenities to create new social spaces; and the use of a "captive audience" (i.e., visitors who come to visit the cultural "charms") to contribute to the local economy.

Figure 10.3a "Tailgating as Municipal Housekeeping"

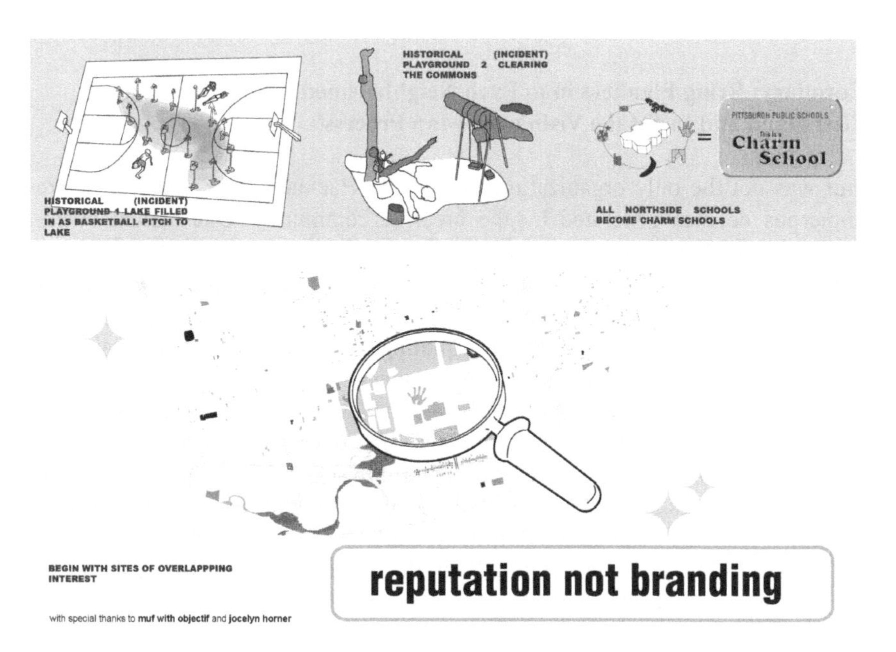

Figure 10.3b "Reputation not branding"

Another important goal was to highlight the resources of each "charm" and help them to see themselves as a part of the neighborhood by adopting a public school and transforming the North Side into an important site for learning and exploration.

Project Example: The Olympic "Fringe"

This idea of suggesting new status for existing spaces and changing their meaning through what happens in a place, is also illustrated in the repurposing models and projects that muf continues to work on near the Olympic site. The building of an Olympic site meant immense change to the built and social environment for surrounding neighborhoods. muf was selected to work on some permanent and temporary projects with a brief and funding headlined with the intention to *lead the visitor through* the neighborhood to the Olympic site. Again, muf began with a mapping of what was in place as a means to tamper with master plans that proposed something more grandiose without seeing grandiose potential in what already exists. One example of a close looking in Hackney Wick and Fish Island revealed many hidden dimensions of the area, including a cadre of creative residents, ecological features, and potential new play spaces. A single proposal among many encouraged collaboration between local artists and businesses over three years. The work gathered authority from mapping to proposing scenarios that could temper development to make space to retain existing tenants and activities. Throughout this process, muf worked closely with a client who supported the expansion of the project beyond its original brief

Corollary: Bring Planners in to Each Neighborhood to Experience them First Hand and to "Make Visible the Map Process"

muf was not the only organization working in Hackney Wick and there were numerous debates generated by the creative community. One annual event, "Legacy Now", was held by Space Studios to highlight the need to safeguard cultural industries as an important element of the Olympic legacy for East London (http://www.spacestudios.org.uk/space/what-we-do/legacy-now). muf used the systematic mapping and collating of data as a conduit, an emissary, between the creative community and the planners.

Mapping is viewed by planners as an empirical activity, and when supplemented by the input of "real people", can have great impact. When muf works with community groups they are careful to schedule meetings in different community venues and to invite planners and city officials to leave their offices and experience the neighborhood directly, sampling food and cultural offerings. In the case of Dalston and Hackney Wick this small move did change the tenor of the conversations, if only because the subject under discussion was fore grounded.

Project: Barking Town Square

The challenge in this project was to design a new square for a town center that had lacked private investment for 60 years. This was an area where the demographic was undergoing enormous change due to immigration and grant-funded development motivated by policies to make this a focus for private housing.

Barking Town Square is made up of four interlocking environments: *The Arcade*, which leads to the Town Hall, celebrates the civic and connects the shopping street to Town Hall; the *Town Hall Square*, which is an open paved area; an extensive *arboretum* of 40 mature trees of 16 different species arranged to create woodland settings of different scale and character, and which invite exploration, performance and play; and the *Folly Wall*, an art commission authored by muf, which encloses the square by introducing a fourth elevation 7-meter-high wall (faux ruin) that employed the skills of a local bricklayer's college to hide the rear of a supermarket and now stands as a memento-mori to the current cycle of regeneration.

muf led the project from concept to completion, including a post-occupancy governance plan by a local city farm that is funded by the developer. The project is one of the Mayor's 100 Public Spaces and won the European Prize for Urban Space in 2008. Critics who admire the art feel it is "rooted in its place" and works well with its surroundings, humanizing "what might otherwise be a harsh new development" (Moore 2010).

Figure 10.4 Barking, Arboretum

Figure 10.5 Barking, Folly Wall

Take Advantage of Friendship Networks. There is a Use/Place for High-end Artists

Project Examples: French Architects – Gillett Square and Dalston Barn

Key to the alternative planning process is the temporary use of space to generate activity. In Dalston this took many different forms. Given the initial premise of the second question of the Dalston map, "Can value be shared?" it was important to test what creative community members could contribute to the public realm and public life. Up to that point, they had been pretty hidden in their studios. Two built projects tested this premise.

muf and J & L Gibbons had proposed to introduce play to animate a small public square, though stakeholders did not want a permanent playground as the space was sometimes used for events. muf proposed a temporary playground housed in a container. The playground gets put away after playing and is easily accessed two or three times a week.

The container is clad in a mirror made by Gary Webb, a local artist whose studio was indicated on the map muf generated, yet up to that point had never made public work. The second collaboration was with the group Exyzt.

Figure 10.6 Gillett Square, play container

muf were approached by the Barbican Centre, who was looking for a site for a commission for this French group of architects and artists. muf proposed one of the sites that appeared in that first mapping exercise, a non-park space identified on the map as a "potential" park. As part of the Radical Nature[1] exhibition Exyzt built a windmill, and for three weeks the site was used as a garden for workshops and making pizzas.

muf was able to move the project along so that local organizations and businesses benefited largely because of the links they had made into local networks gathered through asset mapping and discussions. The 15,000 who visited had their effect in securing £1 million of funding for a number of built projects including a garden built on the same site as the mill. The exhibition drew on ideas that have emerged out of Land Art, environmental activism, experimental architecture and utopianism. The garden, another collaboration with Exyzt, included a permanent pavilion called Dalston Barn within the Eastern Curve Garden (see http://dalstongarden.site11.com/?page_id=6). This is a permanent shelter and base camp for a temporary garden now in its 3rd year of use. Six unemployed young people from the area became paid apprentices to the project and were involved in its construction.

1 Radical Nature – Art and Architecture for a Changing Planet 1969–2009.

Develop "Governance Scenarios" for Spaces to Make Room for Other Voices. Create Possibilities for Encounter among Different Populations while Creating Spaces that Benefit the Indigenous Population.

> "How can public space be made as shared territory for different populations that are there?" "How can two markers on the map come together and in what way might they come together?" (Liza Fior).

muf spends considerable time and effort constructing governance scenarios for the spaces they help to create so that there is room for different voices. In the case of the Eastern Curve Garden there were several markers on the map: the derelict railway line; a host of very different organizations; a theatre; a social enterprise of 60 small businesses; a studios building; the site owners; the local authority; a developer; the local voluntary sector and the campaigning group, "Open Dalston". muf's role was not only to design with Exyzt a big roof pavilion that can host more than one activity at the same time; it was also to make it clear that the governance structure was as needy of structural calculations. This project is still playing out after three successful years in operation. Most recently two of the original organizations, including those involved in Open Dalston, have been given responsibility to run the garden.

Project Example: One of Three Projects High Street 2012 in Whitechapel

There is a repeated theme in muf's work in the attempt to avoid a singular reading of a place and, in turn, a singular meaning and use. Whitechapel has always been a place of radical thought and debate within and across the many different cultures that have occupied and passed through the area in succession. The re-design for Altab Ali Park acknowledges those differences and makes space to accommodate them.

Initial observations saw the park as well used but without the "furniture" to accommodate that use. For example, the large groups of young Bengali men from the language schools in the area congregated on the play equipment or sat on the Shaheed Minar (a shrine to a Bengali Language Martyrs). This gave the appearance that they had "taken over" the space and dissuaded other users. muf's ambition was to make accommodation for all users of the park without prioritizing one over the other. It was also to understand the site as a microcosm of the wider neighbourhood of this part of London where, historically, many different cultural, religious and political influences have shaped the fabric and the people who live here. The design is conceived as a matrix of the religious history of the site, and of the secular, making a setting for the Shaheed Minar, a spatial trace of the realm of faith and an acknowledgment of the attachment to place and landscape.

Figure 10.7 Altab Ali Park

The design provides a setting for the visible remains and traces the invisible but known history of the church and churchyard, which is intersected by a landscape setting for the Shaheed Minar monument. Together these two principles create a place that is a commemorative park with a strong identity, one that offers space for play and repose for all generations.

"Let's Make it as Good as it Can Be." Insinuate Change Wherever and Whenever Possible.

Project Example: The British Pavilion at the Venice Architectural Biennale, 2010

In 2010, muf was invited to author the British Pavilion at the Venice Biennale. Under the direction of Kazuko Seijima, the theme was "People meet in Architecture", with the intention of using the exhibitions to "help people relate to architecture, help architecture relate to people and help people relate to themselves". The *invisible footnote was, "if they pay 20 euros"* (http://www.archivenue.com/sejima-unveils-2010-venice-biennale-theme/).

Figure 10.8 Venice Architectural Biennale, 2010: Villa Frankenstein, British Pavilion

On the theory that "obedience to a brief can be a brief obedience" muf took on the challenge to find a way for people to meet within the pavilion and for it to become a building with programme rather than a gallery. An accompanying goal was to invite Venice to breech the fence and "take advantage of the British Pavilion". The project that muf ultimately designed was as much about challenging the logistics to make it possible for Venice to "take advantage of the British Pavilion" and the Biennale itself, as it was about the contents installed in the British Pavilion.

The exhibition entitled "Villa Frankenstein" highlights the history of Britain's fascination with Venice and contains examples of the detailed documentation of, and conversations with, the city by social critic John Ruskin in the late 19th century (http://villafrankenstein.com/close-looking/venice-and-ruskin/). Ruskin saw architecture as a "political art" and sought to recreate in great detail the relationship between the city's struggles, rise to power, and then decline through changes to the built environment. He fought against a "restoration" in 1870s Venice, believing that it would create rather than solve problems (Hewison 2010). Ruskin also denounced the use of faux Venetian Gothic styling to adorn public houses and cheap suburban villas in the UK, calling these structures "Frankenstein monsters" – hence the title of the exhibition.

In designing the exhibit, muf sought to take on the challenge of retaining local culture in the face of large-scale regeneration by utilizing the relationship between Britain and Venice to reflect on current issues in both settings. The process of trying to make it happen became the means to understanding the seepage of culture, whether it is the result of a Biennale, an Olympics (as in London), or interactions with other more everyday institutions such as Yale University's problematic relationship to New Haven. As with all of muf's work, The Pavilion operated on the principles that any truly successful development strategy requires a detailed understanding of a place. In this spirit, muf did whatever it could to be "disobedient" to the brief and to blur the boundaries between The Pavilion, the city and its people. muf used the Biennale and the long history of a two-way exchange of ideas between Britain and the city of Venice as an opportunity to develop further their methodology of "close looking", and to "breech the fence" that runs around the Biennale in order to allow ordinary Venetians to enter.

The centerpiece of *Villa Frankenstein*, an interior *Stadium of Close Looking*, was a 1:10 timber "model" of the Olympic Stadium repurposed as a drawing studio flanked by two explorations of fragile ecologies created with Venetian collaborators (including Gondola builders).

A record of the neighborhood that surrounds the Giardini was also paired with pages from Ruskin's notebooks and a fragment of live salt marsh fed by the lagoon water. Each had its own Biennale afterlife. The salt marsh was relocated by the entrance to the Giardini, and the timber fragment of stadium was moved to a new location for public use. Other parts of the exhibit that were part of a series called "Made in Venice" involved collaborations with environmental scientists, cultural historians, a variety of artists, and teachers and pupils from several Venetian schools. On the last day of the Biennale the pavilion hosted Salutiamo Venezia, a day of discussions about the future of Venice. Although muf did manage to run a programme, it was only on the last day when the Biennale was free to Venetians and the pavilion was able to act as a platform for Venetian concerns. This Biennale muf have managed to secure space within Arsenale for their collaborators to programme their own meetings.

Ongoing Work

Currently muf is developing two pieces of work. One is the "Open Air Art School", a three-year project based in Hackney Wick directed by artist partner Katherine Clarke. This project places children as co-authors with an artist of a piece of creative work that uses the city as site and subject. The Open Air Art School is linked to an earlier piece of work that asked what role, if any, an artist whose studio happens to be in a neighborhood should take and be given.

The second project that muf have just embarked on is a piece of work with the Association of Children's Museums (an international association that includes the 350 museums throughout the U.S.). The invitation asked muf and three other North American practices to "Re-Imagine the Children's Museum for the C21st". muf's approach is to ask if a building can be more than an architectural reaction to the status quo, and whether it can also be a catalyst for change.

The partners rarely teach, and do not receive grant funding, yet through their work muf continues to ask the question "What is the place of culture in the post industrial city"?

References

Hewison, R. (2010), *Venice and Ruskin – A Dialogue across Time and Place*, 25 August. Available at http://villafrankenstein.com/close-looking/venice-and-ruskin/ [accessed April 2010].

Moore, R. (2010), "The 10 best public works of art", *The Observer*, 17 April. Available at http://www.guardian.co.uk/culture/2010/apr/18/10-best-public-artworks-moore [accessed August 2010].

muf. (2001), *This is What We Do: A muf Manual* (London: Ellipsis).

Stern, M. and Seifert, S. (2007), *Cultivating "Natural Cultural Districts"*, a collaboration between the Social Impact of the Arts Project and The Reinvestment Fund of the Rockefeller Foundation. Available at www.trfund.com/resource/downloads/…/NaturalCulturalDistricts.pdf [accessed September 2008].

Chapter 11
Afterword:
Moving towards "Creative Sustainability" in Post-Industrial Communities

Myrna Margulies Breitbart

In 2002, California documentary filmmaker, Nancy Kelly, produced a film about the postindustrial city where she grew up – North Adams, Massachusetts. The film *DownSide Up* focuses on the opening of Mass MoCA, the largest museum of contemporary art in the U.S., and its relationship to the re-birth of North Adams. The story is told through the eyes of Kelly, who is visiting from her current home in San Francisco. She tours the city with family members who still live in town, some of whom worked in the defunct Sprague Electric factory that now houses the new museum.

The film begins with ocean waves crashing and Kelly's recollection of once nearly drowning as she wondered whether to fight to survive. This memory leads her to ponder about whether her own self-doubt and occasional downbeat moods might be the result of growing up in economically depressed post-industrial North Adams.[1] The advertisements for the film ask deliberately provocative questions: "What happens when an impoverished, working-class town decides that its only hope for survival lies within the world of contemporary art? Can these two disparate worlds possibly benefit each other? And why would they even try?" The film begins to answer these questions as it follows Kelly's family's reaction to the new flagship museum. This reaction moves from disbelief and cynicism to a certain fascination with the modern art and the possibilities that taking a very different social turn might present for themselves and their working class community.

The questions raised in this book are somewhat different from those in the film. Like the film, contributors direct attention to new approaches to culture- and creativity-led forms of urban regeneration that are not restricted to iconic global cities. Their interest is in how small- to medium-sized former manufacturing centers are evolving their own unique creative identities and economic opportunities. Unlike the movie, however, most authors begin from the premise that post-industrial cities have a rich history. They contend that art, cultural production and

1 To confirm negative stereotypes early in the film Kelly references a *New York Times* travel article on the beauties of the Mohawk Trail (route 2) that warns readers, "When you get to the sign for North Adams, to avoid industrial decay, reverse direction".

other forms of creative activity have always been, and continue to be, present in these communities and outside officially named "cultural quarters". The focus is less on *why* a creative or cultural approach to regeneration would be embraced than on *how* this approach is being interpreted and what is most promising in these efforts. This final chapter highlights some of the ways that post-industrial communities are evolving alternative forms of cultural production and creative practice that depart from mainstream models in their search for ways to improve the quality of life and create sustainable economic opportunities for residents.

Creating a (*Different*) Scene

Much creative economy planning in the smaller communities featured in this book mimic practices associated with larger cities, e.g. the promotion of cultural tourism or use of the arts to re-image and market downtown. Apart from officially driven policies, however, there are a whole range of unsanctioned, ad hoc creative practices that emerge through the collaborative efforts of residents and neighborhood-based community organizations. These play an important role, enhancing the quality of life through their many unquantifiable social impacts. It is not possible to understand the current or potential role of creative energy in such cities without examining these community-based activities that occur alongside official planning practice. As more cities adopt mainstream strategies to promote a creative economy it is important to examine how these indigenous, often informal, activities challenge narrow definitions of creative work that target only a limited subset of the population. It is also important to consider how such pursuits could provide economic opportunities for residents if they were better supported.

Mainstream creative economy models, whether applied in smaller post-industrial, or larger cosmopolitan settings, are often subject to the same constraints and flawed assumptions. Such strategies can fail to make a significant dent in addressing pervasive poverty and, at times, even exacerbate social inequality and uneven development. A common assumption is that tourists and middle-income people will be drawn to a city center rejuvenated through the arts and cultural production. Presumably new service sector jobs and tax revenues will then provide benefits that trickle down the class hierarchy improving city finances and the overall competitive position of a city in its region. Research suggests, however, that instrumental use of culture or the arts for downtown investment can, in fact, drive up costs and property values, eventually causing the displacement of people and marginal businesses (Donegan and Lowe 2008). Such policies can also divert funds from improvements to the quality of neighborhood life and programs that enhance the income-earning potential of working class residents. While the goal of reinventing the economy may be central to these mainstream practices, important interconnections between social structure and economy go unrecognized. Little consideration is given to how a different set of more spatially equitable policies might engender more positive and sustainable outcomes for a larger segment of the local population.

The creative economy bandwagon is only one of many historical efforts to regenerate cities that have resulted in bifurcated outcomes for residents who vary by social class and race; the early urban ecology movement is another. At the First International Ecocity conference, held in Berkeley, California in 1990, participants professed a common desire to draw attention to the ways that cities had been built in disregard of "natural principles". Clear divisions soon arose, however, between those whose focus was on urban re-design and environmental land use issues (e.g., congestion, sprawl, air pollution from autos, lack of green space etc.) and those whose more expansive definition of "environmental degradation" led to a focus on the eradication of urban poverty and social injustice (e.g., toxic dumping in poor neighborhoods, uneven urban development, lack of jobs and affordable housing etc.). Those environmentalists who emphasized the many "r's" – *re*cycling, *re*zoning, *re*investing and *re*scaling – tended to focus on the built (or physical) environment as both the source of, and solution to, "urban" environmental problems (Register 1987). Largely ignored were the systemic roots of imbalances in both the social and natural worlds – unequal allocations of political and economic power and the complex interplay of decisions made by individuals, governments, and corporations that mapped race and class inequities directly onto the urban landscape. To effect real change in cities, it became apparent that sustainable city practitioners would have to move beyond simplistic "green city" prescriptions for change that are based upon a very circumscribed understanding of what environmental degradation or restoration means (ignoring, for example, the absence of basic jobs, infrastructure, health and educational opportunity for large segments of the urban population). A case was built for incorporating social justice as a core element into sustainable city practice soon after these criticisms of Ecocity emerged. It was also suggested that this movement adopt a more active participatory process involving alliance building (Anthony 1990; Breitbart and Kirk 1991; King 1991; Kirk 1997).

The parallels between the current creative economy discourse and practice and this early ecological cities movement are worth noting. Several case studies in this book suggest the ineffectual or negative consequences of trickle down theories of culture-led urban regeneration based on neoliberal principles and practices. They also underscore how the application of limited notions of creativity fails to address issues of equity and social exclusion by neglecting to build upon the assets of existing residents. Because of their size, small- to medium-sized post-industrial communities enable a scale of inquiry that can result in a better understanding of the positive and negative impacts of current culture-led regeneration policies. We can begin to identify common elements shared by these communities as well as practices that exhibit a certain particularity to place. While limited in scope, these examples complicate and, in some cases, challenge more generic creative economy policies that have attained such popularity.

Many cultural-led planning practices applied in smaller post-industrial communities provoke justifiable critique and do not contain the elements needed to produce sustainable change. There are other examples, however, that are based on more promising practices that have the potential to enrich residents' lives and shift the direction of planning practice. Strategies that merit further attention initiate what Jeremy Till refers to as a "negotiation of hope". They project a city and its residents "ambiguously forward" rather than allowing them to become immobilized by the past (Till 2005). Such approaches balance efforts to use creativity, culture and the arts to improve the overall economic position of the city by simultaneously drawing in new resources (i.e. people and businesses) and also devoting attention to the interests and needs of existing residents. An effort to articulate this agenda follows.

Cultivating Engaged Citizens

Former industrial powerhouses, such as Holyoke or Lowell, Massachusetts, were conceived in the 19th century as models designed to perfect industrial production and worker/management relations. Constructed as utopian responses to the negative environmental and social conditions present in many large industrial conurbations, city plans emphasized both social and physical design to maximize control over the lives of workers (Hall and Ward 1998). Such "planned" centers were never intended to address imbalances of power and privilege. In fact, during times of economic downturn, these imbalances were often exacerbated. Planning was centralized into few hands, and residents who came from many different cultural and racial backgrounds, though asked to make significant personal sacrifice in times of economic downturn, were never invited to become involved in the envisioning of change.

Many of these same post-industrial communities now depart from this civic model by inviting diverse groups of residents to become involved in visioning exercises. These design charettes are often used to uncover challenges, counter resignation and fight media stereotypes. As one-shot events, they can provoke imaginative speculation about the future, with the results then handed over to "experts" for interpretation and implementation.

The more fruitful examples of participatory planning stem from a social justice agenda that asserts citizens' right to "shape, intervene and participate in the unfolding idea of the city" (Chatterton 2010: 235). As Chatterton suggests, this is not so much about what the city currently is or what it was, but more about "what it could become", what it has never been (234). In practice, such examples of active participation must be sustained over the long term to initiate what Jeremy Till calls "transformative participation". This form of engagement acknowledges imbalances of power and the potential contribution of both local and outside "expert" knowledge while "transform[ing] expectations and futures of all participants" (Till 2005: 36). The goal is not to eradicate all difference or come up with the "best" design or plan, but to begin a sustainable process of negotiation that enables residents who represent very different interests, and have very different experiences, to maintain on-going access to the decision-making arena as a planning process evolves. In this context, inclusive participatory events provide an immediate venue, within which residents and policymakers might embark on the imagining of a different kind of community, much as Sandercock does in *Cosmopolis II*,

> I want city where … city planning is a war of liberation fought against dumb, featureless public space; against STARchitecture, speculatorism and benchmarkers; against multiple sources of oppression, domination and violence; where citizens wrest from space new possibilities, and immerse themselves in their cultures … collectively forging hybrid cultures and spaces. I want a city that is run differently from an accounting firm, where planners "plan" by negotiating desires and fears, mediating memories and hopes, facilitating change and transformation (2003a: 208).

Though a tall order, with many possible pitfalls, some smaller post-industrial cities facing dire needs are pursuing this aim.

Talbot and Magnoli discuss the importance of "community learning" and local knowledge production as,

> a process for ensuring that informed decisions for social change are based on the fullest possible public participation and the most effective use of local expertise and knowledge. Design and planning for sustainable communities need to be firmly grounded in the details of place … Traditional knowledge, which is place and culture-specific, is a critical asset, which needs to be acknowledged, preserved, restored and utilized (Talbot and Magnoli 2000: 95).

There are nonetheless real barriers to convincing local governments of the value of the sustained participation of residents in planning. Many long-term residents also harbor memories of struggle against deleterious forms of economic restructuring tried in the past. They face an uphill battle against negative media in the present, and are loath to participate in planning exercises when they do not believe that their local knowledge and opinions will matter. As the cases in this book reveal, cultivating a city of engaged citizens is made even more difficult when residents have been long excluded from arenas of power or invited to planning events merely to legitimate already-conceived plans where neoliberal agendas place the burden for change *on them*, without providing the resources for implementation.

Inclusive planning must examine how particular policies will impact different constituencies. When muf architects employ creative means to provoke critical visioning (Chapter 10), they bring such questions right to the surface. They work with communities to identify hidden local assets and to elevate the differences among residents to a primary position in the deliberation of an area's future. muf do not assume or even seek a unified identity for a neighborhood or city client (as many place-marketers encourage). Rather, they acknowledge diverse interests and place value on the economic and social dimensions of the creative work that residents are involved in. They commit to addressing needs while encouraging the articulation of desires, often through the temporary transformation of space. What makes this approach so effective is the focus on invigorating aspirations through informal interventionist practices that address needs in new ways. These activities and transformations of space draw people in because they challenge existing political and planning agendas as they reflect on future possibilities. With the help of artists and creative designers such interventions reveal what is unique about a place. This is not an exercise in place branding, but rather a way to lay claim to a place and foster a sense of belonging. The process can lead to the "recovery of imagination" so crucial to thinking differently about the future (Sandercock 2003a: 218).

Any long-term strategy to encourage greater citizen involvement in planning must involve a process of building a sense of community from the bottom up. Cultivating relationships within similar demographic groups, and among different resident groupings, might seem easier in a smaller city. History, however, suggests otherwise. Social divisions often run deep in former manufacturing centers where successive in-migrants compete for fewer and fewer jobs and are perceived by old timers to be drawing excessively on diminishing services. More ways must be found to open up the spaces to collectively address problems, identify opportunities and share information across these social and cultural boundaries.

Prioritizing Diversity and Identifying Local Assets

At a planning meeting for an ArtsWalk festival in Holyoke, Massachusetts that I attended several years ago, a conversation ensued about the social and ethnic divisions present in the city and how these continued to present an obstacle to full community participation. Knowing how many residents from diverse backgrounds had already been targeted for their creative talent to participate in this event, the organizer expressed his extreme frustration at a city that, in his mind, refused to recognize *diversity* as its greatest asset.

Most post-industrial cities have lost a large segment of their middle class and gained significant numbers of residents with less wealth as a result of deindustrialization. A high proportion of those in the U.S. have been ethnically diverse from the time immigrants were first recruited to work in factories in the 19th century. Several such cities have more recently become the gateways for new immigrants. Holyoke, MA, for example, now has the largest percentage of Puerto Rican residents of almost any city in the U.S., while Lowell, Massachusetts has one of the largest percentages of Cambodian immigrants. In contrast, several older industrial cities in the UK, once ethnically homogeneous, are now increasingly diverse (e.g., Birmingham, England, in which 33% of the business activity is now within minority ethnic-owned enterprises) (Sandercock 2003a: 173). Such cities are characterized by multiple identities and histories of struggle revolving around prior economic shifts and government regeneration schemes to address them.

Artists and cultural planners often facilitate bridges across these urban social divides through creative programming; rather than merely be *attracted to* tolerant places (Richard Florida's notion) they *foster* tolerance and imaginative collaborations. In contrast, mainstream creative economy practices can fuel social divisions and inhibit intercultural and cross-class exchange by dichotomizing places (e.g., large vs. small; winner vs. loser; downtown "cultural quarter" vs. neighborhood) and people (e.g., young and creative vs. older and working class).

Many chapters in this book suggest that there is no uniform public sphere, even in smaller places. As Osorio points out in Easthampton, Massachusetts (Chapter 7), even the so-called "creative class" is more diverse than it is assumed to be. This cohort is not something "to be found and measured," he suggests. It consists of a set of connections that is constantly evolving and is influenced by the environment within which it resides. It follows, then, that a one-size-fits-all cultural planning agenda that ignores this dynamism and fails to create balanced opportunities to support differently resourced residents, cannot address the needs and potentials presented by each unique setting. Planning and economic development policies that neglect to acknowledge the zero/sum outcomes of programs that privilege one demographic over another, or one set of assumptions about the sources of economic growth, will not contribute to long-term sustainable development.

The elevation of social diversity as a planning asset still eludes many post-industrial cities even though *economic* diversity has been embraced and urban populations reflect an increasingly transcultural world (Rios et al. 2012: 2). The dangers of economic uniformity are well understood and no city wishes to place all of its survival eggs in the basket of one enterprise. Precipitous economic decline in places such as Pittsfield or North Adams, Massachusetts followed the departure of a single manufacturer. The return of economic vitality and development of a real competitive advantage for these cities now rests on the cultivation of a variety of different, generally smaller, economic enterprises. Such an agenda can encourage innovation, entrepreneurial opportunity and small business development if it seeks to attract new talent while also adopting effective strategies to capacity-build among residents with different life experiences and skills. Acknowledgement of a potential role for diversity within a creative economy departs from mainstream models that only see demographic difference and interesting cuisine as instrumental amenities necessary to attract a new "creative class".[2]

---

2 In the industrial city of Wollongong in New South Wales, Australia, creative economy consultants replaced the city's original slogan, "City of Diversity" with a new brand, "City of Innovation". The same city forced out an internationally recognized design company that used visual means to bring attention to "the sense of placelessness generated through the demise of steel employment" because they feared that negative social commentary would damage the city's image (Waitt and Gibson 2009: 1232, 1241).

The need to identify and build upon the assets of their diverse populations is already recognized by many smaller post-industrial communities. Rather than simply focus on attracting a newer, younger and richer demographic, such places are beginning to turn inward to identify and support the creative potential of local residents. The success of these efforts hinges on acceptance of a much-expanded definition of creative practice, and the adoption of imaginative methodologies. Sophie Handler's description of elderly residents in Newham (Chapter 9) who creatively map their daily routines in the built environment and work with artists to design street furniture that enable them to reclaim and expand their access to valuable space in the city, is one example. muf's work to reveal hidden assets and reflect on current spatial practices in Dalston is another (Chapter 10).

Artists and cultural workers in many post-industrial communities challenge the dichotomies between so-called "high" and "low" forms of creative enterprise that emerge in more cosmopolitan settings. As Bain and McLean (Chapter 4) and Gee (Chapter 5) point out, the smaller scale of a city can foster greater collegiality within the cultural sector. It can also create opportunities to use art to promote social dialogue and encourage community access to the arts, which are valued both for their economic and social impacts. However, as Fitzpatrick points out in Chapter 8 on Liverpool, a diversity of cultural expression can often be rendered invisible by large-scale events brought in from the outside that assume they are entering a vacuous cultural landscape. Indigenous community-based organizations work hard to promote diverse cultural expression. Several years ago, Nuestras Raices, a community-based environmental and cultural development corporation in Holyoke, Massachusetts, joined with two local organizations, the Hispanic Family Festival and the Centro de Artes, to form the South Holyoke Community Arts Initiative (SHCAI). The goals were to promote local talent, connect that to cultural heritage, and do collaborative programming focused on cultivating a large regional demographic that includes significant concentrations of Latinos up and down the Connecticut River. There was also a stated intent to use the festivals and cultural events to improve the income-earning potential and quality of life for local artists and artisans, community gardeners, restaurant owners, health advocates and small producers. Given this inclusive agenda and access to key resources, many people who might not have otherwise participated in the SHCAI-planned events came and benefited.

Expanding Notions of Creativity

Identification of talented residents and the creation of opportunities for them to acquire skills within a more inclusive regeneration agenda requires more than an appreciation for local diversity. A much-expanded and more fluid definition of "creativity" is needed. This definition must move beyond popular measurement indices to include those forms of creativity that are currently hidden yet generate economic and social value.

Most culture-led regeneration policies speak of creativity in relation to individual artists and entrepreneurs. They privilege the "genius" of professionally trained individuals who earn the bulk of their income in a limited array of fields recognized by the census. These categories are meant to count only those individuals who earn the bulk of their income in these pursuits. Creative industry growth is thus premised on a "predominantly individualistic notion of creativity" (Wilson 2010: 368). Aside from ignoring the social context for much creative work, this perspective has transformed support for community-based and collective art practices into more individualized and entrepreneurial programming (Chapter 5; McRobbie 2003).

Creative workers in smaller post-industrial communities often occupy interstitial spaces and engage in less commodified and more collaborative creative practices. These are sometimes based on the cultural heritage of residents and can include a range of professionally- and non-professionally sanctioned pursuits, many of which are experimental in nature. The work of those who cater or make crafts for neighborhood celebrations, party planners, DJs, musicians, self-trained web and graphic designers, bakers and recycled furniture makers in Holyoke, Massachusetts (Chapter 3) contributes in a myriad of ways to quality of life. It also increases the economic autonomy of residents who might otherwise be living even further on the margins. Burt Crenca's painting of his plumbing pipes a copper color to pass a building inspection in his performance venue in Providence, Rhode Island (Chapter 2) is one improvised and subversive creative intervention that had the effect of generating new opportunities for aspiring local artists. Once it attained ownership of the property, AS220 was able to make alliances with other neighborhood businesses and act as a "gentrification watchdog". It was also able to create greater access to cultural resources for underserved and incarcerated youth. Other creative pursuits, such as those undertaken in the Thunder Bay and Petersborough, Ontario postering campaigns (Chapter 4), and the imaginative mapping of urban space by the elderly in Newham (Chapter 9), open up the space for protest and a critical examination of approaches to planning policy. In the latter case, an elderly mapping exercise illustrates the unequal access to urban public space that many residents endure, while allowing older citizens to claim space and a voice in regeneration.

These examples share in common Lefebvre's notion of the "right to the city" as a right to *transform* space not merely occupy it. While illustrating the use of art in the service of resistance and subversive practice, they also underscore important ties between creative practice and economic survival. Most definitions of the creative (or knowledge-based) economy utilized in mainstream economic development confine it to the economic activity of a very well educated workforce consisting of professionals who earn the bulk of their money from their creative work (e.g., professionally trained artists, architects and designers, computer specialists, writers and the people who support this creative work such as journalists and financiers). Census categories enable the easy compilation of data that can be used to measure the economic impact of an activity from this sector. Policies such as the subsidization of housing for loft-living artists, presumed to be young and single, or tax breaks for a narrowly defined group of creative entrepreneurs, are then designed to support practitioners in these sectors. Left out of the equation are large numbers of residents who do not fit into these prescribed categories and yet would benefit immeasurably from more access to resources, start-up subsidies, training and the spaces in which to "create".

Promoting Spatial Equity

The ecological concept of balance, in which no one activity dominates an environment, could begin to address the spatial side of this equation if applied to promote more equitable distribution of resources in the realm of cultural planning and economic development. This requires a consideration of where city resources are distributed (and to whom); how residents access spaces for learning and creative production; and where and how exchanges of talent and knowledge are facilitated across the urban landscape.

At present, most spatial planning to support cultural production and the development of creative industries focuses on downtowns that were decimated by the job and income losses accompanying deindustrialization and urban renewal programs that drove highways through the heart of the city or built malls on the urban periphery. Some post-industrial cities have attempted to jump start regeneration through what Evans describes as culture-led regeneration involving the cultivation of large-scale events or the construction of flagship cultural attractions designed to entice residents and tourists back to the city center (2005).[3] More typical approaches to spatial planning in post-industrial cities involve the recasting as valuable assets of such latent amenities as long-ignored (or even buried) canals, rivers and waterways. Once a key source of energy that drove industrial production, these natural resources are now being transformed into smartly paved "river walks" complete with wrought iron lighting and overlooks. Such large-scale redevelopment schemes are designed to attract people to rejuvenated downtowns where they can consume products and services, and are often accompanied by strict regulation on the use of public space. This can have the ironic effect of limiting many creative and spontaneous forms of activity (Waitt and Gibson 2009: 1237).

Factory and mill spaces alongside resurrected waterways, once considered anachronistic remnants of a lost era, are now being recycled into living quarters or homes for artists, new cultural enterprises and creative industries. The expense of this renovation has frequently necessitated the recruitment of more mundane

3 One example is "Auto World", an expensive and short-lived indoor amusement facility built in Flint Michigan in July 1984 at a cost of $80 million. The complex, expected to be the centerpiece of a new cultural economy for the city, was closed a year later after it was clear that it would not attract the expected outpouring of visitors. It was then demolished in 1997.

activities, such as hair salons or the Registries of Motor Vehicles. Industrial structures often attract people with a fascination for the industrial aesthetic or what some now call "ruin porn" (Edensor 2005). This can lead to a kind of fetishization of spaces of industrial decay (Chayka 2011; Greco 2012). On a recent tour that I took through a former rubber plant in Holyoke, Massachusetts, many of the young artists and new residents on the tour cited the appeal and challenge of hanging pipes, peeling walls and broken left over materials. There was less interest in the history behind the huge waterwheel that once powered the facility or the former workers that once occupied the factory. Cities build on this romanticization of industrial life in order to make themselves more attractive to the middle class and prospective young creatives. Stevenson interprets such hyper-urban referential, and the policies that promote this, as "anti-urban". He believes that the "seeds of a fundamentally anti-urbanism" lie in cultural planning's essentializing of industrial sites as "manifestations of urbanism" (Stevenson 1998: 152). Meanwhile the populations that once earned income within these structures are rendered invisible, even though they maintain active cultural and social lives in neighborhood spaces that *should* be considered the essence of "real" urban life. Stevenson suggests that in this mainstream paradigm that promotes "vital" urban spaces, once vacant, simulated or re-imagined cities can easily come to resemble the "safe" suburbs.

The recycling of industrial space is nevertheless critical to both economic regeneration and the invigoration of hope among residents of post-industrial cities who often see the possibility of improving their own lives in the physical renewal of urban space. A more balanced approach to spatial planning, with less emphasis on attracting a new young "creative" or middle class and more on growing the capacities of existing residents from within, could provide greater benefits. This would require the mapping of sites of creative activity beyond downtown and attention to the need that many residents have for space to improve their quality of life and support their creative work. As many asset inventories in post-industrial cities reveal, much of the creative activity engaged in by residents is collaborative and geographically context-driven – something Becker notes of art in general (Becker 1982; Edensor 2010). This work is often undertaken, however, in less than ideal physical circumstances. A high percentage of residents in smaller post-industrial cities rent rather than own their housing. With the increased presence of growing immigrant populations, this creative work is more often than not conducted within cramped living quarters and extended family settings as opposed to refurbished lofts. Residents living and working within such environments would not only benefit from greater access to educational and material resources; they would also benefit from access to more space in which to develop their talents and learn from others. Surveys such as the one conducted in Holyoke (Chapter 3) uncover a strong latent demand for community arts and design facilities, as well as public kitchens and "maker spaces" that could provide access to equipment too expensive for individual low-income households to purchase. If public policies that currently enable the owners of older factories in lower-income, poorly resourced neighborhoods to convert their space into lofts for single adults were also directed to support affordable family housing with large shared artisan workspaces nearby, many talented residents would be able to extend their creative work and income-earning possibilities so crucial to their survival. The opportunity to obtain grants and purchase space from the city at minimal cost, and as a hedge against gentrification, was available to that music venue AS220, and the arts and technical training program Steel Yard, in Providence, Rhode Island (Chapter 2).

Lack of critical analysis of the social and economic outcomes of creative economy planning has lead to an uncritical adoption of other spatial development strategies. One of the most popular is the use of zoning and tax incentives to rebrand whole industrial areas and downtown neighborhoods as "cultural quarters" or "innovation districts". The designation of special spaces in the city for the cultivation of "creativity" is often driven by the observation that even new communications technologies cannot achieve the kinds of personal interchange that face-to-face interaction allows (Pratt 2000). However, the promotion of cultural districts carries some risk (Markusen and Gadwa 2010). Labeling areas of the city in a tourist brochure or providing financial incentives to locate in a clustered enclave can exacerbate inequality and uneven development without necessarily providing a sustainable boost to the local economy. Among many things, designated "innovation districts" (what Edensor calls "authoritative spatialisation" (2005: 21)) direct the attention of city officials away from more decentralized and organically formed creative clusters and discourage more inclusive and geographically balanced distributions of spatial and financial assets to support them.

In addition, the designation of creative enclaves can build on "particular discourses of fear" that aim at making the city "safe" for consumption (Sandercock 2003a). Even when a rare city simultaneously directs resources to promote creative investment in local neighborhoods, the designation of a whole downtown, or a select quarter, as "cultural", "innovative", or "creative" sets up dichotomies. These then invite contrast with what are perceived to be the adverse "lesser" parts of the city construed as obstacles to tourism (Zukin 1995). The enclaves also take on their own mythologies, which often results in a continued one-way distribution of limited local resources.

Notions of "trickle down" are used to justify spatially targeted investment in the arts and creative industries. This is done even though the low pay service sector jobs that result for residents do not compensate for the redirection of resources that might have been distributed more equitably over the landscape. Over time, such "creative" enclaves also risk losing their caché due to constant replication and proliferation (Breitbart 2004). The bounding of creative activities and naming of spaces as "innovative" or not fails to recognize important relationships between the built environment and distributions of power and economic resources.

Such designations also fail to acknowledge the richer dimensions of cultural life that are a means to economic survival in more marginalized neighborhoods.[4]

The city of Holyoke, Massachusetts is one of many smaller post-industrial cities moving forward with plans to designate a part of its downtown as an "Arts and Innovation Overlay District". The new young Mayor supports this effort and has expressed a desire to include the development of new housing to attract middle- and upper class residents to downtown. While several building renovation projects are already underway to anchor this plan, a Holyoke Community Arts Map, produced 10 years ago, indicates a high proportion of arts and cultural activity in the city is located in organizations within surrounding neighborhoods. The question arises as to whether it is possible for the city to support the infusion of resources into the narrow boundaries of the "Innovation District" while also directing needed funds to develop what is now a primarily informal creative economy in these other locations.

Many important spaces of creative enterprise in post-industrial communities like Holyoke support experimentation and less-commodified and informal types of income generation. By providing limited yet important supplemental income to residents, they enable the forging of new social and economic relationships outside of a wholly capitalist framework. The creative and deliberately provocative mapping of space described by Handler in Chapter 9 and muf's spatial interventions in other post-industrial neighborhoods discussed in Chapter 10, suggest the potential of using urban space to extend these alternative frameworks. On occasion, such participatory design, artistic and critical spatial practices challenge notions of ownership and property, generating dialogue about urban regeneration. They can also provide residents hitherto marginalized in formal planning processes with a means to intervene in their everyday environments and communicate alternative "dreams and desires" for their neighborhoods and themselves. Temporary occupations of everyday space and innovative designs prioritize critical dialogue about how space is used over new construction. They can also promote a more subversive and often aesthetic experimentation with land use as decisions are made about the re-purposing of vacant and underutilized land (Yeh 2005; Loftus 2009). That said, few of the cities described here provide examples of vigilant and sustained advocacy on the part of residents for a more decentralized distribution of cultural resources and greater control over how the spaces of the city are designated or used. Nor do most local governments in post-industrial cities accompany the necessary physical and cultural renewal of central areas with an equally vigorous effort to improve the physical and social environments of surrounding neighborhoods.[5] Yet, as Salkind points out in

4 Numerous studies document the myriad ways that art and culture enhance neighborhood life. Among them are Goldbard, 2006 and Seigert and Stern 2008.

5 Bianchini looks back historically at the innovations introduced by the Greater London Council's involvement in creative industries during its existence from 1981–86. The approach was to work to support the financial viability of community enterprises such

Providence, Rhode Island (Chapter 2), local partnerships between neighborhood-based art and training programs can make a real difference.

Investing in Neighborhoods and Residents

Creative economy policies that highlight cultural quarters or "innovation" districts recognize the economic value that can be derived from the presence of cultural activity yet often fail to assign value to cultural activity in neighborhoods outside these boundaries. This represents more than a missed opportunity; it can exacerbate social inequality. Several case studies in this book suggest how essential the rebuilding of the social and physical infrastructure in neighborhoods is to any form of "creative sustainability" in cities hard hit by the effects of deindustrialization. Bianchini made this observation over a decade ago when he argued that,

> ... an explicit commitment to revitalize the cultural, social and political life of local residents should precede and sustain the formulation of physical and economic regeneration strategies (Bianchini 1999: 14).

In Barnsley, England (Chapter 6) the current recession drives this point home particularly well, as it underscores how improvements to the center city and investments in spaces for new creative industries cannot be sustained without attention to neighborhood development and the improvement of "soft" infrastructure.

Policies to promote creative economic development must generate new opportunities for residents in those areas of the economy that are growing and have the potential to provide meaningful work and a decent quality of life. A high proportion of such jobs fall into the ever-expanding creative industry sector, and innovative approaches to education and workforce development could grow a more substantial portion of that workforce from within. Many post-industrial communities currently import a significant proportion of their workforce because of low skill levels and educational attainment among the existing population. This situation arises in part because young people see little advantage to remaining in school when they are faced with the prospect of unemployment or low paid service jobs. As long as workforce development schemes continue to subscribe to the non-aspirational notion that "a job is a job is a job" young people will walk away from such limited and despairing options.

Creativity is perceived in European policy as the prime source for innovation, which, in turn, is acknowledged as the main driver of sustainable economic

---

as recording, film and video studios, radical book distribution cooperatives, and black publishing houses, and to provide these independent cultural enterprises with access to services such as applications of new technologies, management consulting, and marketing and new technology (Bianchini 1999: 5). The GLC was eliminated when Margaret Thatcher became Prime Minister.

development. A study undertaken recently by the European Union confirmed the importance of introducing more creativity into formal schooling. The report argued for major changes with respect to creative learning and innovative teaching (Cachia et al. 2010). Certain "key competencies" deemed necessary for lifelong learning and "creative development and expression" include: critical thinking, creative initiative, problem solving, digital competence, learning to learn, social and civic competence, a sense of initiative and entrepreneurship, and cultural awareness. The report goes on to discuss how culture affects the types of learning considered valuable, and encourages schools to be more "open-minded about trying … different ways of learning and teaching" (41). One recommendation is to "integrate more cross-curricular skills … such as digital competence, collaboration skills and intercultural understanding" (49). If sustainable economic development depends on innovation, and innovation depends on a more effective incorporation of creativity with the above qualities into education, then post-industrial cities could become forerunners in this process through the cultivation of formal and less formal training and educational models that encapsulate these goals.

Infusing more creativity into learning counters government policies that encourage a competitive rather than collaborative learning culture. This is a good thing. Some worry, however, that touting the importance of creativity to attract the interest of youth while simultaneously defunding community-based cultural organizations and school art programs, may be more about promoting consumer culture than creating new career opportunities (McRobbie 2003). Research nonetheless suggests a strong correlation between the cultivation of creativity, completion of education and future income earning capacity. In one recent study, the proliferation of new technologies through various art and media are thought to help young people identify careers and awaken their desire to learn.

> The arts and design are an underused entry or reentry gateway to education … Programs that target the entertainment industries draw younger students, many of whom are rejecting schooling but are attracted back to programs that connect to their interests and environment (Rosenfeld 2006: 35).

Rosenfeld believes that residents can be prepared for new careers within community art spaces and learning centers outside the current public school system if policy makers "think outside the economic development box that places all the emphasis on 'job readiness' and too little on sparking desire" (Rosenfeld 2006: 32). The Village of the Arts and Humanities, started in 1989 by artist Lily Yeh, in Philadelphia, began with the transformation of a single vacant lot into an art park. It has since supported the aspirations of the community and inspired residents to become agents of change by providing multiple opportunities for self-expression through art, much of which is directed at a revitalization of the physical environment (http://villagearts.org/about/mission-values-awards). A question the Rosenfeld study and the case studies in this book raise, concerns the types of educational and career development models that foster the competencies necessary to propel residents forward into new careers. What kinds of programs attract young people and adults and tap into their creative passions and interests? How do such programs motivate young people to stay in school and prepare residents to access jobs with promise? A few examples are worth noting:

Manchester Craftsmen's Guild The Manchester Craftsmen's Guild (MCG) was founded in 1968 by ceramic artist Bill Strickland in the lower income, predominantly African American neighborhood of north Pittsburgh, Pennsylvania. The program and Strickland have been formally recognized for their success at providing after-school and out-of-school mentored arts and career training for youth and adults.

Starting in a small building with just a few students drawn from the neighborhood, the MCG now occupies 180,000 square feet and maintains a year-round program in photography, film, digital arts, ceramics, design, fiber arts and many other creative areas. With a $1.2 million operating budget, MCG works with over 230 self-selected public school students in grades 9–12, who arrive each day after school on buses provided by the MCG. In a project-based learning format, students acquire many skills that support their interests and create new opportunities. A recent study found that 97% of the participants graduate from high school on time and 86% go on to post-secondary matriculation in addition to performing 2,500 hours of service learning in the community (www.ne-cat.org/replication.html).

Shortly after the founding of the MCG, Strickland was asked to take over the Bidwell Training Center, which originally targeted displaced steel workers and now provides training in such occupational categories as chemical laboratory, pharmaceutical, medical technician, horticulture, and culinary arts. In 1986, a $6.5 million capital campaign allowed MCG to construct a new 62,000-square-foot arts and career training center that added educational programming in other creative areas. MCG has since worked with several national foundations to promote the replication of the Manchester Bidwell model in 200 cities in the U.S. and abroad. This is accomplished through the National Center for Arts and Technology (NCAT). Similar centers have already opened in San Francisco, Cincinnati, Grand Rapids, Michigan, and Cleveland. The New England Center for Arts and Technology (NECAT), based in Boston, MA, recently received an award from the Boston Foundation to work with Strickland to explore the feasibility of developing job training programs for un- and under-employed adults there, along with an after-school program for urban youth.

Artists for Humanity: Creative Jobs for Creative Youth On a smaller scale, artist and teacher, Susan Rodgerson, started Artists for Humanity (AFH) in Boston in 1991. The purpose is to provide underserved youth with an opportunity to develop skills and eventual jobs in many areas of the creative economy. Like MCG, youth aged 14–18, are drawn to AFH from Boston public schools where no art is offered. At AFH, they receive art, technology and business training.

Professional artists, some of whom received their initial training through AFH, run all programs. Besides photography, video, sculpture, painting, screen printing, and graphic, motion, and web design, AFH offers fashion design, and there is a youth-run micro-enterprise development operation. After an initial commitment of 72 hours and community service, participants become eligible for a wage to produce art that is sold to commercial customers (e.g., postcard invitations to events or posters for businesses and local government, community non-profits and schools). The general process involves the youth in a negotiation with prospective clients, which culminates in one student and one mentor being picked to complete the project. Back in 2004, the wage was up to $10/hour but this did not count the additional money youth could receive directly from the proceeds of selling their own work. When I visited AFH several years ago, they were initiating a holiday card business with a $100,000 goal for sales. A piece by Fleet bank, commissioned for $40,000, and inspired by the work of Gauguin, had just hung at the Museum of Fine Arts in Boston, eventually making its way to Logan airport's international terminal.

After an extensive capital campaign to move from cramped quarters, AFH purchased a site and constructed a $6.8 million green building that opened in 2004 called the EpiCenter. Of this amount, $5.3 million came in the form of grants, leaving a $1.5 million mortgage. Part of the building design process began as an architectural studio for students. Three of the youth who participated went on to study architecture with scholarships at Wentworth and Northeastern University. The organization was able to access several sources of funding to pay for light fixture upgrades and solar panels that provide more than 80% of the electricity necessary to run the operation because of the attention given by the designers to green materials and energy efficiency. With their event coordinator, AFH is also able to rent space to for-profit clients in Boston for meetings and celebrations, and to private clients for personal events or nonprofit fundraisers.

Sculpture students built moveable walls with hinges for art work that enable quite a large space on the ground floor to serve the needs of these different functions. The money raised from this outside rental of space and the sale of the art works produced at AFH helps to maintain the programming and pay the mortgage (interview).

Figure 11.1 Interior gallery, Artists for Humanity, Boston, MA

Our House (Nuestra Casa) for Design and Technology While educational models such as AFH and MCG are located in larger cities, there are some notable programs in smaller communities. One of these, Our House for Design and Technology, is in Lawrence, Massachusetts, one of the oldest mill cities in the U.S. and the site of the famous Bread and Roses strike of 1912. Our House is run by Lawrence CommunityWorks (LCW), a membership-only community development organization that focuses on creating affordable housing, family asset building and youth development in a city that currently has a large immigrant population, about 70% Latino.

Our House for Design and Technology is an outgrowth of Movement City, a youth empowerment program of LCW that initially focused on involving young people in the performing arts. It has since expanded to incorporate instruction in design and technology-related creative fields. As the story was relayed to me by the organization, an MIT student intern was working with LCW to evaluate the conditions of some neighborhood buildings. Young people attending Movement City spotted this designer walking the neighborhood with a clipboard and showed up the next day with their own clipboards to follow the architects around and observe. Their curiosity was recognized and rewarded when a Young Architects Program was set up to teach design principles. An MIT post-graduate rounded up eight drafting tables and materials and began to teach Computer Assisted

Design (CAD) in interest areas expressed by the attending youth: web, graphic, fashion and interior design. MIT eventually received a COPC (Community Outreach Partnership Center) grant from the federal office of Housing and Urban Development to combine homework mentoring with training in the creative realm. Partnerships eventually extended to the Massachusetts College of Art and the Berklee School of Music in Boston, as well as to the nearby University of Massachusetts in Lowell (http://www.lawrencecommunityworks.org/real-estate/our-house).

In 1999, LCW began discussions with residents of the North Common neighborhood about the kinds of changes they would like see. Many talked about an abandoned elementary school that they wanted to convert to a meeting space for children and adults – something they described as "another big room in *our house*" to help Lawrence families. LCW conducted a massive capital campaign that culminated in the total rehab of the school. Our House for Design and Technology opened in October 2007 as a "green" building, equipped with solar and geothermal technology. Along with the Hennigan Center, a rehabbed historic Victorian mansion, the two-building campus now houses LCW's youth network and Family Asset Building initiatives, as well as Movement City, a youth empowerment program for ages 10–18 that encourages young people to explore their potential to access "high-skill and creative careers" through design, technology and the performing arts. Youth are brought together here with professionals and volunteers drawn from institutions of higher education who continue to run programs in the fields mentioned above, as well as digital music and video production, architecture, dance, creative writing and voice. The youth pay a membership fee of $60/year to access the programs in keeping with other membership policies of LCW designed to encourage residents' commitment (interview).

The larger MIT/Lawrence partnership that sustains much of the operation is one of the most impressive aspects of the Our House operation. This partnership between the city and an institution of higher education nearly 30 miles away began in 1999. Each year a practicum is offered at MIT that focuses on a priority issue as defined by partner organizations in Lawrence The smaller scale of the city is one factor that enables service learning projects such as this to evolve into more long term institutions. If sustainable over the long term, such partnerships have the potential to provide community-based organizations in smaller post-industrial cities with access to valuable resources that would otherwise reside outside city limits. This depends on a continued valuation of out-of-classroom learning and a commitment on the part of academic partners to allow community organizations to define the issues and terms of the relationship.

Lessons to take forward Many workforce development programs currently focus on necessary skills such as resume preparation and interviewing. What programs such as MCG, AFH and Our House offer beyond the practicalities of job readiness are the ingredients for creative learning and an aspirational hands-on road map for career development. The best programs provide youth with a reason to stay in school, and adults with training in occupations that present the possibility of more meaningful jobs. Instructors who often come from similar under-resourced environments share their own life trajectories and transfer skills in apprentice-like settings.

While it is difficult to generalize from a few programs, certain characteristics seem to contribute to success. MCG and AFH engage in the active and targeted recruitment of young participants. They are also sensitive to the obstacles that inhibit participation (e.g., lack of transportation or the need to be mentored in other more basic areas of education). These programs inspire participants through project-based learning that appeals to young people's interests in a variety of creative areas, and that provide several points of entry into the learning process through collaborative, cross-disciplinary work. Most programs encourage long-term commitment through reward structures (e.g., a reduction in membership fees, pay for work, and opportunities to assume more responsibility over time). Each program also helps participants develop a road map to higher education through portfolio development and access to apprenticeships that build social capital through long-term personal relationships. All three programs received the support of outside funding to renovate or design new neighborhood spaces that are welcoming to residents and have the flexibility to accommodate a range of uses at different times of the day or week. On occasion this flexibility enables outside users to provide a source of revenue to offset expenses.[6]

Creativity and its benefits cannot be confined to a few select cities or individuals. The case studies in this book illustrate that individual capacities are maximized in a social context that supports experimentation and the collaborative exchange of knowledge. A radical revisioning of creative economy planning practice necessitates the adoption of the kind of experiential educational models described above, along with the creation of vibrant mentoring spaces that awaken and support a desire to learn. Cities can begin to promote this agenda through linkage models that require private enterprises that receive financial or other tax incentives, to contribute to a pool of funding that supports apprenticeship and workforce development opportunities for residents at large.

Encouraging Creative Collaborations and Social Networking

The creative economy has sometimes been referred to as a "networked economy" because it "flourishes through connectivity" (Fleming 2004). Focused on larger cities and new technologies, mainstream practice emphasizes the importance of geographic proximity and entrepreneurial "eco-systems" that facilitate a fast-paced

6 If there are local colleges and universities nearby, faculty and students can become teachers within these programs, and green-building techniques can reduce operating expenses further, as illustrated by the partnership in Lawrence, Massachusetts with Our House for Design and Technology.

serendipitous exchange of ideas among proximate inventors, investors and creative industry workers. Many participants are presumed to be young, with little experience in finance and marketing or the communication skills necessary to translate their own creativity into the market. Lacking these skills, and yet thriving on the exchange of ideas, several entrepreneurs that I have spoken with talk about making regular forays into larger cities where there are concentrations of technology-focused peers and venture capital firms.[7]

Jane Jacobs recognized the attributes of urban economic diversity in *The Economy of Cities* written in 1969. Desrochers and Lepälä (2010) carry this research forward in their exploration of the "spillovers" that occur when knowledge and materials move among different occupations and production processes, and when people with different skill sets come together to create new forms of production. They suggest that their findings confirm an earlier contention by Glaeser (1999: 255) that "the primary role of cities may not be the creation of cutting edge technologies, but rather the provision of learning opportunities for "everyday people" (Desrochers and Lepälä 2010: 293). How are the networks we see emerging in post-industrial settings similar to, or different from, these active exchange environments in larger cities, and what potential do they hold for urban regeneration?

In Lawrence, Massachusetts, Lawrence CommunityWorks (LCW) functions more like an open network of residents than a traditional community development corporation as it uses networking to build community and create new leaders who craft specific actions to improve life in some of the city's most impoverished neighborhoods (Plastrik and Taylor 2004). Rather than emphasize an organizational structure, LCW supports existing linkages and loose affiliations among people all over the city; it then uses these to build a network that the former director describes as,

> Not an organization, but … a bundle of thinking, language, habits, value propositions, space and practice – all designed to comprise an environment that more effectively meets people where they are and offers myriad opportunities and levels of engagement (Traynor 2012: 217).

LCW also creates "clusters" among resident groups, but not in the creative economy sense of the arbitrary geographic designation of places for creative or cultural activity. LCW clusters have more to do with the coalescing of members around specific projects (e.g. augmenting family assets) and the sharing of skills and information to address needs or resist plans that are not seen to be in the interest of members. Emphasis is placed on pursuing "real social change" and providing a welcoming environment that enables residents to build power

---

7 Greenhorn Connect is just one example of a network that exists in Boston for the purposes of sharing knowledge and bringing resources and people involved in creative industries together (http://greenhornconnect.com/events/calendar/2012-W10).

through relationships. "Community building", Traynor suggests, "does not start in meetings. It starts – typically – with eating and talking" (Traynor 2012: 217). A premium is placed on flexibility, fluidity, informality, and real accomplishments.

While community-focused networking is a special feature of Lawrence, many smaller cities evidence unique alliances that often arise organically, much as the relationships among artists at One Cottage Street in Easthampton (Chapter 7). Here graduates of a Woodworking School "organically seeded" an artisans community locally because they wanted to maintain their collaborative learning environment after their formal course was complete. Creative networking events, such as Barcamp Barnsley (Chapter 6) also promote an active interchange of ideas, even if they are on a smaller scale. The creative networks established among artists and creative entrepreneurs in Holyoke and Easthampton, Massachusetts (Chapters 3 and 7), Thunder Bay and Petersborough, Ontario (Chapter 4), and smaller northern cities in the UK (Chapter 5) are often less formal, more democratically structured and horizontally integrated than some of the networks in larger cities. Reports from several post-industrial cities also suggest that there is a less competitive environment and a lower barrier to inclusion for individuals within the cultural or creative sector. The synergies forthcoming from smaller networking events help establish informal alliances that generate business and create economies of scale. These enable local artists and smaller creative enterprises to survive and grow. Often these same networks extend out to the larger region and even across international borders. This was evidenced in Holyoke when a more expansive network of Latino performers and craftspeople drew participants to cultural events from as far away as Hartford, New York, and even Puerto Rico.

Besides marketing events, informal networks can encourage backward and forward linkages and cooperative purchases of products or services among members. Barnsley's Design and Media Center and Hudderfield's Media Centre in the UK encourage small start-ups with offices in proximity to one another to learn about and support each other's work through informal weekly gatherings in deliberately designed interior spaces. Oftentimes this leads to the exchange or purchase of services and products among resident businesses.

While networking among a traditionally defined creative "class" holds much potential, the exchange of ideas and skills across social and cultural boundaries presents a further challenge. Some of these divisions may be overcome through approaches to education and training mentioned earlier. Others require planners to incorporate a more inclusive definition of "knowledge" and creativity into public policy, directing some resources away from bounded "innovation districts" and "cultural quarters" and towards more organic creative enclaves emerging in everyday spaces appropriated by residents.

New Forms of "Making"

"Maker spaces" are popular sites for networking and production that have emerged in connection with the new creative economy. These spaces are the result of a slow and growing movement to revive invention, production and entrepreneurship through smaller-scale manufacturing in new collaborative settings. According to a recent Huffington Post article (Chun 2012), the basis for this thinking rests in part in the belief that more Americans want to buy American-made products and that new technologies enable great efficiency. Another impetus is presumed to come from rising fuel prices and the flattening of the "global cost curve" in terms of wages and production methods. Mark Perry, professor of economics at the University of Michigan-Flint, is quoted in this article as foreseeing a "pending renaissance in U.S. manufacturing" due primarily to advances in innovation and entrepreneurship.

Discussion of the impossibility of maintaining a robust economy without production has also captured attention abroad. A recent article from the *Guardian* (Chakrabortty 2011) describes the precipitous decline of the manufacturing sector in the UK over the last 30 years, and recounts how various conservative and liberal governments have responded. Thatcher conservatives suggested austerity and the elimination of jobs to create leaner, more competitive firms. Additional claims were made about workers retooling and moving into new higher skill jobs, even though "the middle-aged engineers who were laid off didn't go away and become software engineers – they largely landed up in worse jobs or on the scrapheap." In contrast, the author presents Tony Blair's prescription for moving into a service economy, which was based on an optimistic view about the potentials of the new knowledge economy and the "creative class". This view, he suggests, was largely driven by the American champion of the "the creative class", Richard Florida. The article critiques the canonizing of technology and the distain with which manufacturing and factory workers are often treated within both paradigms.

Recent articles in *The Economist* and *FastCompany* see a "maker movement" "heralding a new industrial revolution" that would presumably operate parallel to global capitalism (http://makelondon.wordpress.com/). While not focused exclusively on smaller post-industrial cities, the discussion highlights the importance of such attributes as direct place-related personal relationships among people and businesses, and abundant space and machinery. Also cited is the somewhat contradictory role of digital technology, where, it is argued, the boredom of sitting in front of a screen all day fuels a desire for hands-on making, while the technology itself contributes new modes of production. The article suggests that more consideration be given to how "maker spaces" might support inner cities and help to form mixed-use communities that relieve unemployment and foster innovation.

In spite of such pleas, the bulk of current "maker spaces" (e.g., Techshops and do-it-yourself "hacker spaces" that provide users with access to expensive machinery, tools and software) remain in larger cities. Training is provided on sophisticated industrial machines, and members pay sliding scale fees to build prototypes or launch small manufacturing businesses. According to some observers, these large "maker spaces" tend to have a "powertool-welding-boys feel" and exude "a fascination with rockets and robots" (Madden 2012). There are examples, however, of spaces that foster a different and more inclusive environment. SF Made is a non-profit umbrella organization in San Francisco that supports entrepreneurship and the creation of new manufacturing options in low-income neighborhoods. SF tries to build local infrastructure and provide access to capital to support local manufacturing, especially to "green-leaning" enterprises. The Greenpoint Manufacturing and Design Centre, in Brooklyn, New York (http://www.gmdconline.org), and The Crucible, in Oakland, California (http://thecrucible.org) represent other successful non-profits that are accessible to neighborhood residents and bring together industrial making, the arts and training. The Crucible attracts over 5,000 adults and youth each year into project-based classes that use recycled materials and innovative design to produce new items. Once courses are complete, Studio Access Labs and the Crucible's Expanded Access to Tools & Equipment program (CREATE) provide continued access to machinery and tools in an environment that encourages the sharing of ideas and skills. Several for-profit companies also participate in building teams and provide jobs to some of the graduates (http://thecrucible.org/).

While such programs are emerging in larger cities, "maker spaces" are also being cultivated in smaller post-industrial communities. Newport, New Hampshire, is a former textile and shoe manufacturing town that moved several decades ago to the production of advanced precision machine tools and still produces high quality tools and parts in over 60 very small shops, often located in converted garages. The products, including those destined for high-speed trains, aeronautics, and medical industries, are sold all over the world. In the early 1990s, Patryc Wiggins, a French tapestry weaver and daughter of a machinist, became the town's first economic development coordinator. She initiated a number of humanities and arts projects to develop Newport citizens' consciousness about its industrial strengths, and to build local capacity to sustain this into the future.[8]

8 The projects include her own weaving of a 6" x 13" tapestry that draws on her family's history and her memories as a child of Finnish immigrants, the mills, waterwheels, and community life; SPIN, a literary project facilitated by a Vermont poet that includes intergenerational self-portraits of residents; and a machine tool exhibit using photography to interpret the regional and international economy. Patryc also supervised several historic building projects, including the outdoor lighting of the old Opera House in town, mural paintings on buildings with site specific themes, and the total renovation of Eagle Block, a historic building that now houses the machine tool exhibit, a microbrewery and marketplace, a locally owned restaurant, Girl's Incorporated, and public space (Wiggins 2002).

Figure 11.2 Patryc Wiggins at Machine Tool Exhibit, Newport, New Hampshire

Since many precision tool workers were getting old, and the advanced skill set required for precision work did not seem to be passing down to a younger generation, Wiggins organized a New England-wide Artists' Congress to consider innovative methods that combine the arts and community economic development. Another intent was to draw low-income residents, especially girls, into metal working fields in need of skilled workers. Wiggins saw the problem and solution as one in the same: education.

> The coming economy requires that everybody be learning all the time. Right now learning is done by those engaged in interesting, livable wage work. Those in the service industries primarily are under-educated in the mode of the mules of old industry. The new society needs educated people who see education as dynamic and part and parcel to survival. Not necessarily ph.d educated, but really educated in all dimensions ... This education is how the machinists think and live and it is not acknowledged by society (Wiggins 2002).

Wiggins believed in the importance of reducing dependence on fossil fuels and thought precision metal manufacturing could be key. The humanities were used to keep this manufacturing sector alive and generate an appreciation for local cultural enterprise through industry-led educational programs that included a summer camp for elementary-aged youth and a technical curriculum with Girl's Inc.

In a similar effort to build upon and recover the industrial heritage of Providence, Rhode Island, the Steel Yard (Chapter 2), transformed a brownfield into an exceptional space that offers arts and technical training in welding, blacksmithing, ceramics, jewelry, found object sculpture and other metal-related arts to adults and young people. Its founders, who include a Rhode Island School of Design graduate and the great-great grandson of John D. Rockefeller, drew inspiration from The Crucible in Oakland, California, and envision Steel Yard as the "next successful model for industry". They use earned income from higher-end rental housing, the rental of space for events at the foundry, and products made at the Yard and sold to the city (e.g., street furniture, bicycle racks, tree guards, public sculptures etc.) to fund classes and HUD-subsidized live/work units for lower income residents. Steel Yard also includes a workforce development program and has secured contracts with the Providence Downtown Improvement District.

The idea of invigorating an economy by supporting a "spirit of invention" through "industrial laboratories open to all inquirers", where participants can "work out their dreams ... make their experiments ... find .... experts in other branches of industry ... study some difficult problem ... and enlighten each other" is not new (Kropotkin 1892). Nor is the idea of blending art and industry, and making the associations that do so accessible to all. Kropotkin's prescient work, *The Conquest of Bread*, spoke of these ideas yet warned of their impossibility within societies with vast differences in wealth and power.[9] Today, the vision of building the capacity of residents who do not fall within the current definition of the "creative class" has yet to effectively challenge the "one size fits all" (often exclusionary) creative economy models that continue to seduce local planners. Among many topics to explore further is the potential of linking more inclusive and collaborative creative learning and "maker spaces" to educational and training models that combine art and industry and lead to job creation.

9 "Art, in order to develop, must be bound up with industry by a thousand intermediate degrees, blended, so to say, as Ruskin and the great Socialist poet Morris have proved so often so well. Everything that surrounds man, in the street, in the interior and exterior of public monuments, must be of pure artistic form. But this can only be realized in a society in which all enjoy comfort and leisure. Then only shall we see art associations, of which each member will find room for his capacity ... What is now the privilege of an insignificant minority would be accessible to all (Kropotkin 1892)."

"Sustainable Creativity"

Current policies that support a creative economy in the regeneration of smaller post-industrial communities necessarily raise larger concerns about social justice and urban sustainability broadly defined.[10] Early ecologists defined a "sustainable city" as one that could distribute necessities for survival equitably while fostering a sense of collective responsibility and opportunities for discovery and personal growth. The barriers to achieving this balance, they claimed, were not technical but *social*. They rested first and foremost in a reformulation of human relationships and priorities (Kropotkin 1902; Reclus 1905–08; Clark 2004).

Franco Bianchini has taken similar ideas about urban sustainability and applied them directly to the field of cultural planning (1999). He argued that "sustainable creativity" could bring regeneration and social inclusion together in a forward-looking agenda that builds on local assets and encourages shared knowledge and resources. Some post-industrial cities are trying on a small scale to incorporate ecological concepts such as balance, diversified local production, cooperative exchange and active citizen involvement into development planning. How, then, is it possible to prioritize social inclusion and equity within a broader sustainable *creative city* agenda?

Confirming what Kropotkin suggested a century before, an Independent Commission on Sustainable Development in 1997 found that high levels of social exclusion was a real barrier to environmental sustainability (Talbot and Magnoli 2000: 99). Echoing the early social ecologists, Talbot and Magnoli argue that the whole sustainability agenda cannot move forward until "we learn how to plan, build and live equitably within the sustainable city (2000: 93)". How willing are localities to extend definitions of cultural and creative practices to embrace a wider spectrum of their populations? How willing and prepared are they to support creative economy initiatives that increase the income earning capacity of low-wage service and cultural workers (Donegan and Lowe 2008: 58)?

Some believe that post-industrial cities have an important role to play in a "post-carbon" world, and that a more creative and "greener" economy does not have to by-pass the needs of communities of color (Shiffman 2010). Catherine Tumber argues that a reduction in our reliance on expensive environmentally damaging supply chains involves a number of steps that post-industrial cities are well suited to take on – e.g., reducing transportation of people and products, growing local bio-diverse food systems, using local sources of clean energy and creating jobs that "cannot be outsourced". Citing the fragility of the current grid system as an example, she describes how a modern decentralized transmission network could be more easily met in smaller cities with large amounts of space:

---

10 A critique of mainstream models that ignore issues of exclusion, neoliberalism and global competition was, in fact, the focus of a recent conference entitled "Creating Cities, Culture, Space, and Sustainability", held in Munich, Germany in February 2010.

> In New England, a number of projects are underway that will generate three megawatts or less, enough to power a hospital, large shopping center, or small factory. As ideal sites for new energy industries, smaller cities would in turn gain from job creation (Tumber 2010: 224).

This is the great hope in Holyoke, Massachusetts, as the new Massachusetts Green High Performance Computing Center makes use of land along the canals and abundant waterpower. Tumber also notes that renewable energy and energy-efficient industries have already created more than 8.5 million jobs in the U.S. and could produce millions more high-quality jobs in the next decade (225). She expresses the hope that the growing cultural diversity in smaller post-industrial cities will continue to push communities to see the decline of traditional industry "as a framework for creative reinvention" along more sustainable lines. Tumber imagines that a "template that is *de*centralized, *de*concentrated", and "*re*localized" could be created if former centers of industry were placed at the center of a sustainable cities agenda that repurposes manufacturing infrastructure and workforce skills for the production of renewable-energy technology (231).

This potential has not escaped the attention Nuestras Raices in Holyoke. This program supports a community farm, neighborhood gardens, toxic environment remediation, a food and fitness program and a number of sustainable Latino-owned small businesses. Training is provided for local residents in residential and business energy audits, and cultural events are connected to these businesses. Recent reports in the U.S. indicate that stateside Puerto Ricans are continuing to move out of large cities into mid-size and smaller cities. Torres-Vélez argues that the gap between demand and supply in "green" jobs presents a particular opportunity for community-based organizations like Nuestras Raices that have been at the forefront of environmental justice movements in many Puerto Rican neighborhoods (Torres-Vélez 2011). He suggests community-based organizations become more involved in appropriate workforce development to take advantage of these opportunities.

A recent comparison of the workforce development programs in New York City, Hartford, Connecticut, and Springfield/Holyoke, Massachusetts reveal that community-based organizations (CBOs) are playing a bigger role in the latter two cities. The programs are less focused on "work-first" than on sectorial and career-ladder approaches to help residents secure higher paid and longer-term employment (Borges-Méndez 2011). The sectors targeted in the Springfield/Holyoke region are health, advanced manufacturing (especially precision machining in the Pioneer Valley) and early childhood development. Organizations such as Career Point in Holyoke now participate in regional economic forums, including the New England Springfield-Hartford Knowledge Corridor. They function much like a community development organization and do not see themselves as a "conventional service delivery agent" (Borges-Méndez 2011: 81). In contrast, workforce development programs in large cities are reported to be far more distant from neighborhood life and utilize strategies geared more toward placing workers in low-wage jobs, without any real attention to career planning or long-term capital. Borges-Méndez worries about the incentive to participate in such programs and supports the creation of "gateway programs" to facilitate resident movement into promising career paths in areas such as health and environmental preservation (Borges-Méndez 2011: 87). Might the cultural and creative sector be included as well?

At the Centre for Local Economic Strategies (CLES) in the UK there is an on-going effort to apply resilience theory (generally used to bolster communities affected by environmental disasters) to urban regeneration (www.cles.org.uk/). The assumption is that ecosystems thrive because of the interaction of many elements, including "interconnected systems of relationships" between people and their built and natural environments. CLES is working to challenge mechanistic, linear approaches to place making and to address problems of inequality in the face of massive cuts in public spending. Fostering a culture of innovation and creativity is one piece of this agenda.

CLES assumes that resilient places are not static and that there is a need to plan for constant change in the social and environmental as well as economic realms. With respect to the economic there is a concern for the "more qualitative aspects of place development".

> Too many of the strategies we examined focused on "hard" economics – small business start-ups, inward investment, availability of land and premises for business – rather than "softer" aspects of place, such as neighbourhood renewal, environmental sustainability, and levels of community empowerment and participation (McInroy and Longlands 2010).

CLES sees greater interaction between the commercial, public and social economy,[11] and greater attention to the health needs and aspirations of residents through networking and participative practices, as essential stimuli for the creative thinking and action that will result in stronger local economies (16–17).

Concluding Remarks

It is easy to see why a city suffering the long-term effects of deindustrialization, coupled with the current recession, would adopt civic boostering agendas and look to larger urban centers for "easy fix" cultural regeneration formulas (Breitbart

---

11 CLES defines the *commercial* economy as "wealth creation generated by businesses that are privately owned and profit motivated"; the *public* economy as "services delivered on behalf of government organisations whether national, regional or local, and funded by the public purse"; and the *social* economy as a "wide range of community, voluntary and not-for-profit activities that try to bring about positive local change". Each level is seen to interact with the other through partnerships (McInroy and Longlands 2010: 18).

and Stanton 2007). Art, cultural production and creative industries can boost economies and contribute to the quality of life. Many politicians and planners see recommendations to re-use mill spaces and promote their liberal political climates and alternative lifestyles, as attainable goals. However, the essays in this book suggest that there is no simple panacea for the economic and social challenges that post-industrial cities face. Indeed, there is a danger in overlooking the complexity and uniqueness associated with different urban contexts by seeking one formula for success (Waitt and Gibson 2009). Just as post-industrial cities recognize the importance of *economic* diversification, there is a parallel need to recognize the importance of a diversity of creative economy initiatives, and to see the creative economy as only one component of community regeneration within a larger sustainable cities agenda. Good ideas deserve to be replicated, but no models should be imported uncritically from larger cities or even from one small city to another.

The study of the creative economy is an "industry" in and of itself now with new material added on a daily basis. To date, little of this massive research has focused on smaller post-industrial cities facing uncertain economic futures. If these cities are to effectively "negotiate hope" future research must address the inequality of access to resources and opportunity promulgated by current models. The scope of this research must widen to include consideration of the impacts of creative economy policies on distributions of wealth, and the role of low-wage service workers in supporting the lifestyles of a more skilled creative workforce (Donegan and Lowe 2008). The research must also take account of on-going cultural, ethnic and racial transformations of post-industrial cities, and begin to explore more fully the differing impacts of creative economy policies on residents. Which policies bolster support for individual entrepreneurs and traditionally defined artists but ignore the important contributions of everyday, community-based creative and cultural producers, leaving them "to their own devices" in securing meaningful livelihoods (McRobbie 2003)? Whether some post-industrial cities can become models for "sustainable creativity" in their quest to adapt to changes at the local and global scale depends in part on addressing the critical health and educational needs of residents, and on providing them with access to needed capital and training programs that build on their assets. Also key is a willingness to assure that city "rebranding" and marketing strategies do not become substitutes for more substantive and meaningful forms of long-term planning, and that creative city marketing take wider account of residents who stand outside current definitions of the "creative class".

At the end of the film *DownSide Up*, Nancy Kelly reflects on the changes in her hometown as she participates in a community event organized by a local artist who temporarily transforms a street into a "beach" filled with sand, shovels and children playing. While it is impossible for Kelly to speculate as to the long-term advantages of Mass MoCA, or this new cultural turn for her family and other residents, she is certain that the experience of growing up in North Adams today will be very different from her own. The same can be said of all of the post-industrial communities explored in this book. In the end, success will be measured by more than the number of visitors who call in to a flagship museum or the number of new cafes that open to serve young entrepreneurs. It will be measured by the numbers of young people and displaced workers who are introduced to career trajectories and productive enterprises that take advantage of their potential and provide stable high quality jobs. In the end, a sense of place will result not from a marketing campaign at all but from the planned and unplanned actions needed to accomplish these goals.

This book was never intended to generalize from examples in a few post-industrial cities that are struggling to define their futures. It was to complicate the picture presented by generic creative economy planning and to reflect on the potential of practices that take account of the challenges such cities face, without denying residents' aspirations or needs. More close research will undoubtedly add to this picture and expand the notion of "sustainable creativity" in such contexts. The question is not whether cities rooted in an industrial past can "afford" to be creative in reanimating their economies; it is whether they can afford not to.

References

Anthony, C. (1990), *Social Justice and the Sustainable City*, First International Ecocity Conference, Proceedings, edited by Chris Canfield (Berkeley, Ca, May).

Baker, K. (2011), "Making space for makers". Available at http://makelondon.wordpress.com/ [accessed 7 December 2011].

Becker, H. (1982), *Art Worlds* (Berkeley, CA: University of California Press).

Bell, D. and Jayne, M. (eds) (2004), *City of Quarters: Urban Villages in the Contemporary City* (London: Ashgate).

Bianchini, F. (1999), "Cultural planning for urban sustainability", in L. Nystrom (ed.), *City and Culture: Cultural Processes and Urban Sustainability* (Karlskrona, Swedish Urban Environment Council).

Blundell-Jones, P., Petrescu, D. and Till, J. (eds) (2005), *Architecture and Participation* (London: Routledge).

Borges-Méndez, R. (2011), "Stateside Puerto Ricans and the public workforce development system: New York City, Hartford, Springfield/Holyoke". *CENTRO Journal of the Center for Puerto Rican Studies* 23:11, 65–92.

Breitbart, M. (2004), "Blueprinting Soho: the geographic life of the idea to transform space and economy through artists and the arts". Paper presented at the Association of American Geographers Annual Meeting, 14–19 March (Philadelphia, Pennsylvania).

Breitbart, M. and Kirk, G. (1991), "Ecological urbanism: women charting a course for change-work in progress", The Scholar and the Feminist XVIII: Women the Environment and Grassroots Movements Conference paper, 13 April (New York: Barnard College).

Breitbart, M. and Stanton, C. (2007), "Touring templates: cultural workers and regeneration in small New England cities", in M. Smith (ed.), *Tourism, Culture and Regeneration* (Oxfordshire, UK: CABI).

Cachia, R., Ferrari, A., Ala-Mutka, K. and Punie, Y. (2010) *Creative Learning and Innovative Teaching: Final Report on the Study on Creativity and Innovation in Education in the EU Member States* (Seville, Spain: Institute for Prospective Technological Studies).

Chakrabortty, A. (2011), "Why doesn't Britain make things any more?", *The Guardian*, 16 November. Available at http://www.guardian.co.uk/business/2011/nov/16/why-britain-doesnt-make-things-manufacturing [accessed 16 November 2011].

Chatterton, P. (2010), "The urban impossible: a eulogy for the unfinished city", *City* 14:3, 234–44.

Chayka, K. (2011), "Detroit ruin porn and the fetish for decay", *Hyperallergic: Sensitive to Art & Its Discontents*, January. Available at http://hyperallergic.com/16596/detroit-ruin-porn/ [accessed 13 January 2011].

Chun, J. (2012), "In U.S. manufacturing revival, small businesses could play a crucial role", *Huffington Post*, 9 February.

Clark, J. (2004), *Anarchy, Geography, Modernity: The Radical Social Thought of Elisée Reclus* (Lanham, MD: Lexington Books).

Connolly, J. (ed.) (2010), *After the Factory: Reinventing America's Industrial Small Cities* (Idaho: Lexington Books).

DeFilippis, J. and Saegert, S. (eds) (2012), *The Community Development Reader* (New York: Routledge).

Desrochers, P. and Leppälä, S. (2010), "Rethinking 'Jacob Spillovers,' or how diverse cities actually make individuals more creative and economically successful", in S. Goldsmith and L. Elizabeth (eds), *What We See: Advancing the Observations of Jane Jacobs* (Oakland, CA: New Village Press).

Donegan, M. and Lowe, V. (2008), "Inequality in the creative city", *Economic Development Quarterly* 22:46, 46–62.

Edensor, T. (2005), *Industrial Ruins: Space, Aesthetics and Materiality* (London: Berg Publishers).

Edensor, T. et al. (eds) (2010), *Spaces of Vernacular Creativity: Rethinking the Cultural Economy* (London and New York: Routledge).

Evans, G. (2005), "Measure for measure: evaluating the evidence of culture's contribution to regeneration", *Urban Studies* 42:5, 959–83.

Fleming, T. (2004), "Supporting the cultural quarter? The role of the creative intermediary", in D. Bell and M. Jayne (eds), *City of Quarters: Urban Villages in the Contemporary City* (London: Ashgate).

Fox, W. (ed.) (2000), *Ethics and the Built Environment* (London: Routledge).

Gibson, C. and Kong, L. (2005), "Cultural economy: a critical review". *Progress in Human Geography* 29:5, 541–61.

Goldbard, A. (2006), *New Creative Community: The Art of Cultural Development* (Oakland, CA: New Village Press).

Goldsmith, S. and Elizabeth, L. (2010), *What We See: Advancing the Observations of Jane Jacobs* (Oakland, CA: New Village Press).

Greco, J. (2012), "The psychology of ruin porn", *Atlantic Cities*, January. Available at http://www.theatlanticcities.com/design/2012/01/psychology-ruin-porn/886/ [accessed 6 January 2012].

Hall, P. and Ward, C. (1998), *Sociable Cities: The Legacy of Ebenezer Howard* (New York: Wiley).

Jacobs, J. (1969), *The Economy of Cities* (New York: Random House).

King, Y. (1991), "Ecofeminism: a politics of connection", *The Scholar and the Feminist XVIII: Women the Environment and Grassroots Movements Conference Paper*, 13 April (New York: Barnard College).

Kirk, G. (1997), "Ecofeminism and environmental justice: bridges across gender, race, and class", *Frontiers: A Journal of Women Studies* 18:2, 2–20.

Kropotkin, P. (1892) (1972), *The Conquest of Bread* (New York: NYU Press).

Loftus, A. (2009), "Intervening in the environment of the everyday", *Geoforum*, 40, 326–34.

Madden, P. (2012), "DIY sustainability: an emerging trend from the recession?", *The Guardian*, 9 February.

Markusen, A. and Gadwa, A. (2010), "Arts and culture in urban or regional planning: a review and research agenda", *Journal of Planning Education and Research* 29:3 March, 379–91.

McInroy, N. and Longlands, S. (2010), *Productive Local Economies: Creating Resilient Places*, December (Manchester, UK: Centre for Local Economic Strategies).

McRobbie, A. (2003), "Everyone is creative: artists as pioneers of the new economy?", in E. Silva and T. Bennett (eds), *Contemporary Culture and Everyday Life* (Durham, UK: Sociology Press).

Nystrom, L. (ed.) (1999), *City and Culture: Cultural Processes and Urban Sustainability* (Karlskrona, Swedish Urban Environment Council).

Palpscia, R. (ed.) (2003), *The Contested Metropolis* (Basel: Birkhauser).

Plastrik, P. and Taylor, M. (2004), *Lawrence CommunityWorks: Using the Power of Networks to Restore a City*, March (Boston, Ma: Barr Foundation Report).

Pratt, A. (2000), "New media, the new economy and new spaces", *Geoforum* 31, 425–36.

Register, R. (1987), *Ecocity Berkeley* (Berkeley, CA: N. Atlantic Books).

Reclus, E. (1905–1908), *L'Homme et la Terre*, 6 vol. (Paris: Librairie Universelle).

Rosenfeld, S. (2006), *Cool Community Colleges: Creative Approaches to Economic Development* (Washington, DC: Community College Press).

Sandercock, L. (2003), "Practicing utopia: sustaining cities", in R. Palpscia (ed.), *The Contested Metropolis* (Basel: Birkhauser).

Sandercock, L. (2003a), *Cosmopolis II: Mongrel Cities of the 21st Century* (London: Continuum).

Schiffman, R. (2010), "Beyond green jobs: seeking a new paradigm", in S. Goldmith and L. Elizabeth (eds), *What We See: Advancing the Observations of Jane Jacobs* (Oakland, CA: New Village Press).

Silva, E. and Bennett, T. (eds) (2003), *Contemporary Culture and Everyday Life* (Durham, UK: Sociology Press).

Smith, M. (ed.) (2006), *Tourism, Culture and Regeneration* (Oxfordshire, UK: CABI).

Stern, M. and Seifert, S. (2008), "From creative economy to creative society", Report for the University of Pennsylvania's Social Impact of the Arts Project and the Rockefeller Foundation. Available at www.trfund.com/resource/downloads/creativity/Economy.pdf [accessed 13 August 2012].

Stevenson, D. (1998), *Agendas in Place: Urban and Cultural Planning for Cities and Regions* (Rockhampton: RESRC Central Queensland University Press).

Talbot, R. and Magnoli, G. (2000), "Social inclusion and the sustainable city", in W. Fox (ed.), *Ethics and the Built Environment* (London: Routledge).

Till, J. (2005), "The negotiation of hope", in P. Blundell-Jones, D. Petrescu and J. Till (eds), *Architecture and Participation* (London: Routledge).

Torres-Vélez, V. (2011), "Puerto Ricans and the green jobs gap in New York City". *CENTRO Journal of the Center for Puerto Rican Studies* 23:11, 95–112.

Traynor, B. (2012), "Community building: limitations and promise", in J. DeFilippis and S. Saegert (eds), *The Community Development Reader* (New York: Routledge).

Tumber, C. (2010), "Small, green, & good: the role of smaller industrial cities in a sustainable future", in J.J. Connolly (ed.), *After the Factory: Reinventing America's Industrial Small Cities* (Idaho: Lexington Books).

Waitt, G. and Gibson, C. (2009), "Creative small cities: rethinking the creative economy in place", *Urban Studies* 46:5/6, 1223–46.

Wiggins, P. (2002), Talk given at the Planner's Network conference, panel on "The role of the arts in urban revitalization" (Holyoke, MA).

Wilson, N. (2010), "Social creativity: requalifying the creative economy", *International Journal of Cultural Policy* 16:3, 367–81.

Yeh, L. (2005), "Community building through art and youth participation", *Humanistic Educational Journal* 11, Taiwan.

Zukin, S. (1995), *The Culture of Cities* (Oxford: Blackwell).

Index

Page numbers in **bold** refer to a table or figure, page numbers followed by "n" refer to a note at the bottom of the page.

For Product Safety Concerns and Information please contact our EU representative GPSR@taylorandfrancis.com Taylor & Francis Verlag GmbH, Kaufingerstraße 24, 80331 München, Germany